REPRODUCTIVE TECHNOLOGIES AND BIOBANKING FOR THE **CONSERVATION OF AMPHIBIANS**

A catalogue record for this book is available from the National Library of Australia.

ISBN: 9781486313334 (hbk)
ISBN: 9781486313341 (epdf)
ISBN: 9781486313358 (epub)

Published in print in Australia and New Zealand, and in all other formats throughout the world, by CSIRO Publishing.

CSIRO Publishing
Private Bag 10
Clayton South VIC 3169
Australia

Telephone: +61 3 9545 8400
Email: publishing.sales@csiro.au
Website: www.publish.csiro.au
Sign up to our email alerts: publish.csiro.au/earlyalert

Published in print only, throughout the world (except in Australia and New Zealand), by CRC Press, with ISBN 9781032372075.

CRC Press
6000 Broken Sound Parkway NW, Suite 300, Boca Raton, FL 33487-2742
and
2 Park Square, Milton Park, Abingdon, Oxon, OX14 4RN
Website: www.routledge.com

CRC Press is an imprint of Taylor & Francis Group, LLC

Front cover: (top) *Litoria booroolongensis* (photo by Aimee Silla); (bottom left to right) *Pseudophryne pengilleyi* (photo by Aimee Silla), *Gyrinophilus porphyriticus* (photo by Dustin Smith), *Pseudophryne corroboree* (photo by Aimee Silla); (bottom background) *Anaxyrus fowleri* testis (photo by Allison Julien)
Back cover: (left to right) *Pseudophryne corroboree* (photo by Shannon Kelleher and Aimee Silla), *Ambystoma mexicanum* (photo by Michael Bushnell), *Peltophryne lemur* (photo by Diane Barber)

Edited by Joy Window (Living Language)
Cover design by Cath Pirret
Typeset by Envisage Information Technology
Index by Max McMaster
Printed in Singapore by COS Printers Pte Ltd

CSIRO Publishing publishes and distributes scientific, technical and health science books, magazines and journals from Australia to a worldwide audience and conducts these activities autonomously from the research activities of the Commonwealth Scientific and Industrial Research Organisation (CSIRO). The views expressed in this publication are those of the author(s) and do not necessarily represent those of, and should not be attributed to, the publisher or CSIRO. The copyright owner shall not be liable for technical or other errors or omissions contained herein. The reader/user accepts all risks and responsibility for losses, damages, costs and other consequences resulting directly or indirectly from using this information.

CSIRO acknowledges the Traditional Owners of the lands that we live and work on and pays its respect to Elders past and present. CSIRO recognises that Aboriginal and Torres Strait Islander peoples in Australia and other Indigenous peoples around the world have made and will continue to make extraordinary contributions to all aspects of life including culture, economy and science. CSIRO is committed to reconciliation and demonstrating respect for Indigenous knowledge and science. The use of Western science in this publication should not be interpreted as diminishing the knowledge of plants, animals and environment from Indigenous ecological knowledge systems.

The paper this book is printed on is in accordance with the standards of the Forest Stewardship Council® and other controlled material. The FSC® promotes environmentally responsible, socially beneficial and economically viable management of the world's forests.

MIX
Paper from responsible sources
FSC® C016973
www.fsc.org

REPRODUCTIVE TECHNOLOGIES AND BIOBANKING FOR THE
CONSERVATION OF AMPHIBIANS

EDITORS: AIMEE J. SILLA, ANDY J. KOUBA AND HAROLD HEATWOLE

CSIRO PUBLISHING

CRC Press
Taylor & Francis Group
Boca Raton London New York

CRC Press is an imprint of the
Taylor & Francis Group, an **informa** business

Dedication

This volume is dedicated to Paul Licht, PhD. Paul grew up in St Louis, Missouri and obtained his Bachelor of Arts degree in Zoology from Washington University in St Louis, Missouri in 1959. After obtaining his Master's degree (1961) and PhD at the University of Michigan, Ann Arbor (1964) under the mentorship of Professor William Dawson, Paul immediately accepted a position as Assistant Professor at the University of California, Berkeley. There he would spend his entire career as a Professor in Zoology, Professor in Integrative Biology and Endocrinology, Dean of Biological Sciences (1994–2002), Chair of Deans (1998–2002), and finally as Director of the University of California Berkeley Botanical Garden (2003–16).

Paul is a Fellow of the California Academy of Sciences, has given numerous distinguished lectures and has received many honours and awards, including the 1st Grace Pickford Medal, International Society of Comparative Endocrinology (1981); the Berkeley Citation, University of California (2002); the Distinguished Alumni of the Year Award, University of California, Berkeley (2006); and induction into the Berkeley Fellows Centennial Honorary Society (2012).

The concepts and knowledge on amphibian reproductive biology and endocrinology, described by Paul in his collected works, helped lay the foundation for many of the researchers' advancements in amphibian conservation that are described within the following chapters. It is for this reason that we celebrate his foundational accomplishments and dedicate this book to his achievements.

As a research scientist, Paul focused on physiology, environmental adaptation, and endocrinology. Paul pioneered studies that developed assays to measure hormones that circulated at very low levels and his major contributions were his studies of the role of pituitary hormones in growth and reproduction and environmental regulation of reproductive hormones and seasonal cycles. Paul's interest did not end with amphibians (anurans and urodeles) – he has also published on fish, reptiles (turtles, snakes, and lizards), birds (turkey and ostrich) and mammals (rats, bats, moles, ground squirrels, hamsters, hyenas, baboons, and humans). Paul has published over 300 peer-reviewed papers and, in addition to amphibian reproductive physiology, the topics of his papers ranged quite broadly and included purification, identification, and characterisation of the role of pituitary hormones in growth and reproduction in amphibians, reptiles, and mammals; effects of temperature on cardiac function and physiology; effects of temperature and photoperiod on hormones and reproduction; thermoregulation in amphibians, reptiles, and mammals; discovery of a novel thyroid hormone binding protein in turtles; physiological responses to temperature in bats; sex differentiation, reproduction, and siblicide in hyenas; placental steroidogenesis in mammals; salinity tolerance and osmoregulation in reptiles; reproduction in endangered sea turtles; retention and loss of water in scale-less snakes; reproduction and seasonality in amphibians; development and function of cutaneous glands in amphibians; and the role of steroid hormones in metamorphosis and sexual differentiation in amphibians.

As a researcher, Paul was active in both the laboratory and the field. Most impressive, Paul continued to work at the laboratory bench throughout his career. In addition to advising students' projects in the laboratory, he found time to work on his own projects, purchased new equipment to modernise the laboratory, and developed new physiological techniques. As a student, securing time at the bench meant waking up before dawn to make sure that you were in the laboratory before Paul was already set up and running experiments (including weekends).

Above scientific curiosity, Paul simply loved animals. He visited the university's vivarium daily and spent an inordinate amount of time watching and feeding his turtles … until the Office of Laboratory Animal Care finally asked him to stop (he often fed them even after the staff had already cleaned the tanks). In addition, Paul's enthusiasm for life does not end with animals; he maintains quite an impressive personal garden at his home.

Paul was as rigorous as a teacher in the classroom as he was a researcher in the laboratory. He is known for asking the key questions at conferences and lectures. Students and senior scientists alike had nightmares about watching Paul's hand go up after a talk or lecture. I once witnessed him ask such piercing questions of a speaker a few minutes into his talk that the remainder of the speaker's lecture time turned into discussion and he had to return the following week (after doing more experiments) for a second seminar just to answer Paul's questions.

Paul demanded critical thinking of his students, even when he taught elective courses for non-science majors to fulfil their requirements for breadth. Non-biology majors expecting a fluff course quickly found out that Paul was not the instructor to go to for an easy pass. Even for students who had no plans of pursuing a career in science, Paul felt it was of utmost importance for them to understand heat transfer in a counter-current model, to know why a reindeer's feet melt the snow when they stand in it, to understand partial pressures of gases when thinking about diving marine mammals, and how to scale up an LSD dose for an elephant's metabolism.

Paul is also a deep-thinker with regard to social issues and someone who sought to improve diversity, equity, and inclusion in academia. Even when he was not clear whether or not transgressions or misunderstanding occurred in the laboratory, you were sure to have a conversation the next day to clarify points and reach fairness and understanding. In my recent discussions with Paul on current issues about the Black Lives Matter movement, Paul expressed concern, hoping that he had never done anything to discourage diversity, equity and inclusion. 'I hope I treated you all equally harsh,' he said. I assured him that he had and I also assured him that all of his students and colleagues are the better for it.

Above all else, I have learned in my 32 years of knowing Paul … whether for birthdays or academic celebrations … Paul doesn't like surprises. So he is well aware that this dedication is coming. It is my honour to have authored this dedication to a great mentor, advisor, and friend. The field would not be the same without him.

Tyrone B. Hayes
University of California
3 January 2021

Contents

List of contributors

CO-EDITORS (ALSO AUTHORS)

SILLA, Aimee J., School of Earth, Atmospheric and Life Sciences, University of Wollongong, Wollongong, NSW 2522, Australia. asilla@uow.edu.au

KOUBA, Andy J., Department of Wildlife, Fisheries and Aquaculture, Mississippi State University, Mississippi State, MS 39762, USA. a.kouba@msstate.edu

HEATWOLE, Harold, Department of Zoology, University of New England, Armidale, NSW 2351, Australia, and Department of Biology, North Carolina State University, Raleigh, NC 27695-7617, USA. harold_heatwole@ncsu.edu

AUTHORS

BYRNE, Phillip G., School of Earth, Atmospheric and Life Sciences, University of Wollongong, Wollongong, NSW 2522, Australia. pbyrne@uow.edu.au

CLULOW, John, Conservation Biology Research Group, School of Environmental and Life Sciences, University of Newcastle, Callaghan, NSW 2308, Australia. john.clulow@newcastle.edu.au

CLULOW, Simon, Centre for Conservation Ecology and Genomics, Institute for Applied Ecology, University of Canberra, Bruce, ACT 2617, Australia. simon.clulow@canberra.edu.au

FRASER, Barbara, Priority Research Centre for Reproductive Science, University of Newcastle, Callaghan, NSW 2308, and Australia and Pregnancy and Reproduction Program, Hunter Medical Research Institute, New Lambton Heights, NSW 2305, Australia. barbara.fraser@newcastle.edu.au

GERMANO, Jennifer M., Department of Conservation, Nelson 7045, New Zealand. jgermano@doc.govt.nz

GRAHAM, Katherine M., Anderson Cabot Center for Ocean Life, New England Aquarium, Boston, MA 02110, USA. kgraham@neaq.org

HAYES, Tyrone B., Department of Integrative Biology, University of California, Berkeley, CA 94720-5800, USA. tyrone@berkeley.edu

HOUCK, Marlys L., San Diego Zoo Institute for Conservation Research, 15600 San Pasqual Valley Road, Escondido, CA 92027, USA. mhouck@sdzwa.org

JOHNSON, Kevin, Amphibian Ark, Apple Valley, Minnesota, MN 55124, USA. kevinj@amphibianark.org

JULIEN, Allison R., Department of Ectotherms, Fort Worth Zoo, Fort Worth, TX 76110, USA. Ajulien@fortworthzoo.org

KOUBA, Carrie K., Biochemistry, Molecular Biology, Entomology and Plant Pathology, Mississippi State University, Mississippi State, MS 39762, USA. ckv7@msstate.edu

LANGHORNE, Cecilia J., Conservation Science Network, Glasgow, G12 0HW, UK. cecilia.langhorne@gmail.com

LINHOFF, Luke J., Smithsonian's National Zoo and Conservation Biology Institute, Washington DC 20008, USA. LinhoffL@si.edu

MENDELSON III, Joseph R., Zoo Atlanta, Atlanta, Georgia 30315-1440, USA and School of Biological Sciences, Georgia Institute of Technology, Atlanta, Georgia, USA. jmendelson@zooatlanta.org

MOLINIA, Frank C., Manaaki Whenua – Landcare Research, Auckland 1072, New Zealand. MoliniaF@landcareresearch.co.nz

NARAYAN, Edward J., School of Agriculture and Food Sciences, University of Queensland, Gatton 4343, Australia. e.narayan@uq.edu.au

PAHUJA, Harsh K., School of Agriculture and Food Sciences, University of Queensland, Gatton 4343, Australia. h.pahuja@uq.net.au

RAVEN, Brianna H., Department of Biology, University of Ottawa, Ottawa, ON, K1N 6N5, Canada. brave030@uottawa.ca

STRAND, Julie, Aalborg University, Department of Chemistry and Bioscience, Fredrik Bajers V ej 7K, 9220 Aalborg Øst, Denmark and Randers Regnskov, Tørvebryggen 11, 8900 Randers C, Denmark. js@biosfaeren.dk

TRUDEAU, Vance L., Department of Biology, University of Ottawa, Ottawa, ON, K1N 6N5, Canada. trudeauv@uottawa.ca

UPTON, Rose, Conservation Biology Research Group, School of Environmental and Life Sciences, University of Newcastle, Callaghan, NSW 2308, Australia. rose.upton@uon.edu.au

Preface

This book is Volume 12 in the series *Amphibian Biology*, which covers all aspects of the biology of this fascinating group of vertebrates. Volumes 8 and 10 dealt with conservation of amphibians on a topical basis, and Volumes 9 (Western Hemisphere) and 11 (Eastern Hemisphere) examined the threats to amphibians and the status of their decline and conservation on a country-by-country basis. Different subregions were treated in separate issues (Parts) of Volumes 9 and 11. At present the coverage for South America, Asia, Africa, and Europe is complete and has been published; an issue for the Caribbean islands is in press; two issues, one covering the Near East and the other covering Central America and Mexico are in preparation, leaving only the United States and Canada to be launched in a combined Part of Volume 9.

Amphibians, with their permeable skins, cutaneous respiration and dependence on water are especially sensitive to pollution, desiccation, and the fragmentation and loss of habitat, and the series has highlighted the threats they face in a world in which climatic instability, emerging new diseases, and a burgeoning human population, with its attendant assault on the natural world, threaten the very existence of the global biota. Amphibians arguably have suffered the greatest of any vertebrate taxon, and the one most in need of research dedicated to their conservation and the prevention of massive extinction. The above library has covered most aspects of this topic and has reviewed the progress of legislation protecting amphibians; the setting aside of refuges; the spread of emergent diseases; the employment of captive breeding; assessment of the threats and classifying the degrees of threat to a wide variety of species; as well as the relevant aspects of the taxonomy, natural history, and distribution of amphibians. Several herpetologists, however, have stressed to me the need to provide a guide to the practical methodology for snatching endangered species of amphibians from the brink of extinction. In recent years, research on this topic has accelerated as the need for it has become more urgent, and it became imperative to review this new information for the benefit of scientists entering this field, for public officials involved in conservation, and for the public in general. Captive breeding programs were discussed in some of the chapters in the country-by-country assessments, but biobanking and the establishment of repositories for germplasm have not been covered. As an ever-increasing emerging approach, it is timely to publish a review of this important topic.

The 9th World Congress of Herpetology was convened at the University of Otago in New Zealand in January 2020, by the late Phil Bishop as the last of his many outstanding contributions to herpetology. It provided the opportunity for the fruition of the above approaches. The senior two editors of this volume, Aimee Silla and Andrew Kouba, organised a symposium for the Congress entitled 'Reproductive Biotechnologies and Biobanking for Amphibian Conservation'. Many of the innovators and practitioners of new methods for propagating amphibians in captivity, and successfully returning them to viability in nature, participated in that symposium, and exposed pitfalls to avoid and created guidelines to use for achieving the desired results. The papers that were presented, and the heuristic discussions that followed, constituted the seed that germinated and flowered into the present book. It is hoped that this modest beginning will stimulate further, and more effective, research by increasing the number of institutions involved in this endeavour to those already with active programs (see the Appendix at the end of this book) and stimulate collaboration of traditional and emerging technologies, for a better and more enduring conservation of earth's precious amphibian biota.

Harold Heatwole
Editor, *Amphibian Biology*
23 December 2021

Volumes in the series *Amphibian Biology*

Volume 1: The Integument, 1994. Surrey Beatty & Sons, Chipping-Norton, Australia

Volume 2: Social Behavior, 1995. Surrey Beatty & Sons, Chipping-Norton, Australia.

Volume 3: Sensory Perception, 1998. Surrey Beatty & Sons, Chipping-Norton, Australia.

Volume 4: Paleontology, The Evolutionary History of Amphibians, 2000. Surrey Beatty & Sons, Chipping-Norton, Australia.

Volume 5: Osteology, 2003. Surrey Beatty & Sons, Chipping-Norton, Australia.

Volume 6: Endocrinology, 2005. Surrey Beatty & Sons, Chipping-Norton, Australia.

Volume 7: Systematics, 2007. Surrey Beatty & Sons, Chipping-Norton, Australia.

Volume 8: Amphibian Decline: Diseases, Parasites, Maladies, and Pollution, 2009. Surrey Beatty & Sons, Baulkham Hills, Australia.

Volume 9: Status of Decline of Amphibians: Western Hemisphere

Part 1. Paraguay, Chile and Argentina, 2010. Surrey Beatty & Sons, Baulkham Hills, Australia.

Part 2. Uruguay, Brazil, Colombia and Ecuador, 2011. Surrey Beatty & Sons, Baulkham Hills, Australia.

Part 3. Venezuela, Guyana, Suriname, and French Guiana, 2013. Surrey Beatty & Sons, Baulkham Hills, Australia.

Part 4. Peru and Bolivia, 2014. Herpetological Monographs, Allen Press, Lawrence, Kansas, USA

Part 5. The Caribbean, *in press*. Pelagic Press, Exeter, UK.

Part 6. Canada and the United States, *in progress*.

Part 7. Central America and Mexico, *in progress*.

Volume 10: Conservation and Decline of Amphibians: Ecological Aspects, Effect of Humans, and Management, 2012. Surrey Beatty & Sons, Baulkham Hills, Australia.

Volume 11: Status of Conservation and Decline of Amphibians: Eastern Hemisphere

Part 1. Asia, 2014. Natural History Publications, Kota Kinabalu, Sabah.

Part 2. Mauritania, Morocco, Algeria, Tunisia, Libya, and Egypt, 2013. Basic and Applied Herpetology. Madrid

Part 3. Western Europe, 2013. Pelagic Publishing, Exeter, UK.

Part 4. Southeastern Europe and Turkey, 2015. Pelagic Publishing, Exeter, UK.

Part 5. Northern Europe, 2019. Pelagic Publishing, Exeter, UK.

Part 6. Australia, New Zealand, and the Pacific Islands, 2018. CSIRO Publishing, Melbourne, Australia.

Part 7. Status and Threats of Afrotropical Amphibians, 2021. Chimaira, Frankfurt am Main, Germany.

Part 8. Near East, *in progress*.

Volume 12: Reproductive Technologies and Biobanking for the Conservation of Amphibians, 2022. CSIRO Publishing, Melbourne, Australia.

1 Integrating reproductive technologies into the conservation toolbox for the recovery of amphibian species

Aimee J. Silla and Andy J. Kouba

INTRODUCTION

Globally, anthropogenic environmental change is threatening biodiversity and the functioning of natural ecosystems. In order to curb the unprecedented loss of biodiversity, integrated, multidisciplinary approaches will be needed to support the recovery of species. Traditionally, *in situ* and *ex situ* conservation actions have been viewed as separate (perhaps even mutually exclusive) goals and were rarely integrated. In order to preserve global biota from catastrophic demise, the successes of traditional approaches in conservation will need to be supplemented by cooperative integration of *in situ* and *ex situ* strategies to drive proactive wildlife management at all levels of activity (including planning, monitoring, implementation, and assessment). Within the many strategies available to conservationists, reproductive technologies are valuable for enhancing the reproductive output and genetic management of threatened species and offer several tools for linking *in situ* and *ex situ* strategies for conservation (Lueders and Allen 2020). The application of reproductive technologies has been viewed by some as a rival field competing for limited funding and resources. Furthermore, the biobanking of genetic material has raised concerns from conservation practitioners that this will provide governments with an excuse for complacency – that is, to offer them an excuse to allow species to become extinct with a plan to resurrect them in the future (de-extinction),

rather than addressing the immediate need to invest in recovering species.

Fortunately, these barriers have slowly been breaking down in recent years. The paradigm is shifting, and conservation practitioners are increasingly recognising the need for a greater assortment of tools in order to overcome the complex suite of challenges facing them (Corlett 2017). Reproductive technologies offer several promising approaches for species' recovery that can be used to complement, rather than replace, other strategies to create a more holistic and integrated toolbox for conserving amphibians (Figure 1.1). This chapter highlights the important benefits of incorporating reproductive technologies into the conservationist's toolbox, details the history of their development, and provides a framework for integrating these technologies into conservation breeding programs. Coupled with the detailed reviews of the current state of specific reproductive technologies in the chapters to follow, we provide a comprehensive guide to facilitate the accession of reproductive technologies into the toolbox for recovering threatened amphibian species. This concerted approach of coupling reproductive technologies with more traditional methods provides hope for the future and an incentive to strengthen efforts towards solution of the problem. The task is daunting, and will be difficult, but cooperative application of new technologies opens avenues towards previously unavailable success.

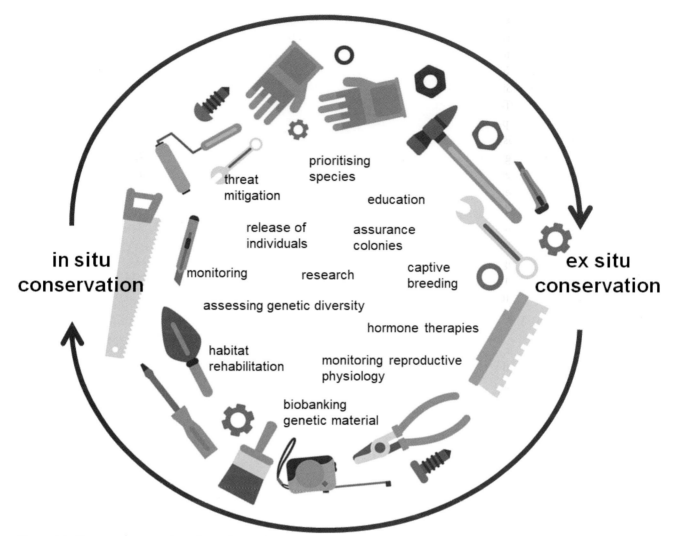

Figure 1.1: The conservationist's toolbox. The rapid worldwide decline in amphibian species illustrates the urgency to increase integrative conservation programs for the benefit of recovering threatened species.

REPRODUCTIVE TECHNOLOGIES: WHAT ARE THEY AND WHAT DO THEY HAVE TO OFFER THE CONSERVATIONIST'S TOOLBOX?

Reproductive technologies (also known as assisted reproductive technologies) encompass a multitude of techniques aimed at understanding and improving reproductive outcomes and enhancing the genetic management of threatened species. Broadly speaking, reproductive technologies include methods for identifying sex (Chapter 3); non-invasive monitoring of reproductive and stress physiology (Chapters 4 and 5); ultrasonography for monitoring reproductive status (Chapter 6); use of hormonal therapies to induce spawning (in male–female pairs), or ovulation (Chapter 6) and spermiation (Chapter 7); protocols for short-term storage, transport, activation, and assessment of sperm (Chapter 7); artificial fertilisation (also known as *in vitro* fertilisation; Chapter 8); and biobanking of amphibian genomes (Chapters 9–12). The potential benefits and application of specific reproductive technologies to amphibian conservation are outlined in the following chapters of this book, in addition to several review articles (Roth and Obringer 2003; Kouba and Vance 2009; Kouba *et al.* 2013; Clulow and Clulow 2016; Vu and Trudeau 2016; Bronson and Vance 2019; Browne *et al.* 2019; Silla and Byrne 2019; Della Togna *et al.* 2020; Strand *et al.* 2020; Silla *et al.* 2021).

Broadly, there are three overarching benefits of incorporating reproductive technologies into existing conservation breeding programs: (1) propagation, (2) genetic management, and (3) safeguarding genetic resources for future applications. First, reproductive technologies can be used to enhance propagation by overcoming reproductive failure and infertility, storing gametes, and synchronising reproductive cycles (Kouba *et al.* 2009; Clulow *et al.* 2014; Silla and Byrne 2019; Silla *et al.* 2021). While simply increasing the number of offspring produced is beneficial to conservation breeding programs (CBPs), especially those that are newly established and aiming to generate offspring for release, the long-term goals of CBPs are more complex and preserving genetic diversity is at the forefront of the objectives of management (Rollinson *et al.* 2014; Russello and Jensen 2018). The indubitable value of reproductive technologies (and the second overarching benefit) lies in the potential to improve the genetic management of threatened species, including the application of methods for assessing and avoiding either inbreeding or outbreeding depression (Chapter 8), in addition to methods facilitating genetic exchange between fragmented *in situ* and *ex situ* populations (Chapter 11). For example, hormonal therapies and techniques for artificial fertilisation can be used to enhance the reproductive output of genetically compatible or genetically valuable individuals (Silla and Byrne 2019). Alternatively, artificial-fertilisation technologies can facilitate quantitative genetic studies that test the genetic compatibility and combining ability of populations (either within *ex situ* populations, or between *ex situ* and *in situ* populations), or explore the potential for assisted gene flow to improve the adaptive potential of wild populations (Chapter 8). Recently, these exciting applications of reproductive technologies have been applied to model amphibian species categorised as Least Concern or Near Threatened, including quacking frogs, *Crinia georgiana* (Dziminski *et al.* 2008; Rudin-Bitterli *et al.* 2018), brown toadlets, *Pseudophryne bibronii* (Byrne *et al.* 2019), red-backed toadlets, *Pseudophryne coriacea* (Byrne *et al.* 2021), Günther's toadlets, *Pseudoprhyne guentheri* (Eads *et al.* 2012), European common frogs, *Rana temporaria* (Laurila *et al.* 2002; Pakkasmaa *et al.* 2003), and moor frogs, *Rana arvalis* (Merilä *et al.* 2004).

Reproductive technologies can also enhance genetic outcomes by facilitating genetic exchange between living and expired generations by conducting artificial fertilisations with freshly obtained eggs and frozen–thawed suspensions of sperm (Chapters 11 and 12). Artificial fertilisations employing frozen–thawed sperm have been achieved in a growing number of species, including the boreal toad, *Anaxyrus boreas boreas* (Langhorne 2016), cane toad, *Rhinella marina* (Proaño

and Pérez 2017), Chiricahua leopard frog, *Rana chirica-huensis* (Chapter 11), dusky gopher frog, *Lithobates sevosus* (Langhorne 2016), eastern dwarf treefrog, *Litoria fallax* (Upton *et al.* 2018), Fowler's toad, *Anaxyrus fowleri* (Poo and Hinkson 2019), green and golden bell frog, *Litoria aurea* (Upton *et al.* 2021), hellbender, *Cryptobranchus alleganiensis* (McGinnity *et al.* 2021), natterjack toad, *Epidalea calamita* (Arregui *et al.* 2020), northern leopard frog, *Lithobates pipiens* (Poo and Hinkson 2019), Puerto Rican crested toad, *Peltophryne lemur* (Chapter 11), Houston toad, *Anaxyrus houstonensis* (Chapter 11), and tiger salamander, *Ambystoma tigrinum* (Marcec 2016). Cryopreservation technologies can allow sperm cells to be stored decades into the future (Kouba *et al.* 2013; Clulow and Clulow 2016; Browne *et al.* 2019; Silla and Byrne 2019), which, through careful management and introgression, can be employed to reinvigorate genetic diversity.

The third overarching benefit of reproductive technologies lies in the ability to establish repositories that safeguard valuable genetic resources for future application, as science and technology progress beyond what is currently possible. For example, biobanking amphibian genetic material (Chapters 9 and 10) may provide future opportunities for reprogramming somatic cells into induced pluripotent stem cells (iPSCs) to derive viable gametes, or the use of techniques such as artificial fertilisation and the editing of genomes (e.g. CRISPR-Cas9 gene-editing approaches), which may allow the production of transgenic offspring that are immune to infectious diseases. Investing resources in the development of reproductive technologies and the biobanking of valuable genetic material now will avoid future regret for having missed opportunities. The banking of amphibian genomes of threatened species, as well as those species in decline that are not yet considered threatened, should be a key conservation strategy moving forward, because once a tipping point of endangerment has been reached, it may already be too late to conserve the populations' historical genetic diversity.

AN HISTORICAL PERSPECTIVE ON THE DEVELOPMENT OF AMPHIBIAN REPRODUCTIVE TECHNOLOGIES FOR CONSERVATION

As we begin, it is appropriate to reflect on and encapsulate some of the landmark contributions over the past

100 years that have advanced the field of reproductive technologies for the conservation of threatened amphibians. Some of the most significant achievements that occurred before 1950 include (1) the use of amphibians in testing for human pregnancy, in which urine from pregnant women could induce ovulation when injected into female frogs (Bellerby 1933; Shapiro and Zwarenstein 1934), or spermiation in male toads (Mainini 1947), and (2) seminal work by Rugh (1934), which described the induction of ovulation in females and amplexus behaviours by males following injections of homoplastic or heteroplastic pituitary extracts. Initial observations summarised by Rugh (1934) revealed the impacts of osmolality on rates of fertilisation and described artificial fertilisation for the first time for several ranid species. The third edition of Rugh's *Experimental Embryology*, first published in the early 1940s, is a must read for anyone entering the field of reproductive technologies (Rugh 1962). One of the most noteworthy accomplishments towards advancing reproductive technologies in amphibians was the work by Gurdon (1962) who demonstrated the developmental capacity of nuclei taken from intestinal epithelial cells of *Xenopus laevis* tadpoles, which were transplanted into enucleated eggs to yield normally feeding tadpoles. Thus, the first example of a cloned species occurred in a frog and was based on the technique of nuclear transplantation originally described by Briggs and King (1952); Gurdon jointly shared the Nobel prize in physiology and medicine in 2012 for this groundbreaking work based on reproductive technologies in amphibians.

Perhaps the most consequential and often cited scientist working in the area of amphibian endocrinology is Paul Licht, to whom this book is dedicated. Licht and colleagues were the first to isolate and characterise amphibian gonadotropic hormones (Papkoff *et al.* 1976), study the role of amphibian gonadotropins on ovulation (Licht *et al.* 1975; Licht and Crews 1976), discover the role gonadotropin-releasing hormone (GnRH) has on regulation of gonadotropin secretion (Licht and Porter 1987), reveal the feedback relationship between gonadotropin secretion and sex steroids (Licht 1979), and evaluate the potency of various isolated preparations of luteinizing hormone (Licht and Papkopff 1976). Licht's contributions to understanding reproductive endocrinology have been instrumental in the application of these hormones for the collection of gametes from various species of amphibians (see Chapters 6 and 7) and other taxa. Seminal papers including Licht's articles on sperm-release in frogs by mammalian gonadotropin-releasing hormone are a must for beginners in this field (Licht 1973, 1974). Conceivably, some of the most influential early works that have impacted the development of amphibian reproductive technologies were several books related to anatomy, physiology, and reproductive ecology of amphibians including Lofts's (1974) chapter on reproduction in *Physiology of the Amphibia Vol. 2*, Duellman's and Trueb's (1994) *Biology of Amphibians*, Norris and Lopez' (2011) book on *Hormones and Reproduction of Vertebrates: Amphibians*, and Wright and Whitaker's (2001) *Amphibian Medicine and Captive Husbandry*. Whitaker's chapter on reproduction was one of the first summaries delineating all the different hormones being used to breed amphibians in conservation breeding programs. These books, and their chapters on reproduction, are still relevant as efforts are made to transfer protocols to novel amphibian species with different reproductive modes.

Early investigations on establishing protocols for artificial fertilisation, which built on the initial work by Rugh (1962), were pioneered by several investigators including Wolf and Hedrick (1971), Cabada (1975), and Hollinger and Corton (1980). All of these researchers established the artificial fertilisation protocols still used today for aquatically breeding species, such as the flooding of dishes with water following a 5-min dry incubation period after combining gametes. Of particular note, studies showing the impact of osmolality on activation and motility of sperm became key to the understanding of the fertilisation processes for many temperate anurans (Hardy and Dent 1986; Inoda and Morisawa 1987; Bernardini *et al.* 1988). The application of reproductive technologies for breeding threatened species first started with the publication of several important papers by Goncharov *et al.* (1989), Waggener and Carroll (1998), Clulow *et al.* (1999), and Obringer *et al.* (2000). Once the applied science of amphibian reproductive technologies began for species' conservation at the end of the 20th century, the number of biologists entering this field grew substantially such that one would be hard pressed to summarise all of the publications or conservation milestones. However, the following chapters provide an excellent background on how the science of amphibian reproductive technologies has evolved and advanced over the past 20 years, allowing real progress to be made in order to ensure that some portion of amphibian biodiversity persists into the future.

HOW TO APPROACH THE INTEGRATION OF REPRODUCTIVE TECHNOLOGIES INTO THE CONSERVATIONIST'S TOOLBOX

The first step towards integrating reproductive technologies into the conservationist's toolbox is to dispel any misconceptions that reproductive technologies will replace other conservation strategies. The development and inclusion of reproductive technologies should always be approached as a complementary strategy rather than an alternative to *in situ*, or other *ex situ*, activities. *Ex situ* conservation breeding programs may serve several purposes in the fight against extinction, including the rescue and maintenance of genetically representative assurance colonies, providing individuals for research (including the development of reproductive technologies), breeding or head-starting offspring for release and reestablishment *in situ*, fulfilling roles as educators and ambassadors for conservation, and fundraising to provide additional financial support for conservation (McFadden *et al.* 2018). *In situ* conservation activities such as conserving habitats, establishing protected areas, the mitigation of threats, monitoring populations, and surveying for additional remnant populations should also occur simultaneously. The conservation of threatened species is a complex and challenging problem that requires a multifaceted and integrated approach to achieve conservation goals, and there are several exemplary amphibian conservation breeding programs setting this new standard (see McFadden *et al.* 2018).

The second step is to identify species to target for the development and application of reproductive technologies. Where *ex situ* conservation is yet to be established, species should first be carefully assessed and prioritised for *ex situ* management (Chapter 2). Once species are brought into captivity (or, if possible, before their collection), individuals will need to be sexed and aspects of captive husbandry carefully considered in order to condition animals for both natural and assisted reproduction (Chapter 3). Prior to the development and application of reproductive technologies, a thorough understanding of the species' reproductive neuroendocrinology is essential (Chapter 4). Non-invasive monitoring of reproductive (Chapter 4) and stress (Chapter 5) physiology are invaluable tools to assess the impacts of particular social and environmental conditions on reproduction and animal welfare. This will facilitate the continued refinement and optimisation of captive husbandry. Ideally, conservation practitioners will have an understanding of the species'

reproductive biology and the specific environmental stimuli required to initiate the cascade of hormonal changes required to achieve reproduction; however, this is not often the case and there have been widespread reproductive failures for many captive colonies. If species-specific requirements are undetermined, replicating and/or cycling environmental conditions reflective of the species' natural environment, or replicating conditions known for a closely related species with a shared breeding biology, may be a good starting point (Pramuk and Gagliardo 2008; Gupta *et al.* 2015). It must be stressed that while reproductive technologies can be used to manipulate reproductive physiology and circumvent the impediments to reproduction that stem from a lack of appropriate stimuli, the use of reproductive technologies does not negate or replace the need for adequate husbandry that provides for different aspects of their life history or that addresses animal welfare (Silla *et al.* 2021). In fact, provision of optimal naturalistic environmental conditions (including climatic cycling, photoperiod, adequate nutrition, and social cues) may improve the efficiency of reproductive technologies and, in turn, enhance the success of amphibian conservation breeding programs (Silla *et al.* 2021).

The development and application of hormonal therapies (Chapters 6 and 7), techniques for artificial fertilisation (Chapter 8), cryopreservation of haploid and diploid genomes (Chapter 9), and establishment and biobanking of cell lines (Chapter 10) should be developed using hypothesis-driven models to the extent possible when working with a limited number of animals for a threatened species. A systematic approach to the refinement of protocols is recommended, using the current knowledge provided in the literature and subsequent chapters of this book as a guide for species-specific refinement and application. Large-scale comparative research, encompassing a diversity of species and reproductive modes, is also recommended in order to investigate patterns in optimal protocols and expedite the application of reproductive technologies to novel threatened species (Silla and Byrne 2019). Once protocols have been established for the application of hormonal therapies, artificial fertilisation, and for the cryopreservation of sperm, these technologies can then be employed to improve the genetic management of threatened species, including use of these tools for the assessment of the genetic consequences of population-mixing (Chapter 8), and for facilitation of genetic exchange between *in situ* and *ex situ* populations

(Chapter 11). Moreover, various 'proofs of concept' for the value of genome resource banks for genetic management have been demonstrated. For example, a threatened Puerto Rican crested toad, *Peltophryne lemur* (Chapter 12), has been produced from frozen–thawed sperm originating from a deceased parent and upon maturity the offspring bred, and progeny were released *in situ*. The development of these reproductive technologies used on *P. lemur* were dependent upon decades of hypothesis-driven research on a common model species, the Fowler's toad, *Anaxyrus fowleri*. Common model species, which can support the large-scale comparative research mentioned above, are often the driving force behind the incremental steps achieved for a wide selection of globally threatened species.

Reproductive technologies have already been applied to a diversity of threatened amphibians globally, including, but not limited to, the black-spotted newt, *Notophthalmus meridionalis* (Guy *et al.* 2020), booroolong frog, *Litoria booroolongensis* (Silla *et al.* 2017; Silla *et al.* 2019), boreal toad, *Anaxyrus boreas boreas* (Calatayud *et al.* 2015; Langhorne *et al.* 2021), dusky gopher frog, *Lithobates sevosus* (Langhorne 2016; Poo and Hinkson 2019), Fijian ground frog, *Platymantis vitiana* (Narayan *et al.* 2010), great crested newt, *Triturus cristatus* (Strand *et al.* 2021), green and golden bell frog, *Litoria aurea* (Upton *et al.* 2021), Hamilton's frog, *Leiopelma hamiltoni* (formally *L. pakeka*; Germano *et al.* 2012), hellbender, *Cryptobranchus alleganiensis* (Unger 2013; McGinnity *et al.* 2021), Mexican axolotl, *Ambystoma mexicanum* (Mansour *et al.* 2011), northern corroboree frog, *Pseudophryne pengilleyi* (Silla *et al.* 2018), Panamanian golden frog, *Atelopus zeteki* (Della Togna *et al.* 2017, 2018), Puerto Rican crested toad, *Peltophryne lemur* (Langhorne 2016), southern corroboree frog, *Pseudophryne corroboree* (Byrne and Silla 2010), and Wyoming toad, *Anaxyrus baxteri* (Browne *et al.* 2006; Poo *et al.* 2019) (Plate 1). The widespread need for amphibian reproductive technologies on a global scale has never been more urgent, with more conservation practitioners now embracing an integrated approach including the application of amphibian reproductive technologies as important tools for conservation. The premise for this book is to encourage an expansion of work in this field by providing a compendium of knowledge on amphibian reproductive technologies that will be useful for the next generation of scientists working to conserve the biodiversity of amphibians. It is hoped that this book will provide an impetus for the expansion of research and development of reproductive technologies and a stimulus for the increased application and integration of these valuable technologies into a greater number of conservation breeding programs into the future.

ACKNOWLEDGEMENTS

Aimee Silla: I acknowledge the traditional custodians of the land on which the research cited in this book was conducted and I pay respect to Elders both past and present. I am grateful for the support and input of Phillip Byrne. The support of the following funding grants is also acknowledged, of which Aimee Silla was in receipt during the preparation of this manuscript: Australian Research Council, Linkage Projects (LP140100808 & LP170100351), Discovery Early Career Researcher Award (DE210100812), and Zoo and Aquarium Association's Wildlife Conservation Fund.

Andy Kouba: I would like to thank the support and input of collaborator Carrie Kouba. Moreover, the following funding grants are also acknowledged, of which Andy Kouba was in receipt during the preparation of this manuscript: the Institute of Museum and Library Services (IMLS) National Leadership Grant (#MG-30–17–0052–17) for sustainability of threatened amphibian collections; an Association of Zoos and Aquariums and Disney Conservation Grant (#CGF-19–1618); the Mississippi Agricultural and Forestry Experiment Station USDA-ARS 58–6402–018; and the material is based upon work supported by the National Institute of Food and Agriculture, USA Department of Agriculture, Hatch project under accession number W3173.

REFERENCES

Arregui L, Bóveda P, Gosálvez J, Kouba AJ (2020) Effect of seasonality on hormonally induced sperm in Epidalea calamita (Amphibia, Anura, Bufonidae) and its refrigerated and cryopreservated storage. *Aquaculture* **529**, 735677. doi:10.1016/j.aquaculture.2020.735677

Bellerby CW (1933) The endocrine factors concerned in the control of the ovarian cycle: *Rana temporaria* as test animal. III. The action of anterior lobe pituitary extracts on the ovary. *The Biochemical Journal* **27**, 2022–2030. doi:10.1042/bj0272022

Bernardini G, Andrietti F, Camatini M, Cosson MP (1988) *Xenopus* spermatozoon: correlation between shape and motility. *Gamete Research* **20**, 165–175. doi:10.1002/mrd.1120200207

Briggs R, King TJ (1952) Transplantation of living nuclei from blastula cells into enucleated frogs' eggs. *Proceedings of the National Academy of Sciences of the United States of America* **38**, 455–463. doi:10.1073/pnas.38.5.455

Bronson E, Vance CK (2019) Anuran reproduction. In *Fowler's Zoo and Wild Animal Medicine Current Therapy*. Volume 9. (Eds RE Miller, N Lamberski and P Calle) pp. 371–379. Elsevier, St Louis MO, USA. doi.org/10.1016/B978-0-323-55228-8.00053-9

Browne R, Seratt J, Vance C, Kouba A (2006) Hormonal priming, induction of ovulation and in-vitro fertilization of the endangered Wyoming toad (*Bufo baxteri*). *Reproductive Biology and Endocrinology* **4**, 34. doi:10.1186/1477-7827-4-34

Browne RK, Silla AJ, Upton R, Della-Togna G, Marcec-Greaves R, Shishova NV, Uteshev VK, Proaño B, Pérez OD, Mansour N (2019) Sperm collection and storage for the sustainable management of amphibian biodiversity. *Theriogenology* **133**, 187–200. doi:10.1016/j.theriogenology.2019.03.035

Byrne PG, Silla AJ (2010) Hormonal induction of gamete release, and *in-vitro* fertilisation, in the critically endangered Southern Corroboree Frog, *Pseudophryne corroboree*. *Reproductive Biology and Endocrinology* **8**, 144. doi:10.1186/1477-7827-8-144

Byrne PG, Gaitan Espitia JD, Silla AJ (2019) Genetic benefits of extreme sequential polyandry in a terrestrial-breeding frog. *Evolution* **73**, 1972–1985. doi:10.1111/evo.13823

Byrne PG, Keogh SJ, O'brien DM, Gaitan Espitia JD, Silla AJ (2021) Evidence that genetic compatability underpins female mate choice in a monandrous amphibian. *Evolution* **75**, 529–541. doi:10.1111/evo.14160

Cabada MO (1975) Sperm concentration and fertilization rate in *Bufo arenarum* (Amphibia: Anura). *The Journal of Experimental Biology* **62**, 481–486. doi:10.1242/jeb.62.2.481

Calatayud NE, Langhorne CJ, Mullen AC, Williams CL, Smith T, Bullock L, Kouba AJ, Willard ST (2015) A hormone priming regimen and hibernation affect oviposition in the boreal toad (*Anaxyrus boreas boreas*). *Theriogenology* **84**, 600–607. doi:10.1016/j.theriogenology.2015.04.017

Clulow J, Clulow S (2016) Cryopreservation and other assisted reproductive technologies for the conservation of threatened amphibians and reptiles: bringing the ARTs up to speed. *Reproduction, Fertility and Development* **28**, 1116–1132. doi:10.1071/RD15466

Clulow J, Mahony M, Browne R, Pomering M, Clark A (1999) Applications of assisted reproductive technologies (ART) to endangered anuran amphibians. In *Declines and Disappearances of Australian Frogs*. (Ed. A Campbell) pp. 219–225 Environment Australia, Canberra.

Clulow J, Trudeau VL, Kouba AJ (2014) Amphibian declines in the twenty-first century: why we need assisted reproductive technologies. In *Reproductive Sciences in Animal Conservation*. (Eds WV Holt, JL Brown and P Comizzoli) pp. 275–316. Springer, New York NY, USA.

Corlett RT (2017) A bigger toolbox: biotechnology in biodiversity conservation. *Trends in Biotechnology* **35**, 55–65. doi:10.1016/j.tibtech.2016.06.009

Della Togna G, Trudeau VL, Gratwicke B, Evans M, Augustine L, Chia H, Bronikowski EJ, Murphy JB, Comizzoli P (2017) Effects of hormonal stimulation on the concentration and quality of excreted spermatozoa in the critically endangered Panamanian golden frog (*Atelopus zeteki*). *Theriogenology* **91**, 27–35. doi:10.1016/j.theriogenology.2016.12.033

Della Togna G, Gratwicke B, Evans M, Augustine L, Chia H, Bronikowski E, Murphy JB, Comizzzoli P (2018) Influence of extracellular environment on motility and structural properties of spermatozoa collected from hormonally stimulated Panamanian Golden Frog (*Atelopus zeteki*). *Theriogenology* **108**, 153–160. doi:10.1016/j.theriogenology.2017.11.032

Della Togna G, Howell LG, Clulow JC, Langhorne CJ, Marcec-Greaves R, Calatayud NE (2020) Evaluating amphibian biobanking and reproduction for captive breeding programs according to the Amphibian Conservation Action Plan objectives. *Theriogenology* **150**, 412–431. doi:10.1016/j.theriogenology.2020.02.024

Duellman WE, Trueb L (1994) *Biology of Amphibians*. Johns Hopkins University Press, Baltimore MD, USA.

Dziminski MA, Roberts JD, Simmons LW (2008) Fitness consequences of parental compatibility in the frog *Crinia georgiana*. *Evolution* **62**, 879–886. doi:10.1111/j.1558-5646.2008.00328.x

Eads AR, Mitchell NJ, Evans JP (2012) Patterns of genetic variation in desiccation tolerance in embryos of the terrestrial-breeding frog, *Pseudophryne guentheri*. *Evolution* **66**, 2865–2877. doi:10.1111/j.1558-5646.2012.01616.x

Germano JM, Molinia FC, Bishop PJ, Bell BD, Cree A (2012) Urinary hormone metabolites identify sex and imply unexpected winter breeding in an endangered, subterranean-nesting frog. *General and Comparative Endocrinology* **175**, 464–472. doi:10.1016/j.ygcen.2011.12.003

Goncharov BF, Shubravy OI, Serbinova IA, Uteshev VK (1989) The USSR programme for breeding amphibians, including rare and endangered species. *International Zoo Yearbook* **28**, 10–21. doi:10.1111/j.1748-1090.1989.tb03248.x

Gupta BK, Tapley B, Vasudevan K, Goetz M (2015) Ex situ *Management of Amphibians*. Assam State Zoo Cum Botanical Garden, Guwahati, Assam, India.

Gurdon JB (1962) Adult frogs derived from the nuclei of single somatic cells. *Developmental Biology* **4**, 256–273. doi:10.1016/0012-1606(62)90043-X

Guy EL, Gillis AB, Kouba AJ, Barber D, Poole V, Marcec-Greaves RM, Kouba CK (2020) Sperm collection and cryopreservation for threatened newt species. *Cryobiology* **94**, 80–88. doi:10.1016/j.cryobiol.2020.04.005

Hardy MP, Dent JN (1986) Regulation of motility in sperm of the red-spotted newt. *The Journal of Experimental Zoology* **240**, 385–396. doi:10.1002/jez.1402400313

Hollinger TG, Corton GL (1980) Artificial fertilization of gametes from the South African clawed frog, Xenopus laevis. *Gamete Research* **3**, 45–57. doi:10.1002/mrd.1120030106

Inoda T, Morisawa M (1987) Effect of osmolality on the initiation of sperm motility in *Xenopus laevis*. *Comparative Biochemistry and Physiology. A. Comparative Physiology* **88**, 539–542. doi:10.1016/0300-9629(87)90077-6

Kouba AJ, Vance CK (2009) Applied reproductive technologies and genetic resource banking for amphibian conservation. *Reproduction, Fertility and Development* **21**, 719–737. doi:10.1071/RD09038

Kouba A, Vance C, Willis E (2009) Artificial fertilization for amphibian conservation: current knowledge and future considerations. *Theriogenology* **71**, 214–227. doi:10.1016/j.theriogenology.2008.09.055

Kouba AJ, Lloyd RE, Houck ML, Silla AJ, Calatayud N, Trudeau VL, Clulow J, Molinia F, Langhorne C, Vance C (2013) Emerging trends for biobanking amphibian genetic resources: the hope, reality and challenges for the next decade. *Biological Conservation* **164**, 10–21. doi:10.1016/j.biocon.2013.03.010

Langhorne CJ (2016) Developing assisted reproductive technologies for endangered North American amphibians. PhD thesis. Mississippi State University, USA.

Langhorne CJ, Calatayud NE, Kouba CK, Willard ST, Smith T, Ryan PL, Kouba AJ (2021) Efficacy of hormone stimulation on sperm production in an alpine amphibian (*Anaxyrus boreas boreas*) and the impact of short-term storage on sperm quality. *Zoology (Jena, Germany)* **146**, 125912. doi:10.1016/j.zool.2021.125912

Laurila A, Karttunen S, Merilä J (2002) Adaptive phenotypic plasticity and genetics of larval life histories in two Rana temporaria populations. *Evolution* **56**, 617–627. doi:10.1111/j.0014-3820.2002.tb01371.x

Licht P (1973) Induction of spermiation in anurans by mammalian pituitary gonadotropins and their subunits. *General and Comparative Endocrinology* **20**, 522–529. doi:10.1016/0016-6480(73)90083-X

Licht P (1974) Induction of sperm release in frogs by mammalian gonadotropin-releasing hormone. *General and Comparative Endocrinology* **23**, 352–354. doi:10.1016/0016-6480(74)90079-3

Licht P (1979) Reproductive endocrinology of reptiles and amphibians: gonadotropins. *Annual Review of Physiology* **41**, 337–351. doi:10.1146/annurev.ph.41.030179.002005

Licht P, Crews D (1976) Gonadotropin stimulation of in vitro progesterone production in reptilian and amphibian ovaries. *General and Comparative Endocrinology* **29**, 141–151. doi:10.1016/0016-6480(76)90015-0

Licht P, Papkoff H (1976) Species specificity in the response of an in vitro amphibian (*Xenopus laevis*) ovulation assay to mammalian luteinizing hormones. *General and Comparative Endocrinology* **29**, 552–555. doi:10.1016/0016-6480(76)90039-3

Licht P, Porter DA (1987) Role of gonadotropin-releasing hormone in regulation of gonadotropin secretion from amphibian and reptilian pituitaries. In *Hormones and Reproduction in Fishes, Amphibians, and Reptiles*. (Eds DO Norris and RE Jones). pp. 61–85X. Springer, Boston MA, USA.

Licht P, Farmer SW, Papkoff H (1975) The nature of the pituitary gonadotropins and their role in ovulation in a urodele amphibian (*Ambystoma tigrinum*). *Life Sciences* **17**, 1049–1054. doi:10.1016/0024-3205(75)90323-9

Lofts B (1974) Reproduction. In *Physiology of the Amphibia*. Volume 2. (Ed. B Lofts) pp. 107–218. Academic Press, New York NY, USA and London, UK.

Lueders I, Allen WT (2020) Managed wildlife breeding – an undervalued conservation tool? *Theriogenology* **150**, 48–54. doi:10.1016/j.theriogenology.2020.01.058

Mainini CG (1947) Pregnancy test using the male toad. *The Journal of Clinical Endocrinology* **7**, 653–658. doi:10.1210/jcem-7-9-653

Mansour N, Lahnsteiner F, Patzner RA (2011) Collection of gametes from live axolotl, *Ambystoma mexicanum*, and standardization of in vitro fertilization. *Theriogenology* **75**, 354–361. doi:10.1016/j.theriogenology.2010.09.006

Marcec R (2016) Development of assisted reproductive technologies for endangered North American salamanders. PhD thesis. Mississippi State University, USA.

McFadden MS, Gilbert D, Bradfield K, Evans M, Marantelli G, Byrne PG (2018) The role of *ex situ* amphibian conservation in Australia. In *Status of Conservation and Decline of Amphibians: Australia, New Zealand, and Pacific Islands*. (Eds H Heatwole and J Rowley) pp. 125–140. CSIRO Publishing, Melbourne.

McGinnity D, Reinsch S, Schwartz H, Trudeau V, Browne R (2021) Semen and oocyte collection, sperm cryopreservation, and in vitro fertilisation with threatened North American giant salamanders (*Cryptobranchus alleganiensis*). *Reproduction, Fertility and Development*. doi:10.1071/RD21035

Merilä J, Söderman F, O'hara R, Räsänen K, Laurila A (2004) Local adaptation and genetics of acid-stress tolerance in the moor frog, *Rana arvalis*. *Conservation Genetics* **5**, 513–527. doi:10.1023/B:COGE.0000041026.71104.0a

Narayan EJ, Molinia FC, Christi KS, Morley CG, Cockrem JF (2010) Annual cycles of urinary reproductive steriod concentrations in wild and captive endangered Fijian ground frogs (*Platymantis vitiana*). *General and Comparative Endocrinology* **166**, 172–179. doi:10.1016/j.ygcen.2009.10.003

Norris DO, Lopez KHRE (Eds) (2011) *Hormones and Reproduction of Vertebrates: Amphibians*. Academic Press, London, UK.

Obringer AR, O'Brien JK, Saunders RL, Yamamoto K, Kikuyama S, Roth TL (2000) Characterization of the spermiation response, luteinizing hormone release and sperm quality in the American toad (*Bufo americanus*) and the endangered Wyoming toad (*Bufo baxteri*). *Reproduction, Fertility and Development* **12**, 51–58. doi:10.1071/RD00056

Pakkasmaa S, Merilä J, O'hara R (2003) Genetic and maternal effect influences on viability of common frog tadpoles under different environmental conditions. *Heredity* **91**, 117–124. doi:10.1038/sj.hdy.6800289

Papkoff H, Farmer SW, Licht P (1976) Isolation and characterization of luteinizing hormone from amphibian (*Rana catesbeiana*) pituitaries. *Life Sciences* **18**, 245–250. doi:10.1016/0024-3205(76)90031-X

Poo S, Hinkson KM (2019) Applying cryopreservation to anuran conservation biology. *Conservation Science and Practice* **1**, e91. doi:10.1111/csp2.91

Poo S, Hinkson KM, Stege E (2019) Sperm output and body condition are maintained independently of hibernation in an endangered temperate amphibian. *Reproduction, Fertility and Development* **31**, 796–804. doi:10.1071/RD18073

Pramuk JB, Gagliardo R (2008) General amphibian husbandry. In *Amphibian Husbandry Resource Guide*. 2nd edn. (Eds VA Poole and S Grow) pp. 4–59, Association of Zoos and Aquariums, Silver Spring MD, USA, <https://saveamphibians.org/wp-content/uploads/2015/09/AmphibianHusbandryResourceGuide2012.pdf>.

Proaño B, Pérez OD (2017) In vitro fertilizations with cryopreserved sperm of *Rhinella marina* (Anura: Bufonidae) in Ecuador. *Amphibian & Reptile Conservation* **11**, 1–6.

Rollinson N, Kieth DM, Houde AL, Debes S, Mcbride MC, Hutchings JA (2014) Risk assessment of inbreeding and outbreeding depression in a captive-breeding program. *Conservation Biology* **28**, 529–540. doi:10.1111/cobi.12188

Roth TL, Obringer AR (2003) Reproductive research and the worldwide amphibian extinction crisis. In *Reproductive Science and Integrated Conservation*. (Eds WV Holt, AR Pickard, JC Rodger and DE Wildt). Cambridge University Press, Cambridge, UK.

Rudin-Bitterli TS, Mitchell NJ, Evans JP (2018) Environmental stress increases the magnitude of nonadditive genetic variation in offspring fitness in the frog *Crinia georgiana*. *American Naturalist* **192**, 461–478. doi:10.1086/699231

Rugh R (1934) Induced ovulation and artificial fertilization in the frog. *The Biological Bulletin* **66**, 22–29. doi:10.2307/1537458

Rugh R (1962) *Experimental Embryology: Techniques and Procedures*. 3rd edn. Burgess Publishing Company, Minneapolis MN, USA.

Russello MA, Jensen EL (2018) *Ex situ* wildlife conservation in the age of population genomics. In *Population Genomics: Wildlife*. (Eds PA Hohenlohe and PO Rajora) pp. 473–492. Springer, Cham, Switzerland.

Shapiro HA, Zwarenstein H (1934) A rapid test for pregnancy on *Xenopus laevis*. *Nature* **133**, 762. doi:10.1038/133762a0

Silla AJ, Byrne PG (2019) The role of reproductive technologies in amphibian conservation breeding programs. *Annual Review of Animal Biosciences* **7**, 499–519. doi:10.1146/annurev-animal-020518-115056

Silla AJ, Keogh LM, Byrne PG (2017) Sperm motility activation in the critically endangered booroolong frog: the effect of medium osmolality and phosphodiesterase inhibitors. *Reproduction, Fertility and Development* **29**, 2277–2283. doi:10.1071/RD17012

Silla AJ, McFadden M, Byrne PG (2018) Hormone-induced spawning of the critically endangered northern corroboree frog *Pseudophryne pengilleyi*. *Reproduction, Fertility and Development* **30**, 1352–1358. doi:10.1071/RD18011

Silla AJ, McFadden MS, Byrne PG (2019) Hormone-induced sperm-release in the critically endangered Booroolong frog (*Litoria booroolongensis*): effects of gonadotropin-releasing hormone and human chorionic gonadotropin. *Conservation Physiology* **7**, coy080. doi:10.1093/conphys/coy080

Silla AJ, Calatayud NE, Trudeau VL (2021) Amphibian reproductive technologies: approaches and welfare considerations. *Conservation Physiology* **9**, coab011. doi:10.1093/conphys/coab011

Strand J, Thomsen H, Jensen JB, Marcussen C, Nicolajsen TB, Skriver MB, Sogaard IM, Ezaz T, Purup S, Callesen H, Pertoldi C (2020) Biobanking in amphibian and reptilian conservation and management: opportunities and challenges. *Conservation Genetics Resources* **12**, 709–725. doi:10.1007/s12686-020-01142-y

Strand J, Callesen H, Pertoldi C, Purup S (2021) Establishing cell lines from fresh or cryopreserved tissue from the great crested newt (*Triturus cristatus*): a preliminary protocol. *Animals (Basel)* **11**, 367. doi:10.3390/ani11020367

Unger S (2013) A comparison of sperm health in declining and stable populations of hellbenders (*Cryptobranchus alleganiensis alleganiensis* and *Ca bishopi*). *American Midland Naturalist* **170**, 382–392. doi:10.1674/0003-0031-170.2.382

Upton R, Clulow S, Mahony MJ, Clulow J (2018) Generation of a sexually mature individual of the Eastern dwarf tree frog, *Litoria fallax*, from cryopreserved testicular macerates: proof of capacity of cryopreserved sperm derived offspring to complete development. *Conservation Physiology* **6**, coy043. doi:10.1093/conphys/coy043

Upton R, Clulow S, Calatayud NE, Colyvas K, Seeto RG, Wong LA, Mahony MJ, Clulow J (2021) Generation of reproductively mature offspring from the endangered green and golden bell frog *Litoria aurea* using cryopreserved spermatozoa. *Reproduction, Fertility and Development* **33**, 562–572. doi:10.1071/RD20296

Vu M, Trudeau VL (2016) Neuroendocrine control of spawning in amphibians and its practical applications. *General and Comparative Endocrinology* **234**, 28–39. doi:10.1016/j.ygcen.2016.03.024

Waggener WL, Carroll EJ, Jr (1998) A method for hormonal induction of sperm release in anurans (eight species) and in vitro fertilization in *Lepidobatrachus* species. *Development, Growth & Differentiation* **40**, 19–25. doi:10.1046/j.1440-169X.1998.t01-5-00003.x

Wolf DP, Hedrick JL (1971) A molecular approach to fertilization: II. Viability and artificial fertilization of *Xenopus laevis* gametes. *Developmental Biology* **25**, 348–359. doi:10.1016/0012-1606(71)90036-4

Wright KM, Whitaker BR (Eds) (2001) *Amphibian Medicine and Captive Husbandry*. Krieger Publishing Company, Malibar FL, USA.

2 Status of global amphibian declines and the prioritisation of species for captive breeding

Kevin Johnson and Joseph R. Mendelson III

INTRODUCTION

The dramatic decline of amphibian populations has been well considered but, unfortunately, not entirely well documented since the mid-1980s, although significant efforts have been made over the past 10–15 years or so to increase knowledge of the status of wild populations. In response to these declines and extinctions, the amphibian conservation community has implemented a range of both *in situ* and *ex situ* conservation actions aimed at reducing the threats to the species in the wild, as well as taking action to protect habitats and to establish captive survival-assurance colonies for species that face imminent risk of extinction. While there is considerable expertise with amphibian-husbandry practices within the *ex situ* conservation community, there often exists a lack of knowledge of the status of wild populations, and the threats they face. The Conservation Needs Assessment was initially developed in 2006, as a decision tree to help the *ex situ* conservation community select and prioritise which taxa are most in need of *ex situ* assistance. In this chapter, we summarise the current status of amphibian declines and outline how Conservation Needs Assessments are a valuable tool to help develop and prioritise a range of conservation actions.

STATUS OF GLOBAL AMPHIBIAN DECLINES

Global amphibian declines and extinctions have been described in detail (e.g. Collins and Crump 2009;

Mendelson and Donnelly 2011; Heatwole 2013; Heatwole and Rowley 2018), international conservation-action plans have been drafted (Wren *et al.* 2015), scientific breakthroughs have emerged (e.g. Longcore *et al.* 1999; Lips *et al.* 2006; James *et al.* 2009; Rosenblum *et al.* 2013; O'Hanlon *et al.* 2018), large-scale intensive conservation programs have been launched (e.g. OEH NSW 2012; Tapley *et al.* 2014; Nahonyo *et al.* 2017), retrospectives have been published (e.g. Lips 2018; Fisher and Garner 2020; Green *et al.* 2020), and the future for amphibians in a changing world has been pondered (e.g. Lips 2013; Lips and Mendelson 2014; Mendelson 2018). Amphibian conservation has received a considerable amount of popular attention and media coverage (Pavajeau *et al.* 2008) and significant, if perpetually insufficient, funding for organisations such as Amphibian Ark (www.amphibianark. org) and the Amphibian Survival Alliance (www. amphibians.org) have been secured and applied towards conservation. Yet we cannot say with any broad level of generality that amphibians are doing better than they were when the amphibian crisis was first articulated in 1989 (Rabb 1990; Vial 1991; Wake 1991). Over 30 years of increased attention to the conservation of amphibians have resulted in significant discoveries too numerous to list here, dire setbacks have occurred including the discovery of additional emerging infectious diseases (Martel *et al.* 2013), and the public is more aware of amphibians in general. The future of amphibians, however, remains uncertain. Nevertheless, some populations have shown unexpected resilience and evidently are recovering, sometimes after having been officially declared to be extinct (e.g. Abarca *et al.* 2010; Newell *et al.* 2013). In the cases involving amphibian chytridiomycosis, such recoveries appear to be remarkable examples of the evolution of tolerance of an invasive pathogen via natural selection. Although such evolution has not been demonstrated specifically, the apparent recoveries certainly represent good news in an otherwise rather bleak landscape of realities. A recent example involved the discovery of an extant population of a micro-endemic, long-missing Mexican fringe-limbed tree frog (*Ecnomiohyla valancifer*) that has already been bred in a captive program in Mexico (V. H. Jiménez-Arcos, *pers. comm.*). Breeding in captivity by species of the genus *Ecnomiohyla*, however, has eluded success in most such programs (e.g. *Ecnomiohyla rabborum*; Mendelson 2011).

Perhaps the most significant publication in the history of global amphibian declines was Stuart *et al.* (2004) which, when combined with the IUCN Red List, presented the results of an impressive campaign to assess the status of the world's amphibians. The Red List data had just been completely updated, so for the first time we had a very good assessment of the status of amphibians in a precise moment of time. The results were shocking: amphibians were far more threatened than either birds or mammals, with 1856 species (32.5%) being globally threatened, and 2468 species (43.2%) experiencing declines in their populations. Many amphibian species were on the brink of extinction. The results from these assessments of threat played a significant role in formalising a global conservation response (Gascon *et al.* 2007) and in mobilising researchers, educators, and conservationists. An interesting historical twist to this story is the fact that, as a result of the timing of the Red List assessments summarised by Stuart *et al.* (2004), the cause of decline of a great many species (*n* = 207) was attributed to 'enigmatic declines'. Amphibian chytridiomycosis was very poorly characterised during the years of the first assessments, but we now know that many of those enigmatic declines were primarily the result of this disease. Scheele *et al.* (2019) estimated that at least 501 amphibian species declined and 90 became extinct in the wild as a result of chytridiomycosis.

A review of how much progress we have made in conservation since the benchmark of Stuart *et al.* (2004) is hampered by unrelated inequities of data. Indeed, in the case of amphibians, conservation efforts have been hampered by lack of data on natural fluctuations of populations, disease ecology, and etiologies in general, natural histories, and basic taxonomy and phylogeny. Stuart *et al.* (2004) derived their primary result of 43.2% of amphibians in decline by simply calculating the ratio of endangered amphibians to the number of known amphibians, fully acknowledging the problem of the many species of amphibians being listed as Data Deficient. Unfortunately, writing the present paper in the year 2021, we continue to have the same deficiencies of data, and thus we cannot realistically calculate a comparative value of the number of threatened amphibians in order to evaluate more recent progress since the report by Stuart *et al.* (2004). This is because of two unrelated issues: (1) the IUCN Red List is expensive and difficult to update, so global amphibian species have not been comprehensively updated since the assessment by Stuart *et al.* (2004) although less complete updates regularly occur; and (2) the rate of discovery of new species of amphibians is not diminishing over time. While numbers of new bird and mammalian species described each year has been declining for many decades, the numbers of new amphibians have remained remarkably steady at ~150–200 species per year (Tapley *et al.* 2018; Streicher *et al.* 2020; Frost 2021). This remarkable level of discovery owes as much to continued field work as it does to major methodological and theoretical advances in taxonomic practices and includes the revalidation or so-called resurrection of species' names previously retired from use (Streicher *et al.* 2020). For example, based on the website Amphibian Species of the World (Frost 2021) 56.4% of the amphibian species recognised in 2010 (7116 species) were not recognised 28 years earlier; in 1985 the number was 4014 species. Among the descriptions are bittersweet discoveries of new species that became extinct before their reality was discovered among historical museum specimens (e.g. Coloma *et al.* 2010; Mendelson 2011). Interestingly, seemingly in response to the slow updates to the IUCN Red List, taxonomists describing new species have started a new tradition of proposing IUCN Red List status for those species in the original descriptions (e.g. Wilson *et al.* 2020).

The result of these inequities in data is that, as of this writing in January 2021, there are Red List assessments that have not been updated in over 15 years, and new species are being described far more frequently than can be assessed by Red List teams; at present over 1000 species have not been assessed by the Red List. These are regrettable realities in the context of an acknowledged major crisis in the scale of biodiversity for amphibians (Wake and Vredenberg 2008). Tapley *et al.* (2018) reviewed this disparity in detail and surmised that financial constraints are a major contributor to the problem. They estimated that keeping just the amphibians on the Red List updated, in the light of continued discoveries of species and changes in conservation status, might require annual investments of approximately US$170 000–320 000. In addition, more than 1200 species currently are listed as Data Deficient (IUCN 2020), which usually indicates that the species are too rare to assess. Many species currently listed as Critically Endangered have not been seen for decades despite repeated visits to sites where they once were common (e.g. *Sarcohyla cyanomma* or *Quilticohyla erythromma* in Mexico; J. R. Mendelson, unpublished data). So, unfortunately, we cannot perform a direct assessment of progress between the data presented by Stuart *et al.* (2004) and now, 17 years later.

THREATS TO AMPHIBIANS

The primary conservation threats to amphibians were nicely articulated by Collins and Storfer (2003) nearly 18 years ago and unfortunately all six of their hypothesised drivers of extinction remain as primary threats to various species globally. Some threats, such as overexploitation, are straightforward in terms of estimations of cause, effect, and pathways towards mitigation. Under such scenarios, mitigation itself also is straightforward in concept; the difficult challenges lie not in the realm of biology, but rather in socioeconomics and policy. Broad-level threats such as contamination, climatic change, or changes in the use of land are not specific to amphibians, so amphibians may benefit from the mitigations brought by a diverse array of conservation stakeholders. The most intractable problem facing amphibians continues to be emerging infectious diseases. Of primary concern are the chytridiomycoses (caused by *Batrachochytrium dendrobatidis*, Bd, and *Batrachochytrium salamandrivorans*, Bsal), ranavirus, and perhaps an enigmatic protist similar to *Perkinsus* (reviewed by Pessier and Mendelson 2017).

As the first serious pathogen to be implicated in amphibian declines, Bd has been the focus of hundreds of technical articles concerning its biology, including the geographical distributions of hosts and pathogens, spread of epidemics, pathogenicity, natural history, evolution, and veterinary treatments. Despite these efforts in the

realm of mitigation (see reviews by Woodhams *et al.* 2011; Scheele *et al.* 2014), we arguably are no closer to a logistically feasible solution to control the disease in the wild than we were upon its discovery in 1999. This is problematic for *ex situ* breeding and release programs (known as conservation breeding programs) because, obviously, the final step of a successful *ex situ* program is the re-establishment of amphibians in the wild. The reality of persistent endemic Bd at virtually every conservation site known for amphibians is a frustratingly real and seemingly intractable problem. We know considerably less about all aspects of the biology and epidemiology of the other pathogens listed, but they appear to present the same level of challenges halting progress towards recovery. Epidemics of fungal pathogens in particular present special challenges for host species and also conservationists (Fisher *et al.* 2012). In our opinion, most progress in conserving amphibians in the context of emerging infectious disease will not be found in attempting to control pathogens (e.g. Woodhams *et al.* 2011; Bletz *et al.* 2013) but rather in incorporating the reality of now-endemic pathogens into planning for conservation (e.g. Sredl *et al.* 2011; Scheele *et al.* 2017; Joseph and Knapp 2018; Mendelson *et al.* 2019). Indeed, 'It's a Bd-world now' (Lips 2013).

Amphibian chytrid fungi may persist at a site, either somehow sustaining themselves in the absence of amphibian hosts or being sustained by local populations of relatively tolerant amphibian species acting as reservoir hosts (Scheele *et al.* 2017; Wilber *et al.* 2019; e.g. Pacific treefrogs, *Pseudacris regilla*, common eastern froglets, *Crinia signifera*, and American bullfrogs, *Lithobates catesbeianus*). These situations represent perhaps the most serious impediment to amphibians and also to completing cycles of released individuals from conservation-breeding programs. A few projects have managed to control Bd, but only in idiosyncratic systems in simple and species-poor environments. For example, Bosch *et al.* (2015) were able to eliminate Bd in a single host–amphibian system by draining the wetland and applying disinfectants to the pond's substrates and immediate environment. Elsewhere, Herculean efforts to remove introduced bullfrogs (*Lithobates catesbeianus*), notorious reservoirs of Bd, at isolated desert wetlands (where recolonisation of bullfrogs would be unlikely, or slow) helped local, native ranid frogs to recover (Sredl *et al.* 2011). Mitigation efforts at these levels are remarkable, but not sustainable on a global scale, and simply not possible in some natural systems. Likewise, such efforts to remove pathogens or reservoir hosts succeeded in very simple ecosystems and are not transferable to complex systems such as cloudforests. The role of conservation breeding programs for amphibians affected by these emerging infectious diseases is compromised because the eventual reintroduction phase is unlikely to succeed if the pathogen persists. The few monitored releases into systems compromised by Bd generally have been unsuccessful, for example with Wyoming toads (*Anaxyrus baxteri*; Polasik *et al.* 2016) or harlequin toads (*Atelopus* spp.; PARC 2018).

When faced with an overall extinction crisis on the scale the world is now facing, and which amphibians exemplify, conservationists quickly become overwhelmed by the enormity of species requiring attention – not to mention the staggering diversity of issues facing them. Thus, prioritisation of species for purposes of conservation becomes a crucial operation in any complex situation. In 2005, one of us (JRM) contributed to an IUCN-sponsored emergency workshop to develop a conservation action plan for harlequin toads (*Atelopus* spp.) in Central and South America as their decimation by Bd was becoming apparent (La Marca *et al.* 2005). Several major zoos from North America and Europe were represented at the workshop. The delegates settled on the establishment of emergency *ex situ* breeding colonies for the most critically endangered species and proceeded to volunteer their institutions as hosts for the breeding colonies on a species-by-species basis. By the end of the workshop, it became clear that the stakeholders had quickly committed their incipient capacities for breeding colonies with the most charismatically coloured species, leaving dozens of drab species unsupported. In the next section we discuss tools for establishing priorities intended to reduce such implicit biases across institutions and programs.

PRIORITISATION OF SPECIES FOR CAPTIVE BREEDING

Given the scope of the amphibian declines in the field, prioritisation for purposes of *ex situ* conservation became a critical issue for biologists. The ideal response to the situation faced by these threatened species is to mitigate threats in the wild, and restore and protect habitat, before populations decrease to unsustainable levels. When it is unlikely that the threats can be mitigated before a species becomes functionally extinct, the establishment of *ex situ* insurance and conservation breeding

programs is considered necessary as part of an integrated conservation response.

Prioritisation algorithms and affinity-group exercises for the purposes of conserving species are not a new concept. There have been many different approaches proposed (Mace *et al.* 2007; Game *et al.* 2013; Le Berre *et al.* 2019) and, at the broadest level, the IUCN Red List acts as a filter for prioritisation. However, what the amphibian crisis showed us was that, regardless of the schema for prioritisation employed, amphibians fell victim to an unspoken value system, driven by the appearances of species. It should be made clear that the institutions that commit to the non-trivial requirements of a proper *ex situ* breeding program extend beyond the more obvious zoos and aquaria to include universities, museums, botanic gardens, and governmental and state agencies, as well as NGOs. The most active group of organisations among these is almost certainly the zoo and aquarium community, but the reality is that the number of amphibian species potentially requiring *ex situ* intervention by far exceeds the current *ex situ* capacity of zoos and aquariums, and, indeed, of the entire *ex situ* community. As the conservation community ramped up efforts on behalf of amphibians, it became clear that not all interested institutions could participate at the same level – for example, maintaining the highest biosecurity levels for endangered species to minimise the risk of transmitting disease, especially when reintroduction of captive animals is proposed. Yet there were much-needed functions that institutions unable to meet certain biosecurity standards could still fulfil, including developing protocols for husbandry using common analogue species, for related and poorly known threatened species. Hence, there was a disconnect between the schema for prioritisation and the culture and logistical realities of the *ex situ* community. The realisation that the *ex situ* community required expert assistance to help select and prioritise the species with which they should be working informed the conceptualisation of the Conservation Needs Assessment tool (https://www.conservation needs.org/AssessmentOverview.aspx) by the Amphibian Ark. While the tool was developed for amphibians, it has shown itself to be customisable for other taxonomic groups as well.

Prioritisation of species for conservation purposes can be contentious because, at its core, it is a subjective topic. Therefore, prioritising which species should be considered for such programs must be effectively and efficiently resolved. The Conservation Needs Assessment tool is designed to reduce subjectivity and also to objectively determine whether *ex situ* breeding might be important for a species' recovery. It is designed to prioritise species more likely to warrant *ex situ* conservation breeding programs (i.e. to ensure that resources are applied where most needed) and is designed to identify species that may not be critically endangered but have something to offer to the broader *ex situ* efforts, for example, as ambassador species for exhibits and outreach education, or as non-endangered proxy species to develop protocols for husbandry of phylogenetically related species that are critically endangered.

Well-established regional coordinating bodies for zoos and aquaria exist in Europe (European Association of Zoos and Aquaria, EAZA), North America (Association of Zoos and Aquariums, AZA) and Australasia (Zoo and Aquarium Association, ZAA), with each of these having some form of prioritisation of species and/or establishing planning processes to guide their conservation of amphibians. The species that their institutional members manage are based to a degree on the geographical focal regions for each association (e.g. AZA focuses its efforts on New World species, primarily those from North America and the Caribbean [Barber and Poole 2014]). In other regions (e.g. Latin America and Asia), priorities for conservation activities are most often guided by national governmental wildlife authorities, local IUCN-SSC amphibian specialist groups (ASGs), universities, and individual field biologists. Within a given geographical scope, priority species for *ex situ* conservation programs are then determined based on various methods, including the five-step assessment process of the IUCN SSC *Guidelines on the Use of Ex Situ Management for Species Conservation* (IUCN SSC 2014) (e.g. European Association of Zoos and Aquaria 2019), and recommendations generated by the Conservation Needs Assessment process. In Australia and New Zealand, ZAA members focus their efforts exclusively on native species, and their priorities closely follow those developed by the wildlife agencies both of state and national governments.

In 2005, the IUCN SSC ASG called for the establishment of an international body to coordinate the *ex situ* response to amphibian conservation, as outlined in the Amphibian Conservation Action Plan (ACAP) (Gascon *et al.* 2007; Wren *et al.* 2015), and the Amphibian Ark was formed to bring together and help coordinate the *ex situ* amphibian conservation community. Prioritising species for conservation breeding programs is a significant aspect

of the Amphibian Ark's activities, as well as helping to raise funds to support conservation programs, and training to ensure that sufficient capacity exists in all regions to manage those programs.

Ex situ management of an amphibian species is considered necessary and appropriate when threats faced by the species in the wild cannot be mitigated in time to prevent extinction, and this is reflected in the Amphibian Ark's mission: 'ensuring the survival and diversity of amphibian species focusing on those that cannot currently be safe-guarded in their natural environments' (Amphibian Ark 2021a). The Amphibian Conservation Action Plan (Gascon *et al.* 2007; Wren *et al.* 2015) and Amphibian Ark (Amphibian Ark 2021b) recommend the establishment of *ex situ* initiatives within the range of ecological origins, to minimise the risk of transmitting diseases, especially when captive animals are likely to be reintroduced into the wild. In line with the ideal of establishing *ex situ* programs within this range, emphasis should be placed on developing species' priorities at the national level (rather than at the global level) because conservation actions are usually developed and implemented at the national level. Some countries with high amphibian biodiversity and high numbers of Critically Endangered species also lack sufficient resources (e.g. funding and the capacity for husbandry) and in these cases it often is essential that *ex situ* conservation programs be established outside of the countries where those resources occur. A well-known example of this was the efforts in 2006–07 when a collective of AZA-member institutions worked with experts in Panama, to collect hundreds of individuals of nearly 30 species and establish captive survival-assurance colonies in those US institutions (Gagliardo *et al.* 2008). Such programs are critical to ensure species' survival, but are not considered optimal, and often have political, social, and legal implications.

In many countries, the *ex situ* conservation community has insufficient knowledge to effectively prioritise species for *ex situ* conservation. Recognising this, a working group for the selection and prioritisation of species was formed during an amphibian *ex situ* conservation planning workshop, which was hosted by the IUCN SSC Conservation Breeding Specialist Group (CBSG, now Conservation Planning Specialist Group, CPSG) and the World Association of Zoos and Aquariums (WAZA) in 2006 (Zippel *et al.* 2006). The working group developed a decision tree that could help to determine whether an *ex situ* conservation program was warranted for each species and, if so, to determine the relevant importance of each program compared with others in the same country or region (Zippel *et al.* 2006). That original decision tree was subsequently expanded and modified, with significantly updated versions in use between 2007 and 2019. The decision tree support tool eventually evolved into the Conservation Needs Assessment process (Amphibian Ark 2019). To date it has been used to complete 4075 assessments of species and generate recommended conservation actions for 3461 species in 45 countries (Figures 2.1 and 2.2), with all assessments created at the national level. *Ex situ* rescue (i.e. conservation breeding programs) has been recommended for 371 (9.1%) of the completed assessments. Here, we review Amphibian Ark's Conservation Needs Assessment in detail.

Crucial to a comprehensive conservation program is full integration between *in situ* and *ex situ* conservation activities, for example following the CPSG One Plan Approach (Byers *et al.* 2013) (also see Chapter 11). This program normally is best highlighted through the establishment of a formal Species Action Plan/Species Recovery Plan that explicitly states the short-term, medium-term, and long-term goals of each component of the initiative (Zippel *et al.* 2006; Amphibian Ark 2021c). When *ex situ* management of an amphibian species is considered necessary and appropriate, the priority should be to establish the initiative within the range of ecological origins. Emphasis should therefore be placed on developing appropriate capacity within that range if such capacity does not exist (Gascon *et al.* 2007; Wren *et al.* 2015). The updated Amphibian Conservation Action Plan (Wren *et al.* 2015)

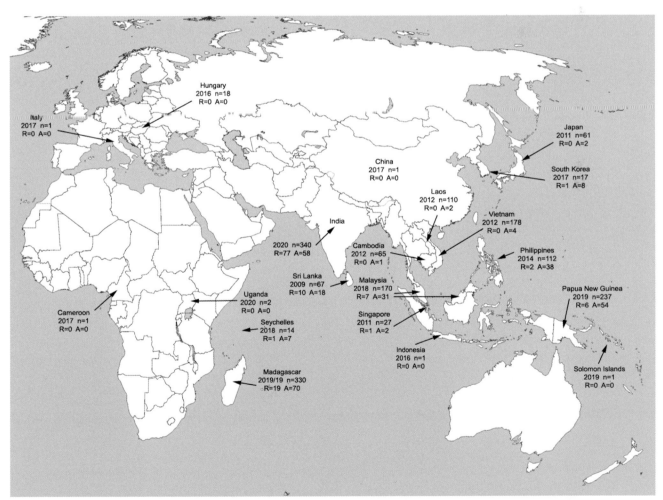

Figure 2.1: Over 1940 Conservation Needs Assessments have been compiled in 19 countries in the Eastern Hemisphere (Conservation Needs Assessments 2021). Recommendations for captive breeding programs, *ex situ* rescue (R) and husbandry analogues (A) are automatically generated from the data within the assessments.

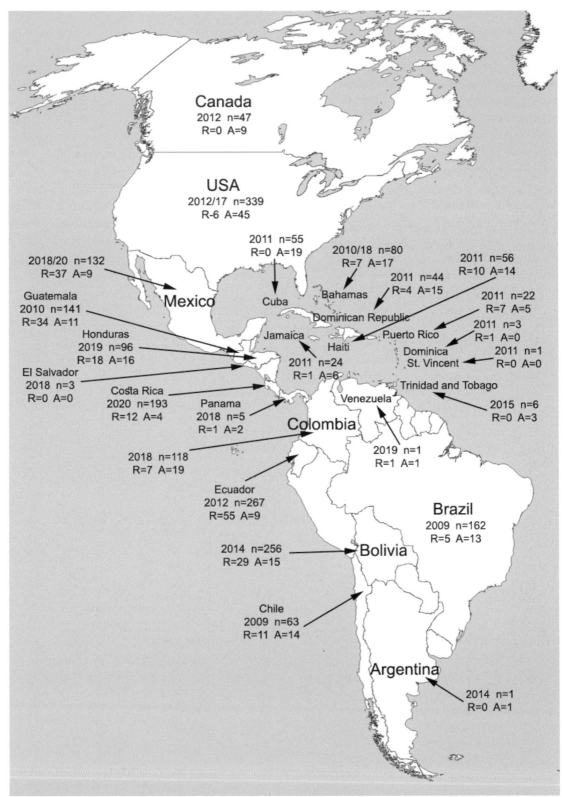

Figure 2.2: In the Western Hemisphere, over 2130 Conservation Needs Assessments have been compiled in 26 countries (Conservation Needs Assessments 2021). Recommendations for captive breeding programs, *ex situ* rescue (R) and husbandry analogues (A) are automatically generated from the data within the assessments.

outlines that the ultimate goal of a prioritisation tool is to develop rigorous criteria to determine whether *ex situ* conservation action is appropriate for a specific taxon, along with evaluation of the prioritisation process and the evaluation of feasibility. Included in the comprehensive evaluation for *ex situ* prioritisation are attentions to ensure that the recommended program leads to the establishment of genetically and demographically viable programs and that the species are prioritised for *ex situ* conservation. Gagliardo *et al.* (2008) also emphasised the importance of administrative and financial stability of such programs – establishing and maintaining high-quality *ex situ* programs is often expensive, and, in many cases, multiple generations will need to be produced and managed until such time that threats in the wild have been mitigated, and secure habitat is available for reintroduction. The high cost of this should never be underestimated. Political instability often can have dire consequences, especially in developing countries.

The entire process and documentation of the Conservation Needs Assessment can be reviewed at https://www.conservationneeds.org. During the Conservation Needs Assessment, review of each species takes the form of a series of questions with weighted scores, along with supporting narrative (Table 2.1). The assessment of each species is assigned to one or more of 10 different conservation actions (see https://conservationneeds.org/Help/EN/ConservationActions.htm). Species are listed according to their priority for the particular conservation action, and the supporting documentation provided during the assessment gives a guide for those species that have the most chance of benefiting from the prescribed action(s) (Johnson *et al.* 2020). Ultimately, priorities are determined by deriving a total score for each species from the responses provided during the assessment. A complete assessment for each species includes current information on the status of the species in the wild; suitable habitat; the threats facing it; cultural, scientific, socioeconomic, and phylogenetic significance; past *ex situ* experience with the species; as well as information about potential authorisation for any proposed *ex situ* conservation programs, and the availability of founder animals. Most of this information is best provided by field biologists and researchers working with the species being reviewed, while some information will be provided by people with relevant experience or knowledge of current and past *ex situ* initiatives in amphibian conservation.

Not all amphibians are suitable candidates for conservation breeding programs. Deciding which species should be established in captivity can be problematic and needs to take into account the geopolitical context and likelihood that both the captive-breeding and release components of the program will succeed. Experts in *ex situ* amphibian husbandry should evaluate the likely success of conservation breeding programs recommended by the Conservation Needs Assessment for the species involved (e.g. are the knowledge and skills, as well as the resources, available to keep animals alive and breeding long-term?). The threats for some species are not currently reversible or may never be reversible. The goal of all *ex situ* conservation programs is, of course, to ensure that conservation resources are directed towards species with the greatest chances of successful reintroduction into threat-free and effectively protected habitats, and that those populations be self-sustaining and thriving in the wild; therefore, limited funding should be allocated to projects that will generate tangible conservation outputs (Wren *et al.* 2015).

Typically, national Chairs of the IUCN SSC ASG will help assemble appropriate experts for a Conservation Needs Assessment. Subsequently, the assembled experts will coordinate the evaluation of all amphibian species in their country over a relatively short time, with the assessments and recommendations for conservation actions being used as the basis for the development of a national amphibian action plan. Delegates at the Conservation Needs Assessment workshops include scientists, field biologists and researchers, experts in animal husbandry, curators, government wildlife authorities, and others considered vital to understanding both the biology and the conservation threats of the species under consideration. It is also important to include representatives from the *ex situ* amphibian conservation community in the assessment process, as they will be responsible for planning and implementing *ex situ* actions. Sharing expertise and experiences enhances the reviews, ensuring that appropriate recommendations for national and global conservation actions are delivered where they are most needed. Assessments can be undertaken in a workshop-based environment, with appropriate experts and other stakeholders present, or can be undertaken online at any time.

Initially, attempts were made to assess all species in each country or region. This level of review was labour-intensive and expensive. In 2018, the decision was made to not assess species listed as Least Concern in the IUCN

Table 2.1: The Conservation Needs Assessment process takes the form of a series of questions with weighted scores, along with supporting narrative.

Section One – Review of external data
1. What is the current IUCN Red List category for the taxon?
2. Is there a strong possibility that this species might be extinct in the wild?
3. The taxon's Evolutionary Distinctiveness (ED) score, as generated by the ZSL EDGE program.
4. Is a population of at least 50% of the individuals of the taxon included within a reliably protected area or areas?

Section Two – Status in the wild
5. Does enough suitable habitat exist, either within or outside of currently protected areas that is suitable for reintroduction or translocation?
6. Have reintroduction or translocation attempts been made in the past for this species?
7. Are any *in situ* conservation actions currently in place for this species?
8. Are additional *in situ* conservation actions (e.g. habitat restoration and/or protection, control of invasive species, national legislation etc.) required to help conserve this species in the wild?
9. Is additional *in situ* research required to better understand the species, e.g. distribution, population trends, natural history, etc.?

Section Three – Threats and recovery
10. Are the threats facing the taxon, including any new and emerging threats not considered in the IUCN Red List, potentially reversible?
11. Is the taxon suffering from unsustainable collection within its natural range, either for food, for the pet trade, or for any other reason, which threatens the species' continued persistence in the wild?
12. Is the known population of this species in the wild large enough to recover naturally, without *ex situ* intervention if threats are mitigated?
13. Does an Action Plan for the species already exist, or is one currently being developed?

Section Four – Significance
14. Does the species exhibit a distinctive reproductive mode, behaviour, aspect of morphology or physiology, **within the Order** to which the species belongs (Anurans, Passeriformes, etc.)?
15. Does the species have a special human cultural value (e.g. as a national or regional symbol, in a historic context, featuring in traditional stories) or economic value (e.g. food, traditional medicine, tourism) within its natural range or in a wider global context?
16. Is the species vital to current or planned research other than species-specific ecology/biology/conservation? (e.g. human medicine, climatic change, environmental pollutants and conservation science), **within the Order** to which the species belongs (Anurans, Passeriformes, etc.).

Section Five – Ex situ activity
17. Does conserving this species (or closely related species) *in situ* depend upon research that can be most easily carried out *ex situ*?
18. Is any *ex situ* research or other *ex situ* conservation action currently in place for this species?
19. If an *ex situ* rescue program is recommended for this species, would an analogue species be required to develop husbandry protocols first?
20. Do the biological and ecological attributes of this species make it suitable for developing husbandry regimes for more threatened related species? That is, could this species be used in captivity to help to develop husbandry and breeding protocols which could be used for a similar, but more endangered species at a later stage?
21. Has this species been successfully maintained and bred in captivity?

Section Six – Education
22. Is the species especially diurnal, active or colourful, or is there an interesting or unusual aspect of its ecology that make it particularly suitable to be an educational ambassador for conservation of the species in the range country, either in zoos or aquariums or within ecotourism activities?

Section Seven – Ex situ program authorisation/availability of animals
23. Is there an existing conservation mandate recommending the *ex situ* conservation of this taxon?
24. Would a proposed *ex situ* initiative for this species be supported (and approved) by the range state (either within the range state or out-of-country *ex situ*)?
25. Are sufficient animals of the taxon available or potentially available (from wild or captive sources) to initiate the specified *ex situ* program?
26. Has a complete taxonomic analysis of the species in the wild been carried out to fully understand the functional unit you wish to conserve (i.e. have species limits been determined)?

Red List unless they were identified as potential husbandry analogues or models for threatened species, which have been identified as requiring urgent *ex situ* rescue. Recently, we have experimented with moving the Conservation Needs Assessment workshops into virtual internet platforms; the results have been productive, with the number of experts able to contribute to the assessments becoming substantially greater because of the reduction in the financial burden of travel and accommodation.

The Conservation Needs Assessment produces recommendations for a range of both *in situ* and *ex situ* conservation actions. Implementation of those recommendations becomes the responsibility of governmental agencies, regional coordinating bodies for zoos and aquaria, as well as individual institutions and organisations. Collectively, Conservation Needs Assessments strive to encourage low-priority species to be phased out of the collections of zoos and aquaria and replaced by species of higher priority (when appropriate), following the seminal recommendations of Conway (2011) to focus collections in zoos nearly exclusively on breeding programs for endangered species. Efforts are made to ensure that recommendations for *ex situ* programs arising from national Conservation Needs Assessments are disseminated appropriately and are readily available to all stakeholders, via the Conservation Needs Assessment website, national ASGs, zoo and aquarium associations, newsletters, and regional and national conferences.

The Conservation Needs Assessment process has been an evolving protocol. The criteria and their rankings have been adjusted as experience with the process was gained, the types of recommended conservation actions have been expanded, and we continue to work with the broader conservation community to identify goals, threats, and conservation options. Assessments and prioritisation of individual species are reviewed and updated as we gain knowledge and as the threats to each species change. Thus, there will be a need to constantly assess species' status and to monitor threats, so that responses to emerging critical situations are sufficiently rapid.

CONCLUSIONS

In summary, conservation managers are faced with a multitude of competing issues that generally require significant resources. The Conservation Needs Assessment process helps to prioritise species for conservation action, ensuring that limited resources can be assigned to species and actions that can most benefit from those resources. This process was first developed in 2006 and has evolved into a transparent and respected method for determining high-level priorities for both *ex situ* and *in situ* conservation action to help save threatened amphibians. Over 4000 sets of recommended conservation actions have been generated to date, with many of the recommended *ex situ* actions being implemented. We encourage program managers to refer to and to utilise these recommendations in their own countries and to seek help in conducting additional Conservation Needs Assessment workshops. In conjunction with data from recent Red-List workshops and other amphibian databases, the assessments are a valuable resource for directing and prioritising planning and action at the national level. We suggest that although the process was initially developed for amphibians, it can be equally useful for other taxonomic groups and has been built to be modified if required.

REFERENCES

Abarca J, Chaves G, García-Rodríguez A, Vargas R (2010) Reconsidering extinction: rediscovery of *Incilius holdridgei* (Anura: Bufonidae) in Costa Rica after 25 years. *Herpetological Review* **41**, 150–152.

Amphibian Ark (2019) *Amphibian Conservation Needs Assessment Process*. <http://www.amphibianark.org/pdf/AArk-Conservation-Needs-Assessment-tool.pdf>.

Amphibian Ark (2021a) *Model Programs*. <https://www.amphibianark.org/conservation-programs/model-facilities/>.

Amphibian Ark (2021b) *Welcome to the Amphibian Ark*. <https://www.amphibianark.org>.

Amphibian Ark (2021c) *Taxon Management*. <https://www.amphibianark.org/conservation-programs/captive-programs/taxon-management/>.

Barber D, Poole V (Eds) (2014) *Association of Zoos and Aquariums Amphibian Taxon Advisory Group Regional Collection Plan*. 3rd edn. Association of Zoos and Aquariums, Silver Spring MD, USA.

Bletz MC, Loudon AH, Becker MH, Bell SC, Woodhams DC, Minbiole KPC, Harris RN (2013) Mitigating amphibian chytridiomycosis with bioaugmentation: characteristics of effective probiotics and strategies for their selection and use. *Ecology Letters* **16**(6), 807–820. doi:10.1111/ele.12099

Bosch J, Sanchez-Tomé E, Fernández-Loras A, Oliver JA, Fisher MC, Garner TWJ (2015) Successful elimination of a lethal wildlife infectious disease in nature. *Biology Letters* **11**(11), 20150874. doi:10.1098/rsbl.2015.0874

Byers O, Lees C, Wilcken J, Schwitzer C (2013) The One Plan approach: the philosophy and implementation of CBSG's approach to integrated species conservation planning. *WAZA Magazine* **14**, 2–5.

Collins JP, Crump ML (2009) *Extinction in Our Times: Global Amphibian Decline.* Oxford University Press, New York NY, USA.

Collins JP, Storfer A (2003) Global amphibian declines: sorting the hypotheses. *Diversity & Distributions* **9**, 89–98. doi:10.1046/j.1472-4642.2003.00012.x

Coloma L, Duellman W, Almendáriz A, Ron S, Teran-Valdez A, Guayasamin J (2010) Five new (extinct?) species of *Atelopus* (Anura: Bufonidae) from Andean Colombia, Ecuador, and Peru. *Zootaxa* **2574**, 1–54. doi:10.11646/zootaxa.2574.1.1

Conservation Needs Assessments (2021) *View Assessments.* <https://conservationneeds.org/master/assessmentsearch>.

Conway W (2011) Buying time for wild animals with zoos. *Zoo Biology* **30**, 1–8. doi:10.1002/zoo.20352.

European Association of Zoos and Aquaria (2019) *EAZA Population Management Manual: Standards, Procedures and Guidelines for Population Management within EAZA.* 3rd edn. European Association of Zoos and Aquaria, Amsterdam, The Netherlands.

Fisher MC, Garner TW (2020) Chytrid fungi and global amphibian declines. *Nature Reviews. Microbiology* **18**, 332–343. doi:10.1038/s41579-020-0335-x

Fisher MC, Henk DA, Briggs CJ, Brownstein JS, Madoff LC, McCraw SL, Gurr SJ (2012) Emerging fungal threats to animal, plant and ecosystem health. *Nature* **484**, 186–194. doi:10.1038/nature10947

Frost DR (2021) *Amphibian Species of the World: An Online Reference.* Version 6.1. <https://amphibiansoftheworld.amnh.org/index.php>. American Museum of Natural History, New York NY, USA. doi.org/10.5531/db.vz.0001

Gagliardo R, Crump P, Griffith E, Mendelson III, Jr, Ross H, Zippel KC (2008) The principles of rapid response for amphibian conservation, using the programmes in Panama as an example. *International Zoo Yearbook* **42**, 125–135. doi:10.1111/j.1748-1090.2008.00043.x

Game ET, Kareiva P, Possingham HP (2013) Six common mistakes in conservation priority setting. *Conservation Biology* **27**, 480–485. doi:10.1111/cobi.12051

Gascon C, Collins JP, Moore RD, Church DR, Mckay JE, Mendelson III, Jr (Eds) (2007) *Amphibian Conservation Action Plan.* IUCN/Species Survival Commission Amphibian Specialist Group, Gland, Switzerland.

Green DM, Lannoo MJ, Lesbarrères D, Muths E (2020) Amphibian population declines: 30 years of progress in confronting a complex problem. *Herpetologica* **76**, 97–100. doi:10.1655/0018-0831-76.2.97

Heatwole H (2013) Worldwide decline and extinction of amphibians. In *The Balance of Nature and Human Impact.* (Ed. K Rohde) pp. 259–278. Cambridge University Press, Cambridge MA, USA.

Heatwole H, Rowley JJL (2018) Introduction. In *Status of Conservation and Decline of Amphibians, Australia, New Zealand, and Pacific Islands.* (Eds H Heatwole and JJL Rowley) pp. 1–4. CSIRO Publishing, Melbourne.

IUCN (2020) *The IUCN Red List of Threatened Species.* Version 2020–3. <https://www.iucnredlist.org>.

IUCN/SSC (2014) *Guidelines on the Use of Ex Situ Management for Species Conservation, Version 2.0.* IUCN Species Survival Commission, Gland, Switzerland.

James TY, Litvintseva AP, Vilgalys R, Morgan JA, Taylor JW, Fisher MC, Berger L, Weldon C, du Preez L, Longcore JE (2009) Rapid global expansion of the fungal disease chytridiomycosis into declining and healthy amphibian populations. *PLoS Pathogens* **5**(5), e1000458.

Johnson K, Baker A, Buley K, Carrillo L, Gibson R, Gillespie G, Lacy RC, Zippel K (2020) A process for assessing and prioritizing species conservation needs: going beyond the Red List. *Oryx* **54**(1), 125–132. doi:10.1017/S0030605317001715

Joseph MB, Knapp RA (2018) Disease and climate effects on individuals drive post-reintroduction population dynamics of an endangered amphibian. *Ecosphere* **9**(11), e02499. doi:10.1002/ecs2.2499>.

La Marca E, Lips K, Lötters S, Puschendorf R, Ibáñez R, Ron S, Rueda-Almonacid J, Schulte R, Marty C, Castro F, *et al.* (2005) Catastrophic population declines and extinctions in Neotropical Harlequin frogs (Bufonidae: *Atelopus*). *Biotropica* **37**, 190–201. doi:10.1111/j.1744-7429.2005.00026.x

Le Berre M, Noble V, Pires M, Frédéric Médail F, Diadema K (2019) How to hierarchise species to determine priorities for conservation action? A critical analysis. *Biodiversity and Conservation* **28**, 3051–3071. doi:10.1007/s10531-019-01820-w

Lips KR (2013) *What if there is No Happy Ending? Science Communication as a Path to Change.* Scientific American. <http://blogs.scientificamerican.com/guest-blog/2013/05/15/what-if-there-is-no-happy-ending-science-communication-as-a-path-to-change/>.

Lips KR (2018) Witnessing extinction in real time. *PLoS Biology* **16**, e2003080. doi:10.1371/journal.pbio.2003080

Lips KR, Brem F, Brenes R, Reeve JD, Alford RA, Voyles J, Carey C, Pessier A, Livo L, Collins JP (2006) Infectious disease and global biodiversity loss: pathogens and enigmatic amphibian extinctions. *Proceedings of the National Academy of Sciences of the United States of America* **103**(9), 3165–3170.

Lips KR, Mendelson III, Jr (2014) Stopping the next amphibian apocalypse. *New York Times.* <https://www.nytimes.com/2014/11/15/opinion/stopping-the-next-amphibian-apocalypse.html>.

Longcore JE, Pessier AP, Nichols DK (1999) *Batrachochytrium dendrobatidis* gen. et sp. nov., a chytrid pathogenic to amphibians. *Mycologia* **91**, 219–227.

Mace G, Possingham H, Leader-Williams N (2007) Prioritizing choices in conservation. In *Key Topics in Conservation Biology.* (Eds D MacDonald and K Service) pp. 17–34. Blackwell Publishing, Oxford, UK.

Martel A, Spitzen-van der Sluijs M, Blooi W, Ducatella B, Fisher M, Woeltjes A, Bosman W, Chiers K, Bossuyt F, Pasmans F (2013) *Batrachochytrium salamandrivorans* sp. nov. causes lethal chytridiomycosis in amphibians. *Proceedings of the National Academy of Sciences* **110**(38), 15325–15329. <www.pnas.org/cgi/doi/10.1073/pnas.1307356110>.

Mendelson III, Jr (2011) Shifted baselines, forensic taxonomy, and Rabbs' fringe-limbed treefrog: the changing role of biologists in an era of amphibian declines and extinctions. *Herpetological Review* **42**, 21–25.

Mendelson III, Jr (2018) Frogs in glass boxes: responses of zoos to global amphibian extinctions. In *The Ark and Beyond: The Evolution of Zoo and Aquarium Conservation.* (Eds B Minteer and JP Collins) pp 298–310. University of Chicago Press, Chicago IL, USA.

Mendelson JR, III, Donnelly R (2011) *The Crisis of Global Amphibian Declines: Causes, Consequences, and Solutions.* Network for Conservation Educators and Practitioners, American Museum of Natural History. <http://research.amnh.org/biodiversity/ncep>.

Mendelson III, Jr, Whitfield SM, Sredl MJ (2019) A recovery engine strategy for amphibian conservation in the context of disease. *Biological Conservation* **236**, 188–191. doi:10.1016/j.biocon.2019.05.025

Nahonyo C, Goboro E, Ngalason W, Mutagwaba S, Ugomba R, Nassoro M, Nkombe E (2017) Conservation efforts of Kihansi spray toad *Nectophrynoides asperginis*: its discovery, captive breeding, extinction in the wild and re-introduction. *Tanzania Journal of Science* **43**, 23–35.

Newell DA, Goldingay RL, Brooks LO (2013) Population recovery following decline in an endangered stream-breeding frog (*Mixophyes fleayi*) from subtropical Australia. *PLoS One* **8**(3), e58559. doi:10.1371/journal.pone.0058559

O'Hanlon SJ, Rieux A, Farrer RA, Rosa GM, Waldman B, Bataille A, Kosch TA, Murray K, Brankovics B, Fumagalli M, *et al.* (2018) Recent Asian origin of chytrid fungi causing global amphibian declines. *Science* **360**, 621–627. doi:10.1126/science.aar1965

OEH NSW (2012) *National Recovery Plan for the Southern Corroboree Frog,* Pseudophryne corroboree, *and the Northern Corroboree Frog,* Pseudophryne pengilleyi. Office of Environment and Heritage (NSW), Hurstville.

PARC [Panama Amphibian Rescue and Conservation Project] (2018) *2018 Annual Report.* <http://amphibianrescue.org/amphibianwordpress/wp-content/upLoads/2019/05/PARCAnnreport2018.pdf>.

Pavajeau L, Zippel K, Gibson R, Johnson K (2008) Amphibian Ark and the 2008 Year of the Frog Campaign. *International Zoo Yearbook* **42**, 24–29. doi:10.1111/j.1748-1090.2007.00038.x

Pessier AP, Mendelson III, Jr (Eds) (2017) *A Manual for Control of Infectious Diseases in Amphibian Survival Assurance Colonies and Reintroduction Programs.* Version 2.0. IUCN/SSC Conservation Breeding Specialist Group, Apple Valley MN, USA.

Polasik JS, Murphy MA, Abbott T, Vincent K (2016) Factors limiting early life stage survival and growth during endangered Wyoming toad reintroductions. *The Journal of Wildlife Management* **80**, 540–552. doi:10.1002/jwmg.1031

Rabb GB (1990) Declining amphibian populations. *Species* **13–14**, 33–34.

Rosenblum E, James T, Zamudio K, Poorten T, Ilut D, Rodriguez D, Eastman J, Richards-Hrdlicka K, Joneson S, Jenkinson T, *et al.* (2013) Complex history of the amphibian killing chytrid fungus revealed with genome resequencing data. *Proceedings of the National Academy of Sciences* **110**(23), 9385–9390. <www.pnas.org/cgi/doi/10.1073/pnas.1300130110>.

Scheele BC, Hunter DA, Grogan LF, Berger L, Kolby JE, McFadden MS, Marantelli G, Skerratt LF, Driscoll DA (2014) Interventions for reducing extinction risk in chytridiomycois-threatened amphibians. *Conservation Biology* **28**, 1195–1205. doi:10.1111/cobi.12322

Scheele BC, Skerratt LF, Grogan LF, Hunter DA, Clemann N, McFadden M, Newell D, Hoskin CJ, Gillespie GR, Heard GW, Brannelly L (2017) After the epidemic: ongoing declines, stabilizations and recoveries in amphibians afflicted by chytridiomycosis. *Biological Conservation* **206**, 37–46. doi:10.1016/j.biocon.2016.12.010

Scheele BC, Pasmans F, Berger L, Skerratt LF, Martel A, Beukema W, Acevedo AA, Burrowes PA, Carvalho T, Catenazzi A, *et al.* (2019) The aftermath of amphibian fungal panzootic reveals unprecedented loss of biodiversity. *Science* **363**, 1459–1463. doi:10.1126/science.aav0379

Sredl MJ, Akins CM, King AD, Sprankle T, Jones TR, Rorabaugh JC, Jennings RD, Painter CW, Christman MR, Christman BL, Crawford C (2011) Re-introductions of Chiricahua leopard frogs (*Lithobates [Rana] chiricahuensis*) in southwestern USA show promise, but highlight problematic threats and knowledge gaps. In *Global Re-introduction Perspectives: 2011. More Case Studies from Around the Globe.* (Ed. PS Soorae) pp. 85–90. IUCN/SSC Re-introduction Specialist Group and Abu Dhabi Environment Agency, Gland, Switzerland.

Streicher JW, Sadler R, Loader SP (2020) Amphibian taxonomy: early 21st century case studies. *Journal of Natural History* **54**, 1–13. doi:10.1080/00222933.2020.1777339

Stuart S, Chanson J, Cox N, Young B, Rodrigues A, Fischman D, Waller R (2004) Status and trends of amphibian declines and extinctions worldwide. *Science* **306**, 1783–1786. doi:10.1126/science.1103538

Tapley B, Harding L, Sulton M, Durand S, Burton M, Spencer J, Thomas R, Douglas T, Andre J, Winston R, George M, *et al.* (2014) An overview of the current conservation efforts for the mountain chicken frog (*Leptodactylus fallax*) on Dominica. *Herpetological Bulletin* **128**, 9–11.

Tapley B, Michaels CJ, Gumbs R, Böhm M, Luedtke J, Pearce-Kelly P, Rowley JJL (2018) The disparity between species descriptions and conservation assessment: a case study in taxa with high rates of species discovery. *Biological Conservation* **220**, 209–214. doi:10.1016/j.biocon.2018.01.022

Vial JL (1991) Declining amphibian populations task force. *Species* **16**, 47–48.

Wake DB (1991) Declining amphibian populations. *Science* **253**, 860. doi:10.1126/science.253.5022.860

Wake DB, Vredenberg VT (2008) Are we in the midst of the sixth mass extinction? A view from the world of amphibians. *Proceedings of the National Academy of Sciences of the United States of America* **105**, 11466–11473. doi:10.1073/pnas.0801921105

Wilber MQ, Johnson PT, Briggs CJ, Wilber M (2019) When chytrid fungus invades: integrating theory and data to understand disease-induced amphibian declines. In *Wildlife Disease Ecology: Linking Theory to Data and Application*. (Eds K Wilson, A Fenton and D Tomkins) pp. 511–543. Cambridge University Press, Cambridge, UK.

Wilson LD, Hofmann EP, Dudek D, Jr, Herrera LA, Luque-Montes IR, Ross AN, Kry C, Duchamp JE, Townsend JH (2020) A critically endangered new species of polymorphic stream frog (Anura: Hylidae: *Atlantihyla*) from the montane forest of Refugio de Vida Silvestre Teguixat, Honduras. *Vertebrate Zoology* **70**, 731–756.

Woodhams DC, Bosch J, Briggs CJ, Cashins S, Davis LR, Lauer A, Muths ER, Puschendorf R, Schmidt BR, Sheafor B, Voyles J (2011) Mitigating amphibian disease: strategies to maintain wild populations and control chytridiomycosis. *Frontiers in Zoology* **8**(1), 1–24. doi:10.1186/1742-9994-8-8

Wren S, Angulo A, Meredith H, Kielgast J, Dos Santos M, Bishop P (Eds) (2015) *Amphibian Conservation Action Plan*. IUCN Species Survival Commission Amphibian Specialist Group. <https://www.iucn-amphibians.org/resources/acap/>.

Zippel K, Lacy RC, Byers O (Eds) (2006) *CBSG/WAZA Amphibian Ex Situ Conservation Planning Workshop Final Report*. IUCN/SSC Conservation Breeding Specialist Group, Apple Valley MN, USA.

3 Methods of identifying the sex of amphibians and of conditioning captive brood stock for assisted reproduction

Luke J. Linhoff, Jennifer M. Germano, and Frank C. Molinia

INTRODUCTION

Breeding amphibians in captivity is a nuanced and highly species-specific process. What works for one species may not work for another. Additionally, most amphibian species have not been kept in captivity, and even fewer have successfully reproduced in captivity. A foundational knowledge of a species' basic natural history, morphology, behaviour, diet, and physiology may be critical to inform the development of husbandry protocols for successful reproduction. Within this chapter, we cover two important topics helpful for successfully employing reproductive technologies. First, identifying the sex of individual amphibians can be surprisingly challenging. It may require the implementation of advanced laboratory techniques or additional novel research into poorly studied species in order to obtain accurate identification. Second, conditioning captive brood stock using advanced husbandry techniques before implementing reproductive technologies may be necessary for successful propagation. These two topics are foundational to implementing a breeding program, and the challenges associated with each should not be underestimated. Both topics continue to be active areas of research and some species may be impossible to breed in captivity using current techniques and technologies. Because of the extensive diversity of behaviours, ecology, and natural histories exhibited by the thousands of extant species of amphibians, exact recommendations are challenging. We hope to provide examples of techniques for identification of sex and conditioning of brood stock that can be tailored for a wide array of amphibian species globally.

IDENTIFICATION OF SEX

Being able to accurately identify the sex of individuals is perhaps one of the most fundamental tenets of establishing a successful conservation breeding program. At the most basic level, captive colonies that lack individuals of both sexes cannot succeed. A heavy weighting away either from females or from males could also lead to failure. This was the case for captive colonies of the threatened and monomorphic New Zealand frogs, which failed for many years and required the use of urinary hormones to compensate for the paucity of males in the captive breeding program (J. M. Germano and F. C. Molinia, unpublished data). Early identification of sex allows institutions to transfer animals at the juvenile or subadult stage between colonies to allow for ideal or balanced ratios. Knowledge of an individual's sex also allows for the separation of males and females for purposes of management and for implementation of reproductive technologies. Additionally, the application of hormonal therapies and protocols for collection of gametes are sex-specific (see Chapters 6 and 7) and so require accurate identification of sex to ensure that the correct doses for priming or induction are administered. Misidentification can result in complete reproductive failure for an entire breeding season.

Fortunately, most amphibian species are sexually dimorphic in their appearance and/or behaviour. These sexual differences develop in response to gonadotropic hormones and often appear only during the breeding season or when reproductive hormones are cycling normally (see Chapter 4), which may not necessarily occur in captivity. For a small proportion of species, sexual dimorphism may be completely absent year-round (sexual monomorphism) as in the leiopelmatid frogs. Furthermore, even in some dimorphic species, sexual differences may not be sufficiently distinct to confidently differentiate between males and females (e.g. an overlap in dimorphisms in size); in others, these differences may not appear until individuals reach sexual maturity.

Phenotype

Most amphibians show external differences between the sexes for at least some part of the year. These dimorphisms can be in physical dimensions, colouration, and skin texture, or in the presence of secondary sexual characteristics (Plate 2). While the evolutionary driver of secondary sexual traits (sexual selection) provides colouration and ornamentation that attracts mates, weaponry for male–male competition, or adaptations that improve grip during mating (either under scenarios of female-choice or forced copulation) (Andersson 1994), they can also be utilised by conservation managers and researchers for discrimination of sex.

Some examples of permanent sexual dimorphisms in amphibians include differences in colour (dichromatism) (Bell and Zamudio 2012; Hoffman and Blouin 2000), elongated digits (Blackburn 2009), spines on the fingers and inguinal regions of males (Blackburn 2009), structure of the larynx or vocal sac (Lowe and Hero 2012; Preininger *et al.* 2016), and cranial shape (Hayes and Licht 1992; Emerson 1996). One of the most common and highly studied differences in frogs and toads is sexual size dimorphism of body-length and weight (Shine 1979; Monnet and Cherry 2002). In 90% of anuran species, females are the larger sex (Shine 1979). The sizes of males and females form a continuum, ranging from complete overlap between the sexes to the extreme case of *Platymantis boulengeri* in which females are on average three times the length of males and 40 times their weight (Kraus 2008). Common seasonal sexual traits include the enlargement of forelimbs (Navas and James 2007; Greene

and Funk 2009; Buzatto *et al.* 2015), enlargement of cloacal glands in salamanders (Whitaker 2001), colouration of the vocal sac (Duellman and Trueb 1994), and the enlargement and darkening of nuptial pads on the thumb and/or forearms in newts and anurans (Cadle 1995; Whitaker 2001; Greene and Funk 2009) that play a role in grasping females during amplexus/oviposition. Development of darkened nuptial pads and the pigmentation of vocal sacs is controlled by testosterone, with these characteristics reaching their maximum state during the breeding season (Greenberg 1942; Epstein and Blackburn 1997; Luna *et al.* 2012). While these traits may be present year-round in some species (although regressed), in others they are absent outside of the breeding season (Luna *et al.* 2012). Further detailed descriptions of amphibian sexual dimorphisms, as well as the significant variation between caecilians, salamanders, and anurans can be found in Duellman and Trueb (1994).

In addition to the more obvious forms of sexual dimorphism, numerous amphibian species have measurable differences in the ratios of bodily features, e.g. in the ratio of length of the tympanic membrane to size of the eye for the American bullfrog, *Rana* (*Aquarana*) *catesbeiana* (Plate 2d) (Asahara *et al.* 2020). Although this ratio is significantly larger in males than in females, there is a small overlap between the sexes, meaning that it would need to be combined with other morphological features for certainty in sex-discrimination (Asahara *et al.* 2020). Another similar measurement that has been shown to have sexual dimorphism in several taxa of birds, mammals, and amphibians is the ratio of lengths of the second-to-fourth digits (2D:4D). This trait is thought to be caused by prenatal steroid hormones. Sexual differences in 2D:4D ratios occur in several anurans (Chang 2008; Beaty *et al.* 2016; Rajabi and Javanbakht 2019), although they are absent in others (Germano *et al.* 2011; Beaty *et al.* 2016). For some species, differences in the 2D:4D ratio may be great enough to use for discrimination of sex. There is potential to incorporate these types of measurements into tools for sexual identification at captive institutions.

Ultrasonography

Ultrasonography is rapidly becoming a valuable tool in amphibian reproductive technologies, although its use is minimal, especially for smaller-sized species due to the difficulty in visualising testes and ovaries (Schildger and Triet 2001; Stetter *et al.* 2001; Roth and Obringer 2003;

Reyer and Bättig 2004). Other imaging techniques for sexing, such as endoscopy and laparoscopy, have been used in amphibians (Roth and Obringer 2003; Kouba and Vance 2009); however, these are invasive and potentially dangerous techniques, thus limiting their use for endangered species. Near-infrared reflectance (NIR) spectroscopy, a technique that measures the characteristic absorption patterns produced by the vibrations of particular chemical bonds, shows promise as a method for identifying the sex of amphibians (Vance *et al.* 2014), but further studies are needed. While both ultrasonography and NIR are advantageous in being non-invasive and requiring only a brief scan of the animal's abdomen, the necessary equipment for these techniques is expensive and trained technicians are required for analysis (Graham *et al.* 2016).

Nevertheless, the use of ultrasonography in amphibians has proven useful for sex determination in several species. For salamanders, this has been limited to a few large species such as the Chinese giant salamander (*Andrias davidianus*) (Li *et al.* 2010) and the hellbender (*Cryptobranchus alleganiensis*) (Miller and Fowler 2012). Recently, ultrasonography proved to be a practical non-invasive tool for precisely determining sex and reproductive status for assessments and for monitoring health in the endangered olm (*Proteus anguinus*) (Holtze *et al.* 2017). ultrasonography allowed detection of the testes in one of 12 male olms, and of different stages of ovarian development in 11 females. However, in females of two species of water frog (*Rana lessonae* and *Rana esculenta*), ultrasonography, compared with dissection and abdominal-incision techniques, was found to be only ~50% effective for determining the presence or absence of eggs (Reyer and Bättig 2004). This method was even less reliable in smaller frogs as was also the case for the electromagnetic scanning of total body electrical conductivity (TOBEC), first described as a rapid non-destructive technique for determining the composition of intact eggs in poultry (Williams *et al.* 1997).

More recently, ultrasonic imaging was 93.4% accurate in correctly identifying the sex of adult, captively reared, endangered dusky gopher frogs (*Lithobates sevosus*). All females were correctly identified as were 88.9% of males; there were three instances of a male being classified as a female with low follicular development (Graham *et al.* 2016). Dusky gopher frogs are weakly dimorphic, but females do produce developing ovarian follicles year-round that are visible and easy to distinguish on ultrasound images, particularly if the female is gravid.

Assessing sex using ultrasonography can be challenging in individuals whose gonads are inactive (e.g. in juveniles or in adults outside of the breeding season); therefore successful imaging of the follicles may be limited by seasonality in many species. However, in the sexually dimorphic African clawed frog (*Xenopus laevis*), in which phenotypic signs of sexual maturation can take 1–2 years to develop, there is one report of the potential use of ultrasonography to sex juveniles (15–22 g) (Brownlee *et al.* 2018).

Beyond the use of ultrasonography for the identification of amphibians' sex, the current focus of ultrasonic imaging in amphibians is to monitor ovarian status in order to improve the success of reproductive technologies, such as the hormonal induction of ovulation (Calatayud *et al.* 2018; Graham *et al.* 2018; Bronson and Vance 2019; Calatayud *et al.* 2019; Ruiz-Fernández *et al.* 2020) (see Chapter 6). Captive amphibian colonies are often housed within or near zoological institutions and universities where the necessary equipment and experienced technicians are available. With ongoing advances in technology, particularly for imaging smaller animals, it is anticipated that the application of ultrasonography as a routine tool for sexing amphibians will become more common in the future.

Quantifying hormones

In amphibians, profiles of androgens, estrogens, and glucocorticoids associated with sex, seasonality, reproduction, sex accessory structures, secondary sexual characteristics, and stress have been reviewed extensively (Rastogi *et al.* 2011; Sever and Staub 2011; Narayan 2013; Graham *et al.* 2016; Calatayud *et al.* 2018). Early studies on amphibian reproductive and stress hormones focused on measuring the levels of hormones in the plasma. In anurans and small salamanders this requires the collection of blood, often by cardiac or aortic puncture, or from the eye, subclavian artery in the arm, or facial maxillary or musculocutaneous vein (Rastogi *et al.* 1986; Delgado *et al.* 1989; Harvey *et al.* 1997; Medina *et al.* 2004). In larger salamanders, samples can be taken from the tail, as detailed by Calatayud *et al.* (2018); also best-practice procedures inclusive of a linked video-guide have been published recently (Calatayud *et al.* 2019). Collection of blood permits real-time measures of physiological parameters. It is, however, highly invasive, and given the risks of injury or death, it is not always feasible, especially in threatened or endangered species. Given the availability of radio-immunoassays and enzyme-immunoassays (Brown *et al.* 2004), there has been a move towards using samples of urine or faeces that can be collected non-invasively or with minimal disturbance from amphibians for measuring reproductive and stress-hormone metabolites (Narayan 2013; also see Chapters 4 and 5). These methods provide ideal tools for identifying the sex of individuals and to ensure that reproductive technologies, including hormonal therapies or the collection of gametes, are best suited for the individual.

The ability to differentiate between males and females initially focused on sexually dimorphic species, so that the technique could be easily validated. In American toads (*Bufo americanus*), concentrations of faecal testosterone metabolites were significantly higher in males than in females (with no overlap in the 95% confidence interval), while for boreal toads (*Bufo boreas boreas*) immunoactivity of oestradiol was significantly higher in females than in males (Szymanski *et al.* 2006). Additionally, the ratio of progesterone metabolites to those of sex hormones were as good a predictor of sex as were concentrations of testosterone or oestradiol metabolites alone in the same species. In southern bell frogs (*Litoria raniformis*), urinary estrone concentrations proved to be a reliable tool for sexing adult frogs, with no overlap between the sexes in 98% of cases, regardless of season (Germano *et al.* 2009). Estrone metabolites were also 80% successful in identifying non-vitellogenic females from males in free-living and captive tropical endangered Fijian ground frogs (*Platymantis vitiana*) (Narayan *et al.* 2010). In the Maud Island frog, *Leiopelma pakeka* (synonymised with and renamed Hamilton's frog, *Leiopelma hamiltoni*, based on recent genetic analyses by Burns *et al.* 2018), sex identification based on levels of urinary estrone metabolite was 94% correct (Germano *et al.* 2012). This was the first report of reliable sexing using this technique in a monomorphic species, in which the sexes overlap in snout-to-vent length for over half of their adult size range and in which no other sexually dimorphic trait is known. This methodology has been found suitable for sexing both Hochstetter's frog (*Leiopelma hochstetteri*) and the critically endangered Archey's frog (*Leiopelma archeyi*) (J. M. Germano and F. C. Molinia, unpublished data). Both species are part of programs with zoos in New Zealand for captive breeding and release. In the threatened monomorphic frogs *Geocrinia alba* and *Geocrinia vitellina*, the ratio of faecal testosterone metabolites to estrone conjugate metabolites was higher in males than in females, and discriminate analysis of data could distinguish (100%) between mature males and females and correctly classify the sex of 75% of the juvenile frogs (Hogan *et al.* 2013). In the weakly dimorphic endangered dusky gopher frog (*Lithobates sevosus*), identification of sex was > 95% accurate, based on the ratio of testosterone to estrone metabolites after urinary hormone analysis (Graham *et al.* 2016).

Sex-based differences have also been reported in measures of metabolites other than those based on reproductive hormones. For example, in a peri-urban population of the common Asian toad (*Duttaphrynus melanostictus*) mean baseline levels of urinary corticosterone metabolites were significantly higher in male toads than in females (Narayan and Gramapurohit 2016). While results vary based on species, season, sample type, and the hormones analysed, these studies show that analyses of non-invasive hormones may be an accurate method for identifying sex in amphibians.

Despite the clear benefits, there are some disadvantages with non-invasive endocrine assays (Brown *et al.* 2004; Narayan 2013). Collection, storage, extraction, and radio-immunoassays or enzyme-immunoassays themselves still take time and require specialised laboratory equipment and experienced personnel. Faecal samples require extraction before analysis and while urine samples can be diluted and assayed directly, data need to be indexed against another assay of creatinine. There are a multitude of assays available and for new species each needs both laboratory and biological validations. For especially small amphibians it is difficult to get sufficient material (e.g. volume of urine) to assay, which often means that repeated sampling is required. Additionally, it can be challenging to prevent contamination in waterborne species.

Several recent studies have measured sex and stress hormones excreted into water (Gabor *et al.* 2013; Baugh *et al.* 2018; Baugh and Gray-Gaillard 2021) as an alternative non-invasive technique suitable for identifying sex. This technique may benefit fully aquatic species, or those individuals that are too small to obtain adequate samples of urine volume, although the technique stills faces some of the same challenges listed above for monitoring urinary and faecal hormones.

Recently, dermal swabbing was developed as an innovative method for collecting and measuring amphibian glucocorticoid secretions in 15 amphibian species, including terrestrial, semi-aquatic, and fully aquatic species

(Santymire *et al.* 2018). Dermal secretions have since been validated as a robust matrix for monitoring glucocorticoid alterations in the edible bullfrog (*Pyxicephalus edulis*) (Scheun *et al.* 2019). Based on the success and innovations above, there is likely to be ongoing use of this technology if it can be validated for measuring reproductive hormones.

Molecular identification of sex

Mammals exhibit XY sex chromosomal systems and determination of sex is controlled by a conserved SRY gene on the degenerate Y chromosome (Kashimada and Koopman 2010). Birds possess ZW sex chromosomal systems, and the DMRT1 gene on the Z chromosome determines sex (Smith *et al.* 2009). Evidence to date suggests that amphibians also have genetically controlled sex determination, with both male heterogamety (XX/XY) and female heterogamety (ZZ/ZW) (Nakamura 2009). However, unlike mammals and birds, sex chromosomes in 96% species of amphibians are homomorphic (undifferentiated) in both sexes (Hayes 1998; Wallace *et al.* 1999). This means that karyotyping as a tool for assigning gender is often inconclusive (Roth and Obringer 2003), except when heteromorphic chromosomes have evolved. For example, the chromosomes of Hochstetter's frog (*Leiopelma hochstetteri*) are cytogenetically distinct with a unique 0W/00 sex-determination system observed in all 11 distinct mainland populations from the North Island, each with different W chromosome morphologies in females. The univalent W chromosome is absent, however, from the Great Barrier Island population (Green 1988, 1994). The pond frog (*Glandirana rugosa*) in Japan consists of five major geographical groups; two groups have homomorphic sex chromosomes with male heterogametic determination of sex, whereas the other three groups have heteromorphic sex chromosomes with male and female heterogametic determination (Miura *et al.* 1998; Miura 2017). The Amazonian frog, *Leptodactylus pentadactylus*, has a karyotype with $2n = 22$ chromosomes and males exhibit an astonishing stable ring-shaped meiotic chain composed of six X and six Y chromosomes (Gazoni *et al.* 2018). Invasive tissue sampling required for such studies often precludes this method from becoming routine for identification of sex in endangered or protected species.

Hypotheses explaining how and why homomorphism of sex chromosomes is maintained, particularly in relation to sex-determining genes in amphibians, are reviewed elsewhere (e.g. Nakamura 2009; Malcom *et al.* 2014; Miura 2017). On the basis of sex linkage, karyological data and a recent genotyping-by-sequencing approach in the torrent frog (*Amolops mantzorum*), eight genes found on up to seven chromosomes are predicted to be candidates for genetic determination of sex in frogs (Miura 2017; Luo *et al.* 2020). Regardless, genetic sex does not necessarily equal phenotypic sex in amphibians, which can be influenced both by temperature and by endocrine-disrupting chemical pollutants causing sex reversal in some species (Flament 2016). For example, in a common frog (*Rana temporaria*) population, 9% of genetically female adults expressed the male phenotype (Alho *et al.* 2010), whereas 2–16% of individuals were sex-reversed in 12 of 16 green frog (*Rana clamitans*) populations (Lambert *et al.* 2019). More recently, in an area associated with anthropogenic use of land, as much as 20% of a wild population of agile frogs (*Rana dalmatina*) that were genotyped as female using single nucleotide polymorphisms (SNPs) showed the male phenotype (Nemesházi *et al.* 2020). In the endangered Chinese giant salamander, *Andrias davidianus*, a female-specific marker, developed using a genome-wide RAD sequencing approach to reveal genetic sex, was used to identify three males reversed from genetic females exposed to high temperature and 13 females reversed from genetic males exposed to 17β-oestradiol (Hu *et al.* 2019a). These results concurred with a previous study in the same species in which sex-reversed individuals were identified by a different female sex-specific marker developed by *de novo* transcriptome assembly (Hu *et al.* 2019b).

Instead of the development of markers based on genotype, a recent study used toe-clip samples from phenotypically sexed *Litoria aurea* individuals to develop sex-linked SNPs and restriction-fragment presence/absence (PA) markers. A male heterogametic XX/XY mode of sex-determination was confirmed in this species through identification of 11 perfectly sex-linked SNP and six strongly sex-linked PA markers (Sopniewski *et al.* 2019). Extending from earlier studies in several species of frogs, toads, newts, and salamanders, vitellogenin proved to be a useful molecular marker for identification of sex and maturation of ovaries in the endangered olm, *Proteus anguinus* (Gredar *et al.* 2019). While vitellogenin was detectable in samples of blood plasma, decreasing over a 10-month period corresponding with degradation of vitellogenic oocytes in adult females, it was not detectable in cutaneous mucus, thereby reducing its potential as a non-invasive sexing tool.

Frequently, the ultimate goal of research on sex identification is to assist conservation efforts and, given the increasing advocacy for non-invasive techniques in wildlife genetics, there is a need to find more non-destructive methods for phenotypic sexing (Zemanova 2019). A relatively novel approach for sex identification is the analysis of genetic markers from a sample taken non-invasively (e.g. a buccal swab), but this method is available for only a few species (Alho et al. 2010; Tamschick et al. 2016; Lambert et al. 2019). All these non-destructive sexing methods can potentially be used on immature animals and sexually monomorphic species, but they are still currently expensive and require specialised equipment and expertise. However, genomic technology is becoming increasingly cheaper and faster as methods are refined. Markers like SNPs now have much shorter reads, meaning they will become increasingly useful on low-template DNA samples. It is likely to be only a matter of time before molecular sexing will become the most efficient way to identify the biological sex of individuals to support amphibian conservation breeding programs globally.

CONDITIONING OF CAPTIVE BROOD STOCK

The importance of good husbandry cannot be understated when attempting to breed amphibians in captivity. In an ideal scenario, animals will breed naturally in captivity without the need for hormonal therapies and associated reproductive technologies (Silla and Byrne 2019; see Chapters 6 and 7). However, many species exhibit reproductive failures, and reproductive technologies have the potential to overcome inhibition of breeding in captivity, offering a valuable tool to improve the success of conservation breeding programs (Kouba et al. 2009; Silla and Byrne 2019). To optimise the efficacy of reproductive technologies, they should be used in conjunction with a high standard of naturalistic husbandry with environmental and social manipulations to condition animals for optimum outcomes (Silla et al. 2021).

Amphibians in the wild often rely on environmental cues for the appropriate timing of breeding (Todd and Winne 2006). While exogenous hormones can immediately stimulate some species to breed in captivity, this can be accompanied by failure or low success rates in artificial propagation if the animal is not naturally conditioned for breeding (e.g. Santana et al. 2015). Additionally, some immediate environmental or conspecific social stimuli act as triggers for the appropriate behaviour and courtship to occur that are either necessary or accompany a higher success rate (Ulloa et al. 2019). Missing environmental cues may cause a variety of reproductive dysfunctions, such as asynchronous release of sperm or eggs, poor amplexus, lack of calling, or low or no fertilisation (e.g. Roth et al. 2010; Santana et al. 2015).

Conditioning triggers can be split into three generalised categories for use on captive amphibians: seasonal cycling, immediate environmental manipulations, and social stimuli. Artificial seasonal cycling is the purposeful manipulation of environmental parameters to mimic the changing of seasons. For example, some species that breed in the spring must first undergo a cooling period in winter before temperatures are elevated to reflect the breeding conditions of spring. This temperature cycling may induce hormone-regulated physiological changes wherein the animal's reproductive system starts preparing for breeding (Zoeller and Moore 1985; Kouba et al. 2012), such as the initiation of maturation of oocytes in females (Minagawa et al. 1993). The use of hormones for breeding may not negate the necessity for pre-breeding conditioning (Lentini 2007; Santana et al. 2015). Seasonal conditioning typically operates on a timescale of weeks or months. In contrast, immediate environmental and social stimuli operate on a timescale of days or hours. Immediate environmental stimuli can act as triggers to make an animal more behaviourally willing to reproduce. For example, immediate environmental manipulations, such as placing an animal in an enclosure that simulates a sudden rainstorm, may promote animals to breed (Plate 3). Social stimuli can also be important for successful breeding. Some species gather in large groups, and individuals may be drawn to conspecific breeding calls though phonotaxis (Farris et al. 2002). Playing recordings of conspecific breeding calls to captive brood stock can simulate a large chorus and stimulate frogs to also start calling and search for a mate, thus prompting higher success in breeding (Birkett et al. 1999; McFadden et al. 2013).

Artificial seasonal manipulations, and immediate environmental and social stimuli can be used in combination and in conjunction with a variety of reproductive technologies (Kouba et al. 2012; Silla et al. 2021). However, making specific recommendations can be problematic because of the extreme diversity in reproductive ecologies exhibited by the thousands of described species of amphibians. Furthermore, only a small fraction of amphibian species have published information on husbandry (Linhoff 2018). When developing protocols for husbandry and making decisions about techniques to use for conditioning, a robust knowledge of the species' natural history and behaviour, in addition to environmental parameters of the species' natural habitat, is helpful (Pramuk and Gagliardo 2012; Michaels *et al.* 2014). We do not provide specific recommendations for every species, but below we discuss a variety of environmental parameters and examples of techniques that may be helpful and can be tailored to a range of species.

Age and body condition

Fundamental to ensuring successful application of reproductive technologies and maximising reproductive fitness, breeding stock must have reached sexual maturity (ideally in their reproductive prime to avoid deleterious effects of parental ageing on the fitness of offspring, e.g. see Schroeder *et al.* (2015)), and be in good breeding condition (Browne and Zippel 2007). The age that amphibians reach sexual maturity is highly species-specific, and it may range from 1 year to over 10 years (Amat and Meiri 2018). The age of sexual maturity may also differ between sexes and populations (Augert and Joly 1993; Sarasola-Puente *et al.* 2011). Second, body condition is an important metric to assess for successful propagation in captivity. Poor nutritional condition may be a contributing source of long-term mortality (e.g. Pessier *et al.* 2014). The conservation breeding program of the Puerto Rican crested toad recommends weighing animals and performing an examination of general health, including an assessment of the reserves of body fat before determining whether an animal is acceptable for captive breeding (Lentini 2007). Body condition may be particularly important for species that require a period of dormancy to prepare for reproduction because animals must have sufficient stores of fat to survive (Seymour 1973; Santana *et al.* 2015), although this may be both species-specific and gender-specific. For example, in a study of boreal toads (*Anaxyrus boreas*), males amplexed more often after undergoing a period of hibernation in captivity (Roth *et al.* 2010). In contrast, female boreal toads that did not undergo a period of hibernation and were treated with GnRHa produced more eggs, probably because they retained more weight. While each species may vary in requirements, all animals must be both sexually mature and in good bodily condition, either for using reproductive technologies or for natural breeding.

Seasonal cycling of environmental conditions

Seasonal cycling mimics natural seasonal variation that a species would normally experience in the wild. While some amphibian species exhibit continuous or year-round breeding, most species are seasonal breeders (Wells 2010). Many amphibians time their breeding to occur during an optimal season based primarily on temperature and rainfall; for example, for most species the wet season is likely to be better for breeding compared to the dry season. Additionally, physiological changes necessary for breeding, such as oogenesis and spermatogenesis, may be triggered by seasonal or environmental changes and are critical for reproduction in captivity (Iturriaga *et al.* 2014). For example, mountain yellow-legged frogs (*Rana muscosa*) that did not undergo a period of hibernation in captivity and were not subjected to hormonal therapies had reduced advertisement calls and amplexus, and less successful breeding (Santana *et al.* 2015). While hormonal therapies may help overcome these barriers, ovulation and spermiation may still be unsuccessful if individuals have not received the appropriate environmental cues to stimulate the maturation of gametes (Whitaker 2001; Kouba *et al.* 2012).

Some amphibians may go into a seasonally dormant state of brumation during cold periods, or they may aestivate and escape periods of dry or hot weather. These processes allow amphibians to survive periods of extreme cold or drought in the wild. Dormancy may last many months. For example, the mountain yellow-legged frog (*Rana muscosa*) undergoes an annual 7-month dormancy period during winter (Bradford 1983). Without a period of brumation, this species typically fails to breed in captivity (Santana *et al.* 2015). Similarly, spadefoot toads (*Scaphiophus* sp.) also undergo a period of dormancy, surviving off of fat stores; ovarian eggs continue to enlarge in preparation for the forthcoming breeding season while the toads are dormant (Seymour 1973). Both brumation and aestivation can be simulated in a captive environment. Artificially creating environmental conditions in captivity to induce brumation is used in various conservation breeding programs. The program for the boreal toad (*Anaxyrus boreas*) recommends using plastic containers containing a disinfected, layered substrate of pea gravel, charcoal, coarse sand, and sphagnum moss (Scherff-Norris *et al.* 2002). A variety of methods to cool animals is available, such as placing them in refrigerators within specially made aestivation chambers (Pramuk and Gagliardo 2012). If animals are kept in a climate-controlled area and all animals within the room or building are to undergo the same environmental manipulations, the entire area can be cooled. Enclosures can also be moved outside to allow the experience of natural changes in temperature if appropriate for the species; however, care must be taken to provide appropriate refugia and avoid the transmission of disease from wild frogs. Hernandez (2017) successfully cycled conditions for mole salamanders (*Ambystoma talpoideum*) by placing them on a porch during colder months before moving them back inside. Of note, some programs such as the Species Survival Plans of the Puerto Rican crested toad and Wyoming toad have recommended prophylactic antifungal baths with itraconazole or eniloconazole before inducing dormancy to reduce chances of fungal infections that are more prevalent at lower temperatures (Lentini 2007). Additionally, animals should be fasted for 1 or 2 weeks before cooling to prevent undigested food from rotting in the animal's stomach during dormancy (Scherff-Norris *et al.* 2002; Lentini 2007; Santana *et al.* 2015). Brumation in amphibians is not without risk and mortalities can occur. To reduce this risk, amphibians should be in good condition before brumation, temperatures should be changed gradually (< 5°C/day) and the duration of brumation should be the same or shorter than natural periods. Programs should also be cautious when developing protocols for poorly studied species. Small trials with few animals initially are recommended. Additionally, brumation may weaken the immune response of amphibians, which highlights the importance of proper biosecurity protocols to limit the impacts of disease on captive populations (Cooper *et al.* 1992; Pessier and Mendelson 2017).

For species that do not undergo dormancy, other seasonal changes in abiotic stimuli are still important and manipulating them in captivity is of benefit. For example, seasonal changes in day length (photoperiod) and light intensity can be simulated in captivity using artificial lighting wired to timers (Browne *et al.* 2007). Regardless of the locality at which the animal is kept in captivity, the lighting system can be designed to simulate the natural light cycle in the animal's native range. Lights that provide appropriate levels of UVA and UVB wavelengths may prevent some nutritional and metabolic disorders. However, too much UV light may be detrimental to animals at any life stage, and embryos in particular may be more sensitive and prone to deformities (Blaustein *et al.* 1997).

Many species time their reproduction to occur during the rainy season. Instead of, or in combination with, strong manipulations in temperature, simulated dry and wet seasons may be necessary for many species, and rainfall can be seasonally simulated. Daily spraying with a hose or increasing the frequency of automatic misting systems wired to timers can be easily performed (Pramuk and Gagliardo 2012). Alternatively, it may also be appropriate to reduce rainfall to simulate a dryer period before starting an artificial rainy season, thus creating a greater environmental gradient (Behr and Roedder 2018). Humidity can also be manipulated in conjunction with rainfall or separately. Increasing or decreasing the frequency of misting, or adding or removing water-retaining substrate and water bowls, can rapidly change humidity in the enclosure. It is important to note that because of their sensitivity to desiccation, a reliable humidity gauge should be kept in the enclosure and animals must be closely monitored when altering any changes in humidity outside of their normal parameters. Rapid or severe reductions in humidity can result in stress or mortality.

For optimum breeding success, animals must be in good bodily condition and receive proper nutrition in captivity. Cycling the quantity and nutrient variety of feed (e.g. increasing feeding for several weeks before the breeding season) may be important for animals to sequester critical reserves of energy and nutrients (Girish and Saidapur 2000). For example, better nutrition may improve the provisioning of yolk and the maturation of eggs (Prasadmurthy and Saidapur 1987). In a study of boreal toads, females with greater weight laid more eggs after being treated with GnRHa in captivity (Roth *et al.* 2010). Increased feeding before a period of hibernation or aestivation may also help animals retain a higher weight that is helpful for breeding immediately afterwards. Each species, and even different population, may require differing husbandry, and protocols successfully used with one species or population may not work for another. Rather, established protocols should just be used as general starting points when developing new species-specific husbandry.

Immediate environmental manipulations

In contrast to seasonal cycling, immediate environmental manipulations are short-term and typically occur immediately before and/or during attempts at breeding. These manipulations may simulate local weather conditions, such as rainstorms. Alternatively, by moving animals into enclosures that mimic the species' preferred type of microhabitat for breeding, it is possible to simulate an animal migrating to a specific habitat for breeding, such as a stream or pond, in contrast to an upland habitat.

Simulating rainfall

Simulated rainfall frequently is used for captive-breeding amphibians. Rain may be a strong immediate environmental cue to induce breeding. For example, a species that breeds in ephemeral bodies of water may rely on associated rare rainfall to inform them of optimal timing for reproduction. Animals can be placed in a specially designed rain chamber that is separate from their regular enclosure. Rain chambers can be partially flooded and contain continuously dripping or flowing water from above (Plate 3). Simulated rain can be used in conjunction with hormonal stimulation. For example, Lentini (2007) recommended placing the Puerto Rican crested toad in a rain chamber for several hours or days before utilising hormonal therapies. Successful captive breeding of Klappenbach's red-bellied frog (*Melanophryniscus klappenbachi*) utilised a 5-month simulated dry season immediately before placing animals into a continuously running rain chamber (Behr and Roedder 2018). A circulating water system or a drain should be installed to prevent the enclosure from flooding. Fluctuations in barometric pressure typically accompany approaching storm systems, which may also be an important trigger for some species to breed in captivity. For example, Díaz *et al.* (2014) found that Hispaniolan yellow tree frogs (*Osteopilus pulchrilineatus*) frequently reproduced in captivity when natural storms occurred. However, manipulation of barometric pressure can be challenging in captivity, and has not been well studied. If flexibility is allowed, it is possible to time captive breeding with the use of reproductive technologies on captive stock when naturally occurring storm systems are approaching. This technique has been used in conjunction with artificial rainfall to simulate storms in captivity (L. J. Linhoff, unpublished data).

Provision of sites for advertisement and oviposition

When hormonal therapy is to be used in isolated males and females for subsequent artificial fertilisation (AF, also referred to as *in vitro* fertilisation, IVF) and/or biobanking, the provision of suitable sites for advertisement and oviposition may not be necessary. However, providing suitable habitat for courtship, oviposition, and embryonic development is of utmost importance when individuals are placed together following hormonal therapy as using hormonal manipulations and seasonal cycling alone does not negate the need for a species' required breeding habitat. For example, moving water typically has higher dissolved oxygen levels, and some species are specifically adapted for reproduction and growth under these conditions. Without moving water to oxygenate eggs and larvae, reproduction may fail in these species. For example, Panamanian golden frogs (*Atelopus zeteki*) require a stream to breed. Adults are normally kept in captivity in an enclosure without flowing water, which is more similar to an upland habitat (Poole 2006). However, before the application of hormonal therapies to induce spawning the animals are transferred to a breeding enclosure with flowing water (Della Togna *et al.* 2017). A partially filled aquarium with a submerged pump and bubblers can be used to cycle water and simulate a flowing stream (Plate 4) (Pramuk and Gagliardo 2012). Care should be taken to make sure that eggs or larvae cannot be sucked into the pumps or filters after reproduction occurs. A variety of microsites in the enclosure that experience different rates of flow is ideal because it provides the animals with a choice of potential oviposition sites.

Some species of amphibians seek very specific microhabitats to breed in the wild, such as a den, phytotelma, burrow, nest site, or other specific substrate for deposition of their eggs. Providing the required microhabitats or resources before breeding helps stimulate reproductive receptivity of brood stock, thus allowing animals time to explore and become comfortable with their altered enclosures and locate possible sites for advertisement and/or breeding. For example, in the northern corroboree frog (*Pseudophryne pengilleyi*), males are placed into enclosures for breeding 1 week before the administration of exogenous hormones to induce spawning in male–female pairs, so that males can first establish a nest site and begin vocal advertisement (Silla *et al.* 2018). To acquire and defend microhabitats or territories for advertisement and breeding, males of some species compete vocally without physical contact (Byrne and Keogh 2007), whereas others may aggressively engage in physical combat such as wrestling (Byrne and Roberts 2004) (discussed further under social stimuli below). Additionally, females may not lay eggs, or hormone-treated females may be induced to lay eggs without amplexus/fertilisation if specific oviposition sites are not available. For example, male hellbender salamanders (*Cryptobranchus alleganiensis*) seek out underwater dens and attract females into them for mating. Ettling *et al.* (2013) successfully utilised artificial nesting boxes in an artificial stream environment combined with seasonal cycling to encourage breeding in hellbender salamanders. The males defended their den from other conspecific males and predators. Without access to a den, it is highly unlikely that reproduction would be successful in captivity.

When developing breeding protocols, the preferred habitat and microsite required for breeding should be assessed. Practitioners should ask whether there is a specific resource that the animals prefer for oviposition, such as a rock den, limestone crack, specific depth of water, phytotelma, or a large leaf overhanging a body of water. How can these structures and conditions be replicated in captivity? Are the artificial environmental conditions also suitable for the survival of embryos and larvae following fertilisation/oviposition? It is important to make sure that the enclosure provides the correct habitat for breeding and contains any necessary microhabitats critical for conditioning animals and successfully implementing hormonal therapies to induce spawning.

Social stimuli

Amphibians exhibit dozens of described mating systems with vast variation in natural history and behaviours (reviewed by Wells 2010). Animals may gather in huge numbers, or they can avoid all other same-sex individuals. Some species exhibit complex parental care, whereas others vigorously compete for access to oviposition sites and females. Understanding a species' mating system and behaviour may help to replicate appropriate conspecific social interactions that optimise reproductive output. Conspecific interactions can prime animals for breeding and may enhance the performance of reproductive technologies.

Any physical social interactions should be closely monitored. In wild populations of some species, aggression is extreme and may result in physical male–male combat and injury or even death to females during forced copulations with multiple males (Byrne and Roberts 1999). With abundant space, individuals may be able to move away from one another. However, in a confined enclosure, aggression may be amplified, co-housed animals may not be able to escape aggressive conspecifics, and brood stock could subsequently be injured or placed in an environment of high stress that is counterproductive for reproduction. While the impacts of crowding and stress are likely to be highly species-specific and

dependent on their duration, and are poorly studied in amphibians, they warrant consideration (e.g. Cikanek *et al.* 2014). For example, many North American and European ranids and bufonids exhibit scramble competition (Hettyey *et al.* 2005). Usually, the males arrive and form dense choruses at the location for breeding before females arrive. The males search for any potential female approaching the site, and they aggressively compete for the possession of a female. In captivity, simulating a scramble type of competition is possible by placing multiple animals in an enclosure before attempts at captive breeding. However, this may not be necessary or beneficial; in fact even in wild populations simultaneous polyandry can lead to reduced fertilisation (Byrne and Roberts 1999).

For many other amphibian species, male–male competitive interactions may consist of relatively harmless posturing or displays, or acoustic competition. In such species with low or no physical aggression, multiple individuals can be placed in breeding enclosures. Alternatively, individuals can receive social stimuli even from isolated enclosures by the playing of a recording of reproductive calls to create an artificial chorus, and/or allowing animals to view each other in adjacent enclosures, and this is routine in many amphibian captive breeding facilities (detailed below). Furthermore, it is important to consider whether parentage needs to be known for the offspring that are generated. In an analysis of wild moor frogs (*Rana arvalis*) that exhibit scramble competition during spawning, polyandry occurred in 67% of egg clutches (Rausch *et al.* 2014). Combining animals in a tank may be beneficial for improving natural breeding and hormonally induced spawning, but it may be impossible to accurately develop a captive-breeding studbook to control genetics within a captive population.

Manipulating the numbers in a group and the composition of various sex ratios may help simulate what an animal experiences in the wild and allow females to exert mate choice. For example, northern leopard frogs (*Lithobates pipiens*) typically are placed in breeding tanks (containing 200 L of water and submerged branches as a substrate for spawning) at a ratio of three males to two females per enclosure following administration of reproductive hormones to induce spawning (Trudeau *et al.* 2013). Scherff-Norris *et al.* (2002) recommended that when captive-breeding the boreal toad (*Anaxyrus boreas*), three or four males first be placed in a breeding enclosure. Once the males exhibit breeding behaviour, such as calling and amplexing one another, multiple females are then added. If natural reproduction does not occur within 24 h, then injections of hormones are applied.

Instead of directly placing animals into the same enclosure, it is also possible to manipulate visual encounters between animals to stimulate reproductive behaviours. Placing an opaque barrier between enclosures allows visual access to be manipulated. This technique has been used to manipulate visual encounters between various species of male harlequin toads (*Atelopus* spp.) (L. J. Linhoff, unpublished data). Male *Atelopus* frogs may visually signal other males by foot waving, a visual agonistic behaviour (Lindquist and Hetherington 1996). Prior to placing males into tanks for breeding, males can be sequestered in enclosures resembling upland habitat without visual access to other males to avoid animals becoming acclimated to male–male interactions. After several weeks, placing males in tanks that simulate a stream environment with visual access to other males may mimic more natural conspecific interactions.

A chorus of calling conspecifics can easily be replicated in captivity using a recording of breeding calls. Playing recorded calls for captive animals at the appropriate taxon-specific duration and time of day may induce more natural breeding behaviour (Birkett *et al.* 1999; McFadden *et al.* 2013). This technique has been used in conjunction with various reproductive technologies. For example, during captive breeding of the Puerto Rican crested toad (*Peltophryne lemur*), breeding calls are played the previous day and night before injecting males with hCG (Lentini 2007). Conspecific interactions are an integral part of successful captive breeding and can improve the success of hormonal therapies to induce spawning. Manipulation of animals' interactions with one another, such as altering the number of animals in an enclosure, can prime animals for successful breeding. Social interactions that mimic what an animal would experience in the wild should be used synergistically with both seasonal cycling and immediate environmental manipulations for optimum success.

CONCLUSIONS

The ongoing amphibian conservation crisis is likely to necessitate breeding programs to be implemented for an ever wider taxonomic variety of species. A program's success requires foundational research in natural history, reproductive biology, and husbandry paired with novel implementation of reproductive technologies. Both the ability to accurately identify the sex of animals and correctly condition brood stock for reproduction are key starting points for every amphibian conservation breeding program and both are active areas of continued research. In monomorphic species in which sex cannot easily be determined visually, the need to optimise animal welfare will drive the technological development of methods to non-invasively ascertain the sex of individuals more efficiently and cheaply. Advances in internal imaging technologies and improved accessibility of genetically based testing combined with a reduction in costs are likely to provide new solutions. New developments in amphibian husbandry combined with concerted efforts by the amphibian conservation community to provide professional training and research in amphibian husbandry are foundational to implementing reproductive technologies (Zipple *et al.* 2011; Harding *et al.* 2017). Employing more naturalistic husbandry combined with advanced environmental and behavioural enrichment will increase the welfare of captive animals and promote greater success in breeding, with and without the use of reproductive technologies. Future research should emphasise a more holistic approach, improving the effectiveness of conservation breeding programs by integrating systematic management of risk, robust scientific trials, and adaptive management in combination with reproductive technologies that benefit both wild and captive populations. New research and refinement of methods for identifying sex, combined with exciting developments in protocols for husbandry, will open new possibilities in the management of captive amphibians that will provide direct benefits for the conservation of endangered species.

REFERENCES

Alho JS, Matsuba C, Merilä J (2010) Sex reversal and primary sex ratios in the common frog (*Rana temporaria*). *Molecular Ecology* **19**(9), 1763–1773. doi:10.1111/j.1365-294X.2010.04607.x

Amat F, Meiri S (2018) Geographical, climatic and biological constraints on age at sexual maturity in amphibians. *Biological Journal of the Linnean Society. Linnean Society of London* **123**, 34–42. doi:10.1093/biolinnean/blx127

Andersson MB (1994) *Sexual Selection*. Princeton University Press, Princeton NJ, USA.

Asahara M, Obayashi Y, Suzuki A, Kamigaki A, Ikeda T (2020) Sexual dimorphism in external morphology of the American bullfrog *Rana* (*Aquarana*) *catesbeiana* and the possibility of sex determination based on tympanic membrane/eye size ratio. *The Journal of Veterinary Medical Science* **82**, 1160–1164. doi:10.1292/jvms.20-0039

Augert D, Joly P (1993) Plasticity of age at maturity between two neighbouring populations of the common frog (*Rana temporaria L.*). *Canadian Journal of Zoology* **71**, 26–33. doi:10.1139/z93-005

Baugh AT, Bastien B, Still MB, Stowell N (2018) Validation of water-borne steroid hormones in a tropical frog (*Physalaemus pustulosus*). *General and Comparative Endocrinology* **261**, 67–80. doi:10.1016/j.ygcen.2018.01.025

Baugh AT, Gray-Gaillard SL (2021) Excreted testosterone and male sexual proceptivity: a hormone validation and proof-of-concept experiment in túngara frogs. *General and Comparative Endocrinology* **300**, 113638. doi:10.1016/j.ygcen.2020.113638

Beaty LE, Emmering QC, Bernal XE (2016) Mixed sex effects on the second-to-fourth digit ratio of túngara frogs (*Engystomops pustulosus*) and Cane Toads (*Rhinella marina*). *The Anatomical Record* **299**, 421–427. doi:10.1002/ar.23322

Behr N, Roedder D (2018) Captive management, reproduction, and comparative larval development of Klappenbach's Red-bellied Frog, *Melanophryniscus klappenbachi* Prigioni and Langone, 2000. *Amphibian & Reptile Conservation* **12**, 18–26.

Bell RC, Zamudio KR (2012) Sexual dichromatism in frogs: natural selection, sexual selection and unexpected diversity. *Proceedings. Biological Sciences* **279**, 4687–4693. doi:10.1098/rspb.2012.1609

Birkett J, Vincent M, Banks C (1999) Captive management and rearing of the roseate frog, *Geocrinia rosea*, at Melbourne Zoo. *Herpetofauna* **29**, 49–56.

Blackburn DC (2009) Diversity and evolution of male secondary sexual characters in African squeakers and long-fingered frogs. *Biological Journal of the Linnean Society. Linnean Society of London* **96**, 553–573. doi:10.1111/j.1095-8312.2008.01138.x

Blaustein AR, Kiesecker JM, Chivers DP, Anthony RG (1997) Ambient UV-B radiation causes deformities in amphibian embryos. *Proceedings of the National Academy of Sciences of the United States of America* **94**, 13735–13737.

Bradford DF (1983) Winterkill, oxygen relations, and energy metabolism of a submerged dormant amphibian, *Rana muscosa*. *Ecology* **64**, 1171–1183. doi:10.2307/1937827

Bronson V, Vance C (2019) Anuran reproduction. In *Fowler's Zoo and Wild Animal Medicine Current Therapy*. Volume 9. (Eds RE Miller, N Lamberski and P Calle) pp. 371–379. Elsevier, St Louis MO, USA.

Brown J, Walker S, Steinman K (2004) Endocrine manual for the reproductive assessment of domestic and non-domestic species. Smithsonian's National Zoological Park, Conservation and Research Center, Front Royal VA, USA.

Browne RK, Odum RA, Herman T, Zippel K (2007) Facility design and associated services for the study of amphibians. *ILAR Journal* **48**, 188–202. doi:10.1093/ilar.48.3.188

Browne RK, Zippel K (2007) Reproduction and larval rearing of amphibians. *ILAR Journal* **48**, 214–234. doi:10.1093/ilar.48.3.214

Brownlee R, Phillips K, Marsh-Armstrong N (2018) Evaluation of African Clawed Frog (*Xenopus laevis*) anatomy and sexing of juveniles with ultrasound. In *International Association for Aquatic Animal Medicine, Annual Meeting and Conference*. 19–23 May, Long Beach, California. <https://www.vin.com/apputil/content/defaultadv1.aspx?pId=20778&meta=Generic&catId=113374&id=8504986&ind=38&objTypeID=17>

Burns R, Bell B, Haigh A, Bishop P, Easton L, Wren S, Germano J, Hitchmough R, Rolfe R, Makan T (2018) *Conservation Status of New Zealand Amphibians, 2017*. New Zealand Threat Classification Series 25. Department of Conservation, Wellington, NZ.

Buzatto BA, Roberts JD, Simmons LW (2015) Sperm competition and the evolution of precopulatory weapons: increasing male density promotes sperm competition and reduces selection on arm strength in a chorusing frog. *Evolution* **69**, 2613–2624. doi:10.1111/evo.12766

Byrne PG, Keogh JS (2007) Terrestrial toadlets use chemosignals to recognize conspecifics, locate mates and strategically adjust calling behaviour. *Animal Behaviour* **74**, 1155–1162. doi:10.1016/j.anbehav.2006.10.033

Byrne PG, Roberts JD (1999) Simultaneous mating with multiple males reduces fertilization success in the myobatrachid frog *Crinia georgiana*. *Proceedings of the Royal Society of London. Series B, Biological Sciences* **266**, 717–721. doi:10.1098/rspb.1999.0695

Byrne PG, Roberts JD (2004) Intrasexual selection and group spawning in quacking frogs (*Crinia georgiana*). *Behavioral Ecology* **15**, 872–882. doi:10.1093/beheco/arh100

Cadle JE (1995) A new species of *Boophis* (Anura: Rhacophoridae) with unusual skin glands from Madagascar, and a discussion of variation and sexual dimorphism in *Boophis albilabris* (Boulenger). *Zoological Journal of the Linnean Society* **115**, 313–345. doi:10.1111/j.1096-3642.1995.tb01427.x

Calatayud NE, Stoops M, Durrant BS (2018) Ovarian control and monitoring in amphibians. *Theriogenology* **109**, 70–81. doi:10.1016/j.theriogenology.2017.12.005

Calatayud NE, Chai N, Gardner NR, Curtis MJ, Stoops MA (2019) Reproductive techniques for ovarian monitoring and control in amphibians. *Journal of Visualized Experiments* **147**, e58675. doi:10.3791/58675

Chang JL (2008) Sexual dimorphism of the second-to-fourth digit length ratio (2D: 4D) in the strawberry poison dart frog (*Oophaga pumilio*) in Costa Rica. *Journal of Herpetology* **42**, 414–416. doi:10.1670/07-153.1

Cikanek SJ, Nockold S, Brown JL, Carpenter JW, Estrada A, Guerrel J, Hope K, Ibáñez R, Putman SB, Gratwicke B (2014) Evaluating group housing strategies for the ex-situ conservation of harlequin frogs (*Atelopus* spp.) using behavioral and physiological indicators. *PLoS One* **9**, e90218. doi:10.1371/journal.pone.0090218

Cooper EL, Wright RK, Klempau AE, Smith CT (1992) Hibernation alters the frog's immune system. *Cryobiology* **29**, 616–631. doi:10.1016/0011-2240(92)90066-B

Delgado MJ, Gutierrez P, Alonso-Bedate M (1989) Seasonal cycles in testicular activity in the frog, *Rana perezi*. *General and Comparative Endocrinology* **73**, 1–11. doi:10.1016/0016-6480(89)90049-X

Della Togna G, Trudeau VL, Gratwicke B, Evans M, Augustine L, Chia H, Bronikowski EJ, Murphy JB, Comizzoli P (2017) Effects of hormonal stimulation on the concentration and quality of excreted spermatozoa in the critically endangered Panamanian golden frog (*Atelopus zeteki*). *Theriogenology* **91**, 27–35. doi:10.1016/j.theriogenology.2016.12.033

Díaz LM, Incháustegui SJ, Marte C (2014) Preliminary experiences with the husbandry, captive breeding, and development of the Hispaniolan Yellow Tree Frog, *Osteopilus pulchrilineatus* (Amphibia: Anura: Hylidae), with ecological and ethological notes from the wild. *Herpetological Review* **45**, 52–59.

Duellman WE, Trueb L (1994) *Biology of Amphibians.* Johns Hopkins University Press, Baltimore MA, USA.

Emerson SB (1996) Phylogenies and physiological processes – the evolution of sexual dimorphism in Southeast Asian frogs. *Systematic Biology* **45**, 278–289.

Epstein MS, Blackburn DG (1997) Histology and histochemistry of androgen-stimulated nuptial pads in the leopard frog, *Rana pipiens,* with notes on nuptial gland evolution. *Canadian Journal of Zoology* **75**, 472–477. doi:10.1139/z97-057

Ettling JA, Wanner MD, Schuette CD, Armstrong SL, Pedigo AS, Briggler JT (2013) Captive reproduction and husbandry of adult Ozark hellbenders, *Cryptobranchus alleganiensis bishopi. Herpetological Review* **44**, 605–610.

Farris HE, Rand AS, Ryan MJ (2002) The effects of spatially separated call components on phonotaxis in túngara frogs: evidence for auditory grouping. *Brain, Behavior and Evolution* **60**, 181–188. doi:10.1159/000065937

Flament S (2016) Sex reversal in amphibians. *Sexual Development: Genetics, Molecular Biology, Evolution, Endocrinology, Embryology, and Pathology of Sex Determination and Differentiation* **10**, 267–278. doi:10.1159/000448797

Gabor CR, Bosch J, Fries JN, Davis DR (2013) A non-invasive water-borne hormone assay for amphibians. *Amphibia-Reptilia* **34**, 151–162. doi:10.1163/15685381-00002877

Gazoni T, Haddad CFB, Narimatsu H, Cabral-de-Mello DC, Lyra ML, Parise-Maltempi PP (2018) More sex chromosomes than autosomes in the Amazonian frog *Leptodactylus pentadactylus. Chromosoma* **127**, 269–278. doi:10.1007/s00412-018-0663-z

Germano JM, Molinia FC, Bishop PJ, Cree A (2009) Urinary hormone analysis assists reproductive monitoring and sex identification of bell frogs (*Litoria raniformis). Theriogenology* **72**, 663–671. doi:10.1016/j.theriogenology.2009.04.023

Germano JM, Cree A, Bishop PJ (2011) Ruling out the boys from the girls: can subtle morphological differences identify sex of the apparently monomorphic frog, *Leiopelma pakeka? New Zealand Journal of Zoology* **38**, 161–171. doi:10.1080/03014223.2010.548076

Germano JM, Molinia FC, Bishop PJ, Bell BD, Cree A (2012) Urinary hormone metabolites identify sex and imply unexpected winter breeding in an endangered, subterranean-nesting frog. *General and Comparative Endocrinology* **175**, 464–472. doi:10.1016/j.ygcen.2011.12.003

Girish S, Saidapur SK (2000) Interrelationship between food availability, fat body, and ovarian cycles in the frog, *Rana tigrina*, with a discussion on the role of fat body in anuran reproduction. *The Journal of Experimental Zoology* **286**, 487–493.doi:10.1002/(SICI)1097-010X(20000401)286:5<487::AID-JEZ6>3.0.CO;2-Z

Graham KM, Kouba AJ, Langhorne CJ, Marcec RM, Willard ST (2016) Biological sex identification in the endangered dusky gopher frog (*Lithobates sevosa*): a comparison of body size measurements, secondary sex characteristics, ultrasound imaging, and urinary hormone analysis methods. *Reproductive Biology and Endocrinology* **14**(1), 1–14. doi:10.1186/s12958-016-0174-9

Graham KM, Langhorne CJ, Vance CK, Willard ST, Kouba AJ (2018) Ultrasound imaging improves hormone therapy strategies for induction of ovulation and in vitro fertilization in the endangered dusky gopher frog (*Lithobates sevosa*). *Conservation Physiology* **6**, coy020. doi:10.1093/conphys/coy020

Gredar T, Leonardi A, Novak M, Sepčić K, Mali LB, Križaj I (2019) Vitellogenin in the European cave salamander, *Proteus anguinus*: its characterization and dynamics in a captive female as a basis for non-destructive sex identification. *Comparative Biochemistry and Physiology. Part B, Biochemistry & Molecular Biology* **235**, 30–37. doi:10.1016/j.cbpb.2019.05.010

Green DM (1988) Cytogenetics of the endemic New Zealand frog, *Leiopelma hochstetteri:* extraordinary supernumerary chromosome variation and a unique sex-chromosome system. *Chromosoma* **97**, 55–70. doi:10.1007/BF00331795

Green DM (1994) Genetic and cytogenetic diversity in Hochstetter's frog, *Leiopelma hochstetteri*, and its importance for conservation management. *New Zealand Journal of Zoology* **21**, 417–424. doi:10.1080/03014223.1994.9518011

Greenberg B (1942) Some effects of testosterone on the sexual pigmentation and other sex characters of the cricket frog (*Acris gryllus*). *The Journal of Experimental Zoology* **91**, 435–451. doi:10.1002/jez.1400910308

Greene AE, Funk WC (2009) Sexual selection on morphology in an explosive breeding amphibian, the Columbia spotted frog (*Rana luteiventris*). *Journal of Herpetology* **43**, 244–251. doi:10.1670/08-112R.1

Harvey LA, Propper CR, Woodley SK, Moore MC (1997) Reproductive endocrinology of the explosively breeding desert spadefoot toad, *Scaphiopus couchii. General and Comparative Endocrinology* **105**, 102–113. doi:10.1006/gcen.1996.6805

Hayes T, Licht P (1992) Gonadal involvement in sexual size dimorphism in the African bullfrog (*Pyxicephalus adspersus*). *The Journal of Experimental Zoology* **264**, 130–135. doi:10.1002/jez.1402640203

Hayes TB (1998) Sex determination and primary sex differentiation in amphibians: genetic and developmental mechanisms. *The Journal of Experimental Zoology* **281**, 373–399. doi:10.1002/(SICI)1097-010X(19980801)281:5<373::AID-JEZ4>3.0.CO;2-L

Hernandez A (2017) Successful reproduction of the mole salamander *Ambystoma talpoideum in* captivity, with an emphasis on stimuli environmental determinants. *Herpetological Bulletin* **141**, 29.

Hettyey A, Török J, Hévizi G (2005) Male mate choice lacking in the agile frog, *Rana dalmatina*. *Copeia* **2005**, 403–408. doi:10.1643/CE-04-115R2

Hoffman EA, Blouin MS (2000) A review of colour and pattern polymorphisms in anurans. *Biological Journal of the Linnean Society. Linnean Society of London* **70**, 633–665. doi:10.1111/j.1095-8312.2000.tb00221.x

Hogan LA, Lisle AT, Johnston SD, Goad T, Robertston H (2013) Adult and juvenile sex identification in threatened monomorphic *Geocrinia* frogs using fecal steroid analysis. *Journal of Herpetology* **47**, 112–118. doi:10.1670/11-290

Holtze S, Lukač M, Cizelj I, Mutschmann F, Szentiks CA, Jelić D, Hermes R, Göritz F, Braude S, Hildebrandt TB (2017) Monitoring health and reproductive status of olms (*Proteus anguinus*) by ultrasound. *PLoS One* **12**, e0182209. doi:10.1371/journal.pone.0182209

Hu Q, Chang C, Wang Q, Tian H, Qiao Z, Wang L, Meng Y, Xu C, Xiao H (2019a) Genome-wide RAD sequencing to identify a sex-specific marker in Chinese giant salamander *Andrias davidianus*. *BMC Genomics* **20**, 1–8. doi:10.1186/s12864-019-5771-5

Hu Q, Tian H, Li W, Meng Y, Wang Q, Xiao H (2019b) Identification of critical sex-biased genes in *Andrias davidianus* by *de novo* transcriptome. *Molecular Genetics and Genomics* **294**, 287–299. doi:10.1007/s00438-018-1508-4

Iturriaga M, Sanz A, Oliva R (2014) Seasonal reproduction of the greenhouse frog *Eleutherodactylus planirostris* (Anura: Eleutherodactylidae) in Havana, Cuba. *South American Journal of Herpetology* **9**, 142–150. doi:10.2994/SAJH-D-13-00039.1

Kashimada K, Koopman P (2010) Sry: the master switch in mammalian sex determination. *Development* **137**, 3921–3930. doi:10.1242/dev.048983

Kouba AJ, Vance CK (2009) Applied reproductive technologies and genetic resource banking for amphibian conservation. *Reproduction, Fertility and Development* **21**, 719–737. doi:10.1071/RD09038

Kouba AJ, Vance CK, Willis EL (2009) Artificial fertilization for amphibian conservation: current knowledge and future considerations. *Theriogenology* **71**, 214–227. doi:10.1016/j.theriogenology.2008.09.055

Kouba A, Vance C, Calatayud N, Rowlison T, Langhorne C, Willard S (2012) Assisted reproductive technologies (ART) for amphibians. In *Amphibian Husbandry Resource Guide*. 2nd edn. (Eds VA Poole and S Grow) pp. 60–118. Association of Zoos and Aquariums, Silver Spring MA, USA.

Kraus F (2008) Remarkable case of anuran sexual size dimorphism: *Platymantis rhipiphalcus* is a junior synonym of *Platymantis boulengeri*. *Journal of Herpetology* **42**, 637–644. doi:10.1670/07-238Rl.1

Lambert MR, Tran T, Kilian A, Ezaz T, Skelly DK (2019) Molecular evidence for sex reversal in wild populations of green frogs (*Rana clamitans*). *PeerJ* **7**, e6449. doi:10.7717/peerj.6449

Lentini A (2007) *Husbandry Manual Puerto Rican Crested Toad (Peltophryne lemur), 2006/07 update*. Keeper and Curator Edition. Toronto Zoo, Toronto, Canada.

Li P-Q, Zhu B-C, Wang Y-F, Xiang X-J (2010) Sex identification of Chinese giant salamander (*Andrias davidianus*) by Doppler B-ultrasound method. *Journal of Biology* **27**, 94–96.

Lindquist ED, Hetherington TE (1996) Field studies on visual and acoustic signaling in the 'earless' Panamanian golden frog, *Atelopus zeteki*. *Journal of Herpetology* **30**, 347–354. doi:10.2307/1565171

Linhoff LJ (2018) Linking husbandry and behavior to enhance amphibian reintroduction success. PhD thesis. Florida International University, USA.

Lowe K, Hero J (2012) Sexual dimorphism and color polymorphism in the Wallum Sedge Frog (*Litoria olongburensis*). *Herpetological Review* **43**, 236–240.

Luna MC, Taboada C, Baêta D, Faivovich J (2012) Structural diversity of nuptial pads in Phyllomedusinae (Amphibia: Anura: Hylidae). *Journal of Morphology* **273**, 712–724. doi:10.1002/jmor.20016

Luo W, Xia Y, Yue B, Zeng X (2020) Assigning the sex-specific markers via genotyping-by-sequencing onto the Y chromosome for a torrent frog *Amolops mantzorum*. *Genes* **11**, 727. doi:10.3390/genes11070727

Malcom JW, Kudra RS, Malone JH (2014) The sex chromosomes of frogs: variability and tolerance offer clues to genome evolution and function. *Journal of Genomics* **2**, 68. doi:10.7150/jgen.8044

McFadden M, Hobbs R, Marantelli G, Harlow P, Banks C, Hunter D (2013) Captive management and breeding of the critically endangered southern corroboree frog (*Pseudophryne corroboree*) (Moore 1953) at Taronga and Melbourne Zoos. *Amphibian & Reptile Conservation* **5**, 70–87.

Medina MF, Ramos I, Crespo CA, González-Calvar S, Fernández SN (2004) Changes in serum sex steroid levels throughout the reproductive cycle of *Bufo arenarum* females. *General and Comparative Endocrinology* **136**, 143–151. doi:10.1016/j.ygcen.2003.11.013

Michaels CJ, Gini BF, Preziosi RF (2014) The importance of natural history and species-specific approaches in amphibian ex-situ conservation. *The Herpetological Journal* **24**, 135–145.

Miller ER, Fowler ME (2012) *Fowler's Zoo and Wild Animal Medicine Current Therapy*. Volume 7-E-Book. Elsevier Health Sciences, St Louis MO, USA.

Minagawa M, Chiu J-R, Kudo F, Ito F, Takashima F (1993) Female reproductive biology and oocyte development of the red frog crab, *Ranina ranina*, off Hachijojima, Izu Islands, Japan. *Marine Biology* **115**, 613–623. doi:10.1007/BF00349369

Miura I (2017) Sex determination and sex chromosomes in Amphibia. *Sexual Development: Genetics, Molecular Biology, Evolution, Endocrinology, Embryology, and Pathology of Sex Determination and Differentiation* **11**, 298 306. doi:10.1159/0004852/0

Miura I, Ohtani H, Nakamura M, Ichikawa Y, Saitoh K (1998) The origin and differentiation of the heteromorphic sex chromosomes Z, W, X, and Y in the frog *Rana rugosa*, inferred from the sequences of a sex-linked gene, ADP/ATP translocase. *Molecular Biology and Evolution* **15**, 1612–1619. doi:10.1093/oxfordjournals.molbev.a025889

Monnet J-M, Cherry MI (2002) Sexual size dimorphism in anurans. *Proceedings of the Royal Society of London. Series B, Biological Sciences* **269**, 2301–2307. doi:10.1098/rspb.2002.2170

Nakamura M (2009) Sex determination in amphibians. *Seminars in Cell & Developmental Biology* **20**, 271–282. doi:10.1016/j.semcdb.2008.10.003

Narayan EJ (2013) Non-invasive reproductive and stress endocrinology in amphibian conservation physiology. *Conservation Physiology* **1**, cot011. doi:10.1093/conphys/cot011

Narayan EJ, Gramapurohit NP (2016) Sexual dimorphism in baseline urinary corticosterone metabolites and their association with body-condition indices in a peri-urban population of the common Asian toad *(Duttaphrynus melanostictus)*. *Comparative Biochemistry and Physiology. Part A, Molecular & Integrative Physiology* **191**, 174–179. doi:10.1016/j.cbpa.2015.10.016

Narayan EJ, Molinia FC, Christi KS, Morley CG, Cockrem JF (2010) Annual cycles of urinary reproductive steroid concentrations in wild and captive endangered Fijian ground frogs *(Platymantis vitiana)*. *General and Comparative Endocrinology* **166**, 172–179. doi:10.1016/j.ygcen.2009.10.003

Navas CA, James RS (2007) Sexual dimorphism of extensor carpi radialis muscle size, isometric force, relaxation rate and stamina during the breeding season of the frog *Rana temporaria* Linnaeus 1758. *The Journal of Experimental Biology* **210**, 715–721. doi:10.1242/jeb.000646

Nemesházi E, Gál Z, Újhegyi N, Verebélyi V, Mikó Z, Üveges B, Lefler KK, Jeffries DL, Hoffmann OI, Bókony V (2020) Novel genetic sex markers reveal high frequency of sex reversal in wild populations of the agile frog *(Rana dalmatina)* associated with anthropogenic land use. *Molecular Ecology* **29**, 3607–3621. doi:10.1111/mec.15596

Pessier A, Mendelson J, III (Eds) (2017) *A Manual for Control of Infectious Diseases in Amphibian Survival Assurance Colonies and Reintroduction Programs*. Version 2.0. IUCN/SSC Conservation Breeding Specialist Group, Apple Valley MN, USA.

Pessier AP, Baitchman EJ, Crump P, Wilson B, Griffith E, Ross H (2014) Causes of mortality in anuran amphibians from an *ex situ* survival assurance colony in Panama. *Zoo iBology* **33**, 516–526. doi:10.1002/zoo.21166

Poole V (2006) *Panamanian Golden Frog Husbandry Manual*. 2nd edn, <http://www.atelopus.com/uploads/pdf/Husbandry Manual.pdf>.

Pramuk J, Gagliardo R (2012) General amphibian husbandry. In *Amphibian Husbandry Resource Guide*. 2nd edn. (Eds A Poole and S Grow) pp. 4–59. Association of Zoos and Aquariums, Silver Spring MD, USA.

Prasadmurthy YS, Saidapur SK (1987) Role of fat bodies in oocyte growth and recruitment in the frog *Rana cyanophlyctis* (Schn.). *The Journal of Experimental Zoology* **243**, 153–162. doi:10.1002/jez.1402430117

Preininger D, Handschuh S, Boeckle M, Sztatecsny M, Hödl W (2016) Comparison of female and male vocalisation and larynx morphology in the size dimorphic foot-flagging frog species *Staurois guttatus*. *The Herpetological Journal* **26**, 187–197.

Rajabi F, Javanbakht H (2019) Sexual dimorphism in digit length ratios in Marsh Frog, *Pelophylax ridibundus* (Ranidae) from Iran. *Journal of Applied Biological Sciences* **13**, 33–36.

Rastogi RK, Iela L, Delrio G, Bagnara JT (1986) Reproduction in the Mexican leaf frog, *Pachymedusa dacnicolor*: II. The male. *General and \Comparative Endocrinology* **62**, 23–35. doi:10.1016/0016-6480(86)90090-0

Rastogi RK, Pinelli C, Polese G, D'Aniello B, Chieffi-Baccari G (2011) Hormones and reproductive cycles in anuran amphibians. In *Hormones and Reproduction of Vertebrates*. (Eds D Norris and K Lopez) pp. 171–186. Elsevier, Denver CO, USA.

Rausch AM, Sztatecsny M, Jehle R, Ringler E, Hödl W (2014) Male body size and parental relatedness but not nuptial colouration influence paternity success during scramble competition in *Rana arvalis*. *Behaviour* **151**, 1869–1884. doi:10.1163/1568539X-00003220

Reyer H-U, Bättig I (2004) Identification of reproductive status in female frogs – a quantitative comparison of nine methods. *Herpetologica* **60**, 349–357. doi:10.1655/03-77

Roth T, Obringer A (2003) Reproductive research and the worldwide amphibian extinction crisis. In *Reproductive Science and Integrated Conservation*. (Eds W Holt, A Pickard, J Roger and D Wildt) pp. 359–374. Cambridge University Press, Cambridge, UK.

Roth TL, Szymanski DC, Keyster ED (2010) Effects of age, weight, hormones, and hibernation on breeding success in boreal toads *(Bufo boreas boreas)*. *Theriogenology* **73**, 501–511. doi:10.1016/j.theriogenology.2009.09.033

Ruiz-Fernández MJ, Jiménez S, Fernández-Valle E, García-Real MI, Castejón D, Moreno N, Ardiaca M, Montesinos A, Ariza S, González-Soriano J (2020) Sex determination in two species of anuran amphibians by magnetic resonance imaging and ultrasound techniques. *Animals (Basel)* **10**, 2142. doi:10.3390/ani10112142

Santana FE, Swaisgood RR, Lemm JM, Fisher RN, Clark RW (2015) Chilled frogs are hot: hibernation and reproduction of the endangered mountain yellow-legged frog *Rana muscosa*. *Endangered Species Research* **27**, 43–51. doi:10.3354/esr00648

Santymire RM, Manjerovic MB, Sacerdote-Velat A (2018) A novel method for the measurement of glucocorticoids in dermal secretions of amphibians. *Conservation Physiology* **6**, coy008. doi:10.1093/conphys/coy008

Sarasola-Puente V, Gosá A, Oromí N, Madeira MJ, Lizana M (2011) Growth, size and age at maturity of the agile frog *(Rana dalmatina)* in an Iberian Peninsula population. *Zoology (Jena, Germany)* **114**, 150–154. doi:10.1016/j.zool.2010.11.009

Scherff-Norris KL, Livo LJ, Pessier A, Fetkavich C, Jones M, Kombert M, Goebel A, Spencer B (2002) *Boreal Toad Husbandry Manual*. Colorado Division of Wildlife, Fort Collins CO, USA. <https://cpw.state.co.us/Documents/Research/Aquatic/pdf/FinalHatcheryManual12-24-02.pdf>.

Scheun J, Greeff D, Medger K, Ganswindt A (2019) Validating the use of dermal secretion as a matrix for monitoring glucocorticoid concentrations in African amphibian species.

Conservation Physiology 7, coz022. doi:10.1093/conphys/coz022

Schildger B, Triet H (2001) Ultrasonography in amphibians. *Seminars in Avian and Exotic Pet Medicine* 10, 169–173. doi:10.1053/saep.2001.24673

Schroeder J, Nakagawa S, Rees M, Mannarelli M-E, Burke T (2015) Reduced fitness in progeny from old parents in a natural population. *Proceedings of the National Academy of Sciences of the United States of America* 112, 4021–4025. doi:10.1073/pnas.1422715112

Sever DM, Staub NL (2011) Hormones, sex accessory structures, and secondary sexual characteristics in amphibians. In *Hormones and Reproduction of Vertebrates*. (Eds D Norris and K Lopez) pp. 83–98. Elsevier, Denver CO, USA, 83–98.

Seymour RS (1973) Energy metabolism of dormant spadefoot toads (*Scaphiopus*). *Copeia* 1973, 435–445. doi:10.2307/1443107

Shine R (1979) Sexual selection and sexual dimorphism in the Amphibia. *Copeia* 1979, 297–306. doi:10.2307/1443418

Silla AJ, Byrne PG (2019) The role of reproductive technologies in amphibian conservation breeding programs. *Annual Review of Animal Biosciences* 7, 499–519. doi:10.1146/annurev-animal-020518-115056

Silla AJ, McFadden M, Byrne PG (2018) Hormone-induced spawning of the critically endangered northern corroboree frog *Pseudophryne pengilleyi*. *Reproduction, Fertility and Development* 30, 1352–1358. doi:10.1071/RD18011

Silla AJ, Calatayud NE, Trudeau VL (2021) Amphibian reproductive technologies: approaches and welfare considerations. *Conservation Physiology* 9, coab011. doi:10.1093/conphys/coab011

Smith CA, Roeszler KN, Ohnesorg T, Cummins DM, Farlie PG, Doran TJ, Sinclair AH (2009) The avian Z-linked gene DMRT1 is required for male sex determination in the chicken. *Nature* 461, 267–271. doi:10.1038/nature08298

Sopniewski J, Shams F, Scheele BC, Kefford BJ, Ezaz T (2019) Identifying sex-linked markers in *Litoria aurea*: a novel approach to understanding sex chromosome evolution in an amphibian. *Scientific Reports* 9, 1–10. doi:10.1038/s41598-019-52970-4

Stetter MD, Wright KM, Whitaker BR (2001) Diagnostic imaging of amphibians. In *Amphibian Medicine and Captive Husbandry*. (Eds K Wright and W Whitaker) pp. 253–272. Krieger Publishing Company, Malabar FL, USA.

Szymanski DC, Gist DH, Roth TL (2006) Anuran gender identification by fecal steroid analysis. *Zoo Biology* 25, 35–46.

Tamschick S, Rozenblut-Kościsty B, Ogielska M, Lehmann A, Lymberakis P, Hoffmann F, Lutz I, Kloas W, Stöck M (2016) Sex reversal assessments reveal different vulnerability to endocrine disruption between deeply diverged anuran lineages. *Scientific Reports* 6, 1–8. doi:10.1038/srep23825

Todd BD, Winne CT (2006) Ontogenetic and interspecific variation in timing of movement and responses to climatic factors during migrations by pond-breeding amphibians. *Canadian Journal of Zoology* 84, 715–722. doi:10.1139/z06-054

Trudeau VL, Schueler FW, Navarro-Martin L, Hamilton CK, Bulaeva E, Bennett A, Fletcher W, Taylor L (2013) Efficient induction of spawning of northern leopard frogs (*Lithobates pipiens*) during and outside the natural breeding season. *Reproductive Biology and Endocrinology* 11, 1–9. doi:10.1186/1477-7827-11-14

Ulloa JS, Aubin T, Llusia D, Courtois ÉA, Fouquet A, Gaucher P, Pavoine S, Sueur J (2019) Explosive breeding in tropical anurans: environmental triggers, community composition and acoustic structure. *BMC Ecology* 19, 1–17. doi:10.1186/s12898-019-0243-y

Vance CK, Kouba AJ, Willard ST (2014) Near infrared spectroscopy applications in amphibian ecology and conservation: gender and species identification. *NIR News* 25, 10–15. doi:10.1255/nirn.1444

Wallace H, Badawy GMI, Wallace BMN (1999) Amphibian sex determination and sex reversal. *Cellular and Molecular Life Sciences CMLS* 55, 901–909. doi:10.1007/s000180050343

Wells KD (2010) *The Ecology and Behavior of Amphibians*. University of Chicago Press, Chicago IL, USA.

Whitaker B (2001) Reproduction. In *Amphibian Medicine and Captive Husbandry*. (Eds K Wright and B Whitaker) pp. 285–307. Krieger, Malabar FL, USA.

Williams TD, Monaghan P, Mitchell PI, Scott I, Houston DG, Ramsey S, Ensor K (1997) Evaluation of a non destructive method for determining egg composition using total body electrical conductivity (TOBEC) measurements. *Journal of Zoology* 243, 611–622. doi:10.1111/j.1469-7998.1997.tb02805.x

Zemanova MA (2019) Poor implementation of non-invasive sampling in wildlife genetics studies. *Rethinking Ecology* 4, 119. doi:10.3897/rethinkingecology.4.32751

Zipple K, Johnson K, Gagliardo R, Gibson R, McFadden M, Browne R, Martinez C, Townsend E (2011) The Amphibian Ark: a global community for ex situ conservation of amphibians. *Herpetological Conservation Biology* 6, 340–352.

Zoeller RT, Moore FL (1985) Seasonal changes in luteinizing hormone-releasing hormone concentrations in microdissected brain regions of male rough-skinned newts (*Taricha granulosa*). *General and Comparative Endocrinology* 58, 222–230. doi:10.1016/0016-6480(85)90338-7

Plate 1. Reproductive technologies have been applied to a diversity of threatened amphibian species, including the (A) booroolong frog, *Litoria booroolongensis*, (B) boreal toad, *Anaxyrus boreas boreas*, (C) dusky gopher frog, *Lithobates sevosus*, (D) Fijian ground frog, *Platymantis vitiana*, (E) great crested newt, *Triturus cristatus*, (F) green and golden bell frog, *Litoria aurea*, (G) Hamilton's frog, *Leiopelma hamiltoni*, (H) hellbender, *Cryptobranchus alleganiensis*, (I) Mexican axolotl, *Ambystoma mexicanum*, (J) northern corroboree frog, *Pseudophryne pengilleyi*, (K) Panamanian golden frog, *Atelopus zeteki*, (L) Puerto Rican crested toad, *Peltophryne lemur*, (M) southern corroboree frog, *Pseudophryne corroboree*, and (N) Wyoming toad, *Anaxyrus baxteri*. Photographs A, J, and M courtesy of Aimee Silla, B courtesy of Natalie Calatayud, C courtesy of Sinlan Poo, D courtesy of Edward Narayan, E courtesy of David O'Brien, F courtesy of Cassandra Bugir, G courtesy of Ken Miller and Jennifer Germano, H courtesy of Brian Gratwicke, Smithsonian Conservation Biology Institute, I courtesy of Jonathan Lai, K courtesy of Gina Della Togna, L courtesy of Dustin Smith, and N courtesy of Kristin Hinkson, all used with permission.

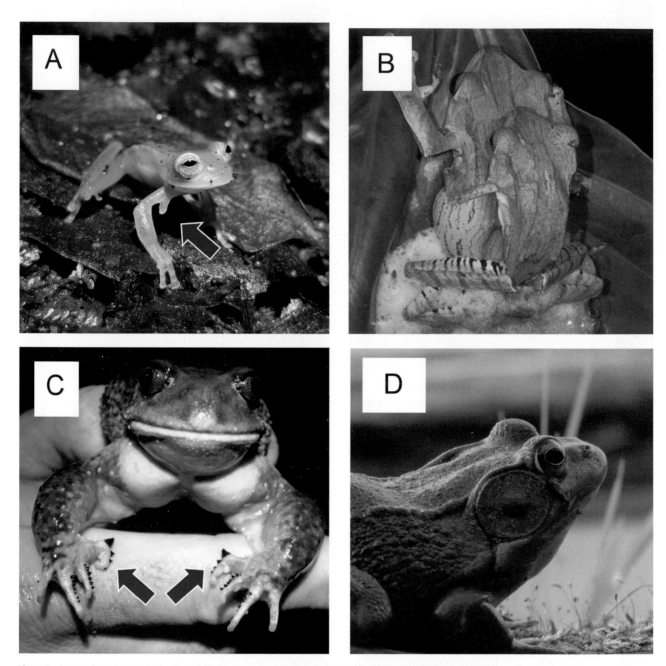

Plate 2. Examples of morphological differences used for the identification of sex: (A) males of the glass frog *Espadarana prosoblepon* have a protruding humoral spine; (B) females of *Polypedates otilophus* are much larger than males, making sexual differentiation of adults possible using size alone; (C) male giant burrowing frogs, *Heleioporus australiacus*, exhibit large nuptial spines on their fore fingers during the breeding season; and (D) a male bullfrog, *Lithobates catesbeianus*, has a tympanum much larger in diameter than that of a female. The large size does not change seasonally, allowing for year-round identification of sex. Photo credits: E. Lassiter (A); J. Reardon (B); L. Linhoff (C–D).

Plate 3. A rain chamber used for amphibian captive breeding. Frequency of rain can be set using a timer and the enclosure includes a variety of oviposition sites. A standing drainpipe slightly raised above the floor of the enclosure prevents the water level from rising and, upon draining, circulates the water into the lower sump reservoir to be pumped back into the enclosure as rain. Photo credit: B. Gratwicke, Smithsonian Conservation Biology Institute.

Plate 4. A harlequin toad (*Atelopus* ssp.) breeding enclosure that mimics a flowing stream. The stacked valves allow husbandry workers to easily adjust water levels. The variety of large and small rocks can provide a variety of oviposition sites. Multiple bubblers help oxygenate the water and its rate of flow to mimic a moving stream, which is the species' natural habitat. Photo credit: J. Guerrel.

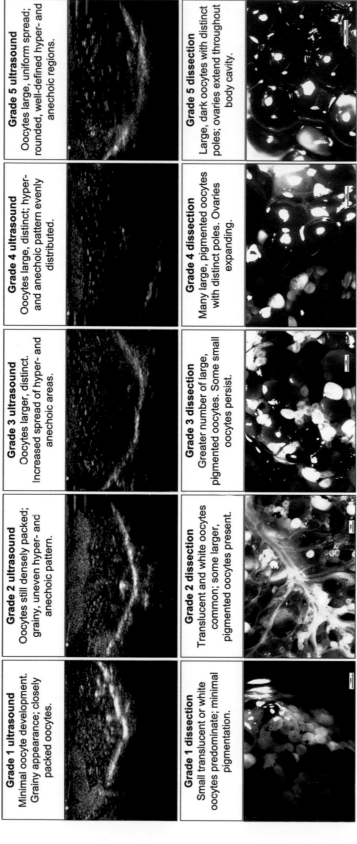

Plate 5. Top row: example of ultrasound grading scale with five grades showing low (Grade 1) to high (Grade 5) oocyte development on sonograms of the ovaries of Mississippi gopher frog (*Lithobates sevosus*). Bottom row: dissected ovaries (matched to sonogram in top row) from ultrasound validation experiment demonstrating different ultrasound grades are representative of physical changes to the oocytes and ovaries. Ultrasound images were taken just before the dissection of the ovary of each female (see Graham *et al.* 2018 for additional data). Blue scale-bar on dissection images represents 1 mm. Images by Katherine Graham, Mississippi State University, Starkville MS, USA.

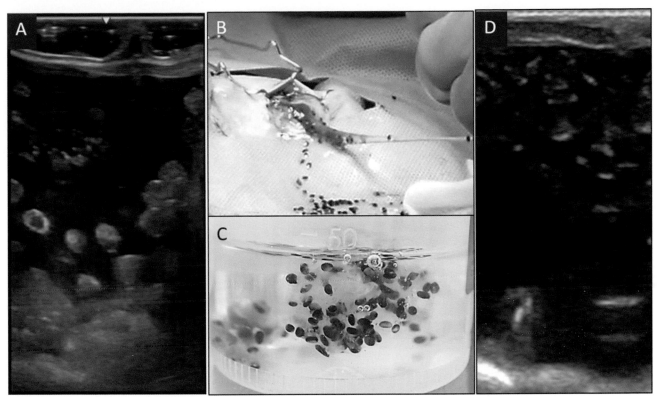

Plate 6. Post-ovulatory eggs surrounded by fluid and free in the coelomic cavity in an egg-bound female Chiricahua leopard frog (*Lithobates chiricahuensis*). (A) Coelomic ultrasound with a GE LOGIQTM e ultrasound system and L10–22-RS wide-band high frequency linear array transducer showing uneven bound eggs aggregating in 0.14–0.25 cm diameter clumps. (B) Surgical removal of eggs by veterinarian. (C) Degrading eggs in thick, abnormal jelly masses. (D) Normal egg development seen in ultrasound of recovered female 3 months after surgery. Images: Panels A–C by Kimberly Rainwater DVM, Fort Worth Zoo, Fort Worth TX, USA. Panel D by Shaina Lampert Mississippi State University, Starkville MS, USA.

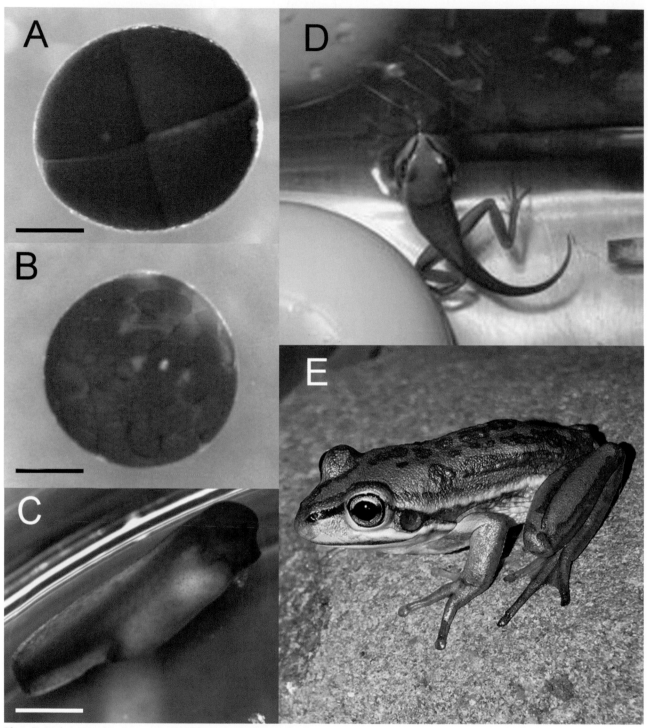

Plate 7. Embryonic development of *Litoria aurea* following fertilisation with cryopreserved spermatozoa (a) Four-cell embryo, Gosner Stage 4. (b) Blastula, Gosner Stage 8. (c). Embryo ready to hatch, Gosner Stage 19. (d) Metamorphosed tadpole, ~40–50 mm long from the snout to tip of the tail. (e) Subadult frog, 3 months after metamorphosis, 50 –60 mm long. Scale bar = 0.5 mm. Reproduced with permission from Upton R, Clulow S, Calatayud NE, Colyvas K, Seeto RG, Wong LA, Mahony MJ, Clulow J (2021). Generation of reproductively mature offspring from the endangered green and golden bell frog *Litoria aurea* using cryopreserved spermatozoa. *Reproduction, Fertility and Development* **33**, 562–572.

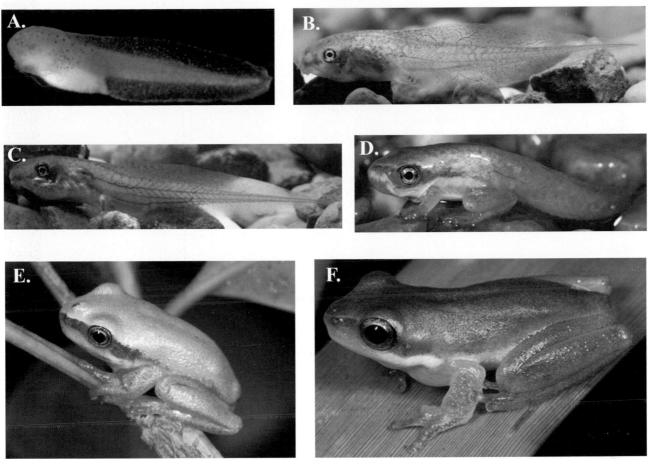

Plate 8. Images of *Litoria fallax* produced from cryopreserved testicular macerates (Gosner Stages 20–46). (A) Freshly hatched, free-swimming tadpole; 5 mm, Gosner Stage 20. (B) Tadpole nearing metamorphosis, hind legs beginning to develop; Gosner Stage 35, 20–30 mm. (C) Tadpole with fully developed hind legs; Gosner Stage 40, 30–40 mm. (D) Metamorph, eruption of the forelimb and resorption of the tail commenced; Gosner Stage 42, 30–40 mm. (E) Juvenile frog; tail resorbed; Gosner Stage 46, 10–20 mm. (F) Sexually mature male, 20–30 mm. Reproduced with permission from Upton R, Clulow S, Mahony MJ, Clulow J (2018). Generation of a sexually mature individual of the Eastern dwarf tree frog, *Litoria fallax*, from cryopreserved testicular macerates: proof of capacity of cryopreserved sperm derived offspring to complete development. *Conservation Physiology* **6**, coy043.

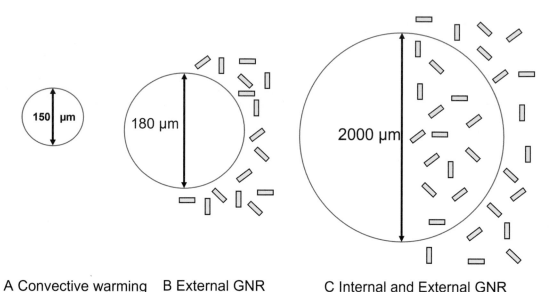

A Convective warming   B External GNR   C Internal and External GNR

Plate 9. Observed and predicted limits of the sizes of oocytes and embryos during various types of thawing post-vitrification. **A.** Maximum diameters of mammalian oocytes cryopreserved by vitrification and convective warming. **B.** Predicted limits for volumes of biomaterials thawed by laser-warming of external GNRs (Khosla *et al.* 2019). **C.** Predicted limits for the volumes of biomaterials thawed by laser-warming of internal and external GNRs (Khosla *et al.* 2019). Reproduced with permission from Clulow J, Upton R, Trudeau V, Clulow S (2019). Amphibian assisted reproductive technologies: moving from technology to application. In *Reproductive Sciences in Animal Conservation.* 2nd edn. (Eds P Comizzoli, JL Brown and WV Holt) pp. 413–463. Springer, Cham, Switzerland.

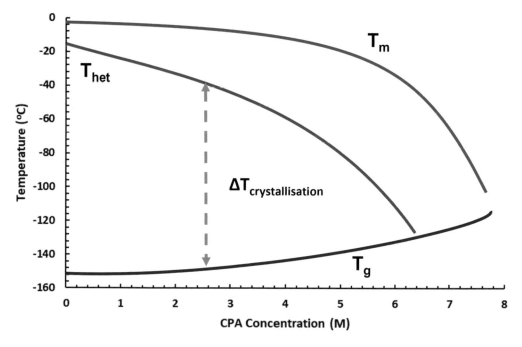

Plate 10. Generalised diagram of the relationship between cryoprotectant concentration and T_g (glass-transition temperature), T_{het} (heterogeneous ice-nucleation temperature) and $\Delta T_{crystallisation}$ (zone of formation of ice crystals between T_{het} and T_g. T_m (melting temperature) of cryoprotectant solution. Higher levels of cryoprotectant reduce $\Delta T_{crystallisation}$ during both freezing and thawing, which in combination with high rates of vitrification and thawing (laser warming) allow intracellular water to make the transition between the liquid and glass states without formation of lethal intracellular ice crystals. Adapted with permission from Khosla K, Zhan L, Bhati A, Carley-Clopton A, Hagedorn M, Bischof J (2019). Characterisation of laser gold nanowarming: a platform for millimetre-scale cryopreservation. *Langmuir* 35(23), 7364–7375. Copyright American Chemical Society.

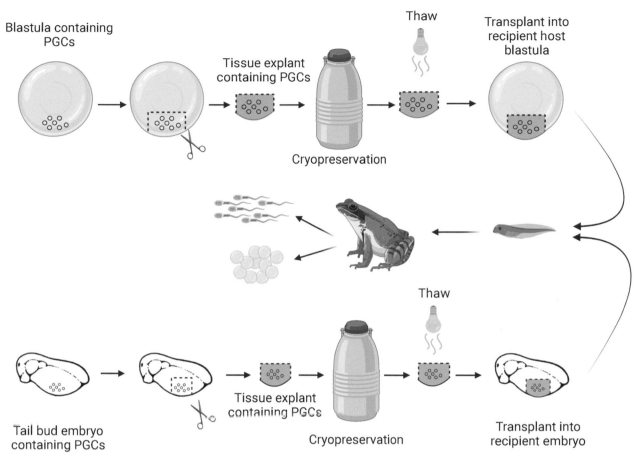

Plate 11. Proposed procedure for the biobanking and recovery of amphibian diploid genomes using cryopreserved primordial germ cells (PGCs). Tissue explants containing PGCs are taken from the vegetal pole of the blastulae or from the abdominal region of the tail-bud of embryos that are cryopreserved in amphibian biobanks. These explants can be thawed when required and transplanted into either blastulae or tail bud of hosts' embryos of the same species, or possibly even of different species. The PGCs are then taken up into the developing gonads of the host, resulting in the host producing either sperm or eggs of the original donor species upon sexual maturity. These could, in turn, generate progeny containing the genome of the donor individual through natural breeding of the host, despite the fact that the donor might have died many decades earlier.

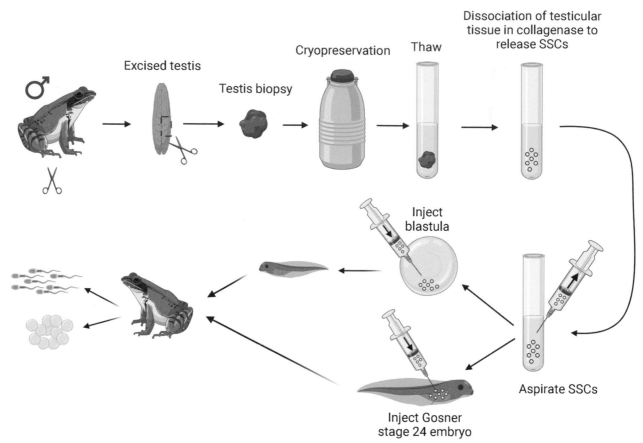

Plate 12. Proposed procedure for the biobanking and recovery of amphibian diploid genomes using spermatogonial stem cells (SSCs). Whole testes or testis biopsies are removed from the donor male and cryopreserved in amphibian biobanks. These tissues can be thawed out when required and dissociated in collagenase to release SSCs. The SSCs can then be aspirated and injected into either a host's blastulae or Gosner Stage 24 embryos of the same, or possibly even different, species. The SSCs are then incorporated into the developing gonads of the host, with the result that at sexual maturity the host produces sperm containing the genome of either the host or the original donor species. These could, in turn, create progeny containing the genome of the donor individual through natural breeding of the host with a female, despite the fact that the donor might have died many decades earlier.

4 Hormonal control of amphibian reproduction

Vance L. Trudeau, Brianna H. Raven, Harsh K. Pahuja, and Edward J. Narayan

INTRODUCTION

A fundamental tenet in reproductive physiology is that hypothalamic neurons at the base of the forebrain control the synthesis and release of the two gonadotropins, luteinizing hormone (LH) and follicle stimulating hormone (FSH), from the pituitary to the blood, in turn controlling gonadal functions in both sexes. Amphibians share the main common reproductive neuropeptides, pituitary hormones, and gonadal steroids with other vertebrates, yet very little is actually known about neuroendocrine regulation of reproduction in all but a few species. This is surprising given their paramount position in vertebrate evolution, ecological significance, and the threat to ecosystem functions by global declines in amphibian populations. This lack of knowledge and slow progress in research continues to hamper the development of a comprehensive hormonal toolkit for efficient propagation of endangered species under captive or semi-captive conditions. The main goal here is to review key aspects of the hypothalamo-pituitary-gonadal (HPG) axis that are relevant to the induction of ovulation (Chapter 6) and spermiation (Chapter 7), or that represent potential research opportunities where work is clearly warranted. Recent reviews covering other aspects (Vu and Trudeau 2016; Silla and Byrne 2019; Di Fiore *et al.* 2020; Silla *et al.* 2021) should also be consulted.

HYPOTHALAMO-PITUITARY ANATOMY

The amphibian hypothalamo-pituitary complex shares structural and functional similarities with those of fish, reptiles, birds, and mammals (Tonon *et al.* 1985; Vu and Trudeau 2016; Trudeau and Somoza 2020). The preoptic area in the hypothalamus, located in the ventral diencephalon, is the central regulatory site for reproduction. Attached to the hypothalamus through the infundibular stalk is the pituitary gland that consists of the adenohypophysis (anterior pituitary), intermediate lobe, and neurohypophysis (posterior pituitary). The pituitary gland of amphibians contains six different hormone-secreting cells that regulate major physiological functions. The cell types (and the hormones they secrete) are the corticotrophs (adrenocorticotropic hormone, ACTH), gonadotrophs (LH and FSH), lactotrophs (prolactin, PRL), melanotrophs (melanocyte stimulating hormone, MSH), somatotrophs (growth hormone, GH), and thyrotrophs (thyroid stimulating hormone, TSH) (Trudeau and Somoza 2020). The pituitary also presents parenchymal, non-endocrine folliculostellate cells, serving to communicate with the hormone-producing cells (Trudeau and Somoza 2020).

The endocrine cells in the amphibian anterior pituitary are assumed to be controlled by numerous hypothalamic neuropeptides and catecholamines as in other tetrapod classes, although very little fundamental research has directly addressed this. The neurohormones are released into a well-developed amphibian median eminence-portal blood system and then transported to the anterior pituitary (Trudeau and Somoza 2020). It is specifically the cells of the *pars distalis* of the anterior pituitary that synthesise and secrete hormones that regulate growth, reproduction, and the endocrine stress response among many other functions (Vu and Trudeau 2016; Trudeau and Somoza 2020). Amphibian pituitary hormone-secreting cells are somewhat regionally distributed, but not as clearly as the highly regionalised distribution of the teleost anterior pituitary (Trudeau and Somoza 2020). Little is known about hypothalamo-pituitary relations in Caudata and Gymnophiona, although the limited available evidence suggests that newts and salamanders (Richerson and DeRoos 1971; Olivereau *et al.* 1987) are similar to Anura (frogs and toads).

The magnocellular and parvocellular neurons of the amphibian preoptic area synthesise mesotocin (MT) and arginine vasotocin (AVT). These neurons send axonal projections to the posterior pituitary, the site of release of these peptides into the systemic circulation.

NEUROENDOCRINE FACTORS CONTROLLING THE RELEASE OF PITUITARY HORMONE

Gonadotropin-releasing hormone (GnRH)

The neuronal system producing the reproductive neuropeptide GnRH is considered the central integrator of other inputs from higher regions of the brain. Pulses of GnRH are delivered to the anterior pituitary to stimulate pulsatile or surge release of LH and FSH. Research to date in amphibians supports this role; however, very little is actually known about the control of endogenous GnRH synthesis and release in this class. Amphibians express both the *gnrh1* and *gnrh2* genes and produce the mature GnRH1 and GnRH2 peptides (Collin *et al.* 1995), formerly known as mammalian GnRH (mGnRH) and type II chicken GnRH (cGnRHII), respectively. The possibility of other amphibian GnRH isoforms exists as well, since a novel and biologically active GnRH1-like form with a Trp8 substitution was identified in the pituitary of the brown frog, *Rana dybowskii* (Yoo *et al.* 2000), and there is some evidence for the presence of GnRH3 (formerly called salmon GnRH) in the European edible frog, *Pelophylax esculentus* (Di Fiore *et al.* 2020).

Neuronal cell bodies for GnRH1 are localised in the well-defined septal-anterior preoptic region and fibres project through the ventral diencephalon and terminate in the median eminence. Cell bodies and fibres for GnRH2 neurons are much more widely distributed. In the telencephalon, GnRH2 cell bodies can be found scattered ventrally from the olfactory bulb to the septal area, a pathway that corresponds to the terminal nerve. Many GnRH2 cell bodies are also found in the septal-anterior preoptic area. These neurons send projections that terminate at the median eminence. Importantly, many neurons and fibres co-express both GnRH1 and GnRH2, indicating that GnRH2 may also have hypophysiotropic functions, and would also be expected to stimulate production and release of gonadotropins. In *Pelophylax esculentus*, GnRH2 levels are higher than GnRH1 (Di Fiore *et al.* 2020). Numerous other GnRH2 cell bodies can be found in the mid-diencephalon, near the third ventricle, and in the posterior preoptic and infundibular areas, sending projections widely throughout the brain. Several hypothalamic peptides are also produced and exert paracrine actions directly in the gonads (Di Fiore *et al.* 2020). For example, *in vivo* and *in vitro* studies have shown that GnRH1 and GnRH2 directly regulate testicular production of androgen in *Pelophylax esculentus* (Pierantoni *et al.* 1984; Fasano *et al.* 1995).

The neuropeptide GnRH acts through membrane-bound G-protein-coupled receptors GnRH receptors (GnRHRs) found on gonadotrophs. Three distinct subtypes of GnRHRs have been identified in the bullfrog *Lithobates catesbeianus* that exhibit differential ligand selectivity and intracellular signalling (Wang *et al.* 2001; Oh *et al.* 2003). For all GnRHRs, GnRH2 exhibits a higher potency than GnRH1. Important here is that GnRHR1 is the prominent subtype in the pituitary, whereas GnRHR2 is in the brain and GnRHR3 is expressed in the brain and pituitary. The presumed pituitary-specific GnRHR1 and also GnRHR3 have the highest affinity for GnRH2 followed by GnRH1 (Wang *et al.* 2001). The pituitary cells exhibiting differential expression of the GnRHRs remain to be determined and will be important if we are to target subtypes of receptors for optimal stimulation of specific pituitary functions, especially the release of gonadotropins. Receptors for GnRH found in the brain suggest additional neuromodulatory roles that control reproductive behaviour.

To our knowledge, only GnRH1 has been tested for LH-releasing ability in amphibians. Additionally, only synthetic analogues of the GnRH1 have ever been tested for their abilities to induce ovulation, spermiation, or spawning in amphibians. Thus, despite clear indications that GnRH2 is found in the median eminence and activates anuran GnRH receptors, its role for production of pituitary hormones remains unknown. The existence of multiple GnRHs and GnRHRs in anurans is highly reminiscent of the situation in teleosts, as many different GnRH analogues for fish that exhibit high specificity of action and potency have been developed (Murthy *et al.* 1994; Lovejoy *et al.* 1995). The exploration of a wider range of GnRH agonists with differing GnRHR modulatory actions and potentially differential effects on the release of LH and FSH is thus of immediate importance. Improved GnRH analogues with targeted abilities to induce the LH surge would improve induced spawning. Identifying GnRH analogues that may preferentially enhance the release of FSH could offer new methods to enhance control of gonadal growth and gametogenesis, leading to an overall advance in amphibian reproductive technologies.

The RF-amides kisspeptin and gonadotropin-inhibitory hormone

Kisspeptin (KISS)

The neuropeptide KISS has been identified in many vertebrate species, yet is absent in birds (Tena-Sempere *et al.* 2012). Kisspeptin is the RF-amide (contains an Arg-Phe-NH$_2$ motif at the C-terminus) product of the gene *kiss1* (Roseweir and Millar 2009). The initial length of the KISS precursor is 145 amino acids, but this is cleaved into several smaller fragments with lengths of 54, 14, 13, and 10 amino acids (Kotani *et al.* 2001). These fragments share the common C-terminal decapeptide referred to as KISS-10, which is the smallest form of KISS that can invoke activation of the G-protein-coupled receptor GPR54, the kisspeptin receptor (Kotani *et al.* 2001; Muir *et al.* 2001). The number of genes encoding for KISS and GPR54 range across vertebrates from as few as zero in birds to as many as three in frogs (*kiss1*, *kiss1b*, *kiss2*; *gpr54–1a*, *gpr54–1b*, *gpr54–2b*) (Tena-Sempere *et al.* 2012). Mammals typically have one of each gene (*kiss1*, *gpr54–1a*), whereas most fish contain two of each (*kiss1*, *kiss2*; *gpr54–1b*, *gpr54–2b*) (Tena-Sempere *et al.* 2012). Mutations in GPR54 lead to hypogonadotropic hypogonadism, a condition that causes improper development of the gonads and infertility in humans (de Roux *et al.* 2003) and rodents (Seminara *et al.* 2003).

In mammalian and non-mammalian vertebrates, KISS-10 has been found to be highly conserved, only varying at position 8 from the C-terminus in fish and in the Western clawed frog *Silurana tropicalis* (Lee *et al.* 2009). In fish, position 8 of KISS contains a leucine residue whereas in *Silurana tropicalis*, this site in KISS-1b contains a valine residue (Lee *et al.* 2009). Kisspeptins and their receptors have been found to be expressed strongly in the hypothalamus and gonads of vertebrates (Lee *et al.* 2009; Tena-Sempere *et al.* 2012; Chianese *et al.* 2017). The role of KISS in reproduction has been clearly established only in mammals, and emerging roles in fishes have been reported (Somoza *et al.* 2020). Kisspeptins stimulate GnRH1 leading to an increase in plasma LH and FSH, driving the onset of puberty and ovulation in mammals (Decourt *et al.* 2016). Specific populations of KISS neurons also mediate feedback actions of sex steroids to regulate GnRH neurons (Smith *et al.* 2005). This is supported by multiple studies that have observed infertility, low levels of gonadotropins and sex steroids, delayed onset of puberty, and decreases in weight and size of gonads of both male and female murine *kiss1* gene knockouts (d'Anglemont de Tassigny *et al.* 2007; Lapatto *et al.* 2007).

Kisspeptin and its receptor have been identified in the testes of numerous mammals and non-mammals including fish and frogs (Meccariello *et al.* 2020). More specifically, both the KISS ligand and receptor have been found to be expressed in Leydig cells in mice (Anjum *et al.* 2012) suggesting an autocrine function, and also in spermatozoa of humans (Pinto *et al.* 2012) suggesting a role in function of sperm. The kisspeptin receptor also has been identified in Sertoli cells of primates (Irfan *et al.* 2016) and frogs (Chianese *et al.* 2013), and in spermatozoa and peritubular myoid cells of *Pelophylax esculentus* (Chianese *et al.* 2013, 2017), possibly contributing to the release of sperm (Sharma *et al.* 2020). Given the location of kisspeptin in Leydig cells and GPR54 in Sertoli cells, kisspeptin is likely acting in a paracrine manner in frogs' testes (Chianese *et al.* 2017). Based on *ex vivo*, *in vivo*, and *in vitro* experiments with KISS-10 (Chianese *et al.* 2013, 2015, 2017) it has been proposed that kisspeptin may be a regulator of intratesticular GnRH, spermatogenesis, proliferation of germ cells, and steroidogenesis in *Pelophylax esculentus*. For example, KISS-10 stimulates intratesticular expression of GnRH1 and all three GnRHRs in the testes of *Pelophylax esculentus*. This is important for the production of androgen and for the progression of germ cells (Chianese *et al.* 2015). The role of KISS and GnRHs in the ovaries of frogs remains to be explored. It is not known whether KISS regulates GnRH in the amphibian brain, or whether it can regulate LH and FSH in the pituitary; however, this is possible given that pituitary *gpr54*-mRNA levels are highest in the breeding phase of *Pelophylax esculentu*s (Chianese *et al.* 2013). The KISS1R of *Lithobates catesbeianus* is a 379-amino G protein-coupled receptor (Moon *et al.* 2009), exhibiting ~45% amino acid identity with mammalian and ~70% identity with mammalian and fish GPR54s, respectively.

Gonadotropin-inhibitory hormone (GnIH)

The discovery and roles of GnIH and orthologs with the defining C-terminal LPXRFamide motifs have recently been reviewed, with a focus on amphibians (Jadhao *et al.* 2017), so will only be briefly summarised here. While GnIH was so named for its clear inhibition of the release of GnRH and gonadotropins in birds and mammals, GnIH orthologs in amphibians and fish have varying hypophysiotropic activities and may even stimulate the release of the gonadotropins in some species (Jadhao *et al.* 2017). In frogs, various GnIH orthologs have been shown to stimulate the release of GH and/or PRL from pituitary cells in culture (Koda *et al.* 2002; Ukena *et al.* 2003) but showed no effects on LH and FSH (Figure 4.1). The role in release of LH, FSH, GH, and PRL has to be more clearly established. Perhaps with this information it may be possible to use GnIH to modulate specific aspects of amphibian reproduction for captive breeding. Further studies of the functions of GnIH and its orthologs in amphibians are clearly warranted.

Mesotocin and arginine vasotocin

The oxytocin/vasopressin family of nonapeptides are conserved across the vertebrates. The homologues in Amphibia are commonly referred to as MT and AVT, differing only at amino acids 3 and 9, and acting on distinct G-protein-coupled receptors. Their precursor proteins are respectively encoded by the oxytocin (*oxt*) and arginine vasopressin (*avp*) genes. Surprisingly, very little is known about the function of MT although there are overlapping functions with AVT for osmoregulation, kidney function, and control of the release of MSH, among others. Relevant here are the roles of the nonapeptides in reproductive function.

In amphibians, AVT neuronal cell bodies largely are found in the basal forebrain from the accumbens–amygdala area to the preoptic area and hypothalamus, projecting widely in the brain with heavy innervation of areas associated with social behaviour, decision-making, social signalling, and aggression (Wilczynski *et al.* 2017). At least some of these roles are probably mediated by the stimulatory effects of AVT and MT on neurosteroid production. Both neuropeptides induce a concentration-dependent stimulation of the formation of P4, 17α-hydroxypregnenolone (OHPG), 17α-hydroxyprogesterone (OHP4), and dehydroepiandrosterone (DHEA) (see Figure 4.2). Using a series of agonists and antagonists selective for mammalian receptors, it was shown that neurosteroid production is via the activation of V1a AVT receptors and MT receptors (Do-Rego *et al.* 2006). There are important season-dependent and androgen-dependent roles for AVT in amplexus in male urodeles and in courtship vocalisations in numerous species of frogs; these have been reviewed recently in detail (Wilczynski *et al.* 2017). Much less is known about AVT in female amphibians, although it may stimulate oviposition in female rough-skinned newts, *Taricha granulosa* (Moore *et al.* 1992).

There is emerging evidence that isotocin, the piscine homologue of MT and oxytocin, stimulates the release of LH in goldfish (Mennigen *et al.* 2017). Both isotocin and AVT have a stimulatory role to play in spawning in zebrafish (Altmieme *et al.* 2019). Little is known about potential mechanisms by which the nonapeptides may stimulate spawning. This could potentially involve actions on GnRH neurons since in a sex-changing grouper, *Epinephelus adscensionis*, V1a2 AVT receptor colocalises with GnRH in preoptic neurons (Kline *et al.* 2016). In the African clawed frog (*Xenopus laevis*), MT immunoreactivity can be localised in proximity to GnRH neurons (Alvarez-Viejo *et al.* 2003), suggesting a link to reproduction.

Arginine vasotocin may have a role in controlling sexual behaviours in male newts. Sexual behaviours increased after injection of AVT into intact male *Taricha granulosa* (Moore and Zoeller 1979). In the Japanese fire belly newt, *Cynops pyrrhogaster*, AVT acts on the brain to increase the incidence and frequency of androgen-induced courtship behaviour (Toyoda *et al.* 2003). It was also reported that AVT induces the discharge of sex pheromone, acting peripherally on a contractile structure of the abdominal gland. Deposition of spermatophores in *Cynops pyrrhogaster* can be induced by AVT even in the absence of the female. Administration of a vasopressin

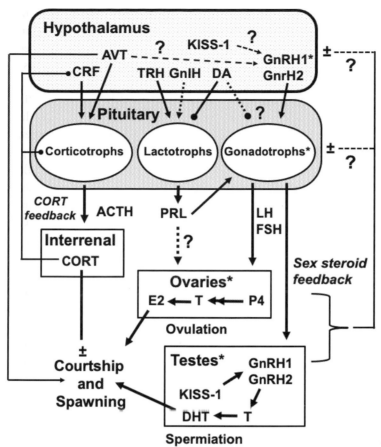

Figure 4.1: Proposed model for hormonal control of reproduction in anurans. The stimulatory neuropeptide gonadotropin-releasing hormone 1 (GnRH1) and perhaps GnRH2 are released from hypothalamic nerve terminals into the median eminence-portal blood system (not illustrated) and transported to the anterior pituitary, where these decapeptides act on G-protein-coupled GnRH receptor subtypes on gonadotrophs to stimulate the synthesis of luteinizing hormone (LH) and follicle-stimulating hormone (FSH). The release of the gonadotropins is enhanced by prolactin (PRL) secreted from pituitary lactotrophs. Prolactin is stimulated by the tripeptide thyrotropin-releasing hormone (TRH), originally named for that action because of its role in mammals. In anurans it does not effectively stimulate thyrotropin but does stimulate PRL and growth hormone. The RF-amide peptide gonadotropin-inhibitory hormone (GnIH) and the catecholamine dopamine (DA) also play respective stimulatory and inhibitory roles in the regulation of PRL. The effects of the GnRHs on gonadotropins are inhibited by DA in some species, but experimental evidence is lacking for many amphibians. The gonadotropins act on their respective G-protein-coupled receptors in the ovaries and testes to drive steroidogenesis and the release of gametes. In the ovaries, progesterone (P4) is converted through multiple enzymatic steps to testosterone (T), which is aromatised to oestradiol (E2). Oestradiol plays an important role in stimulating the hepatic synthesis of the egg-yolk vitellogenins (VTGs) in females (not shown). The synthesis and release of VTGs from the liver (not shown) is also regulated by LH, FSH, and PRL. Testosterone is converted by 5α-reductase to 5α-dihydrotestosterone (DHT), the main androgen in amphibians. These sex steroids generally play various roles in positive and negative feedback (±) regulation of gonadotropin-release, but direct evidence is lacking in many amphibian species. Steroids are involved in gonadal development, development of secondary sex characters, and can modulate courtship behaviours that are controlled by arginine vasotocin (AVT). Note that AVT neurons project to the posterior pituitary for release into the circulation, and also project widely in the brain. For clarity, this is not depicted in the figure. Intratesticular kisspeptin (KISS-1) and the GnRHs regulate steroidogenesis, proliferation of germ cells, and spermatogenesis. No data for intraovarian KISS and GnRH have yet been reported for amphibians. Kisspeptin is also found in hypothalamic neurons but it is not known if, or how, it may regulate GnRH neurons or gonadotropin-release. Corticotropin-releasing factor (CRF) stimulates the synthesis and release of adrenocorticotropic hormone (ACTH) from the pituitary, which in turn stimulates corticosterone (CORT) from the steroidogenic cells of the interrenal (amphibian equivalent to mammalian adrenal cortex embedded within the kidney complex). Corticosteroids negatively feedback at the level of CRF and ACTH to reduce the synthesis of glucocorticoid. This constitutes the hypothalamo-pituitary-interrenal axis. Accumulating evidence indicates that CORT can modulate reproductive processes at multiple sites (see *). Solid lines indicate functions supported by strong evidence and dashed lines with ? indicate very limited evidence or speculations that will need further investigation. Lines terminating in an arrowhead indicate stimulation whereas lines terminating in an ellipse indicate inhibition. The ± sign indicates a modulatory response since both positive and negative effects have been reported. See text for additional details.

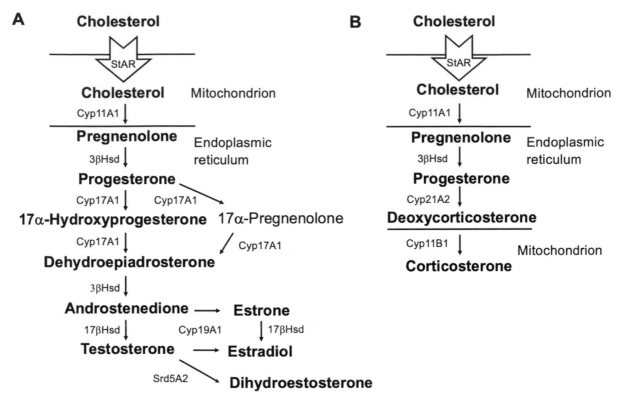

Figure 4.2: Main reproductive and stress-related steroids produced by anurans. (A) Once inside a steroidogenic cell, cholesterol is transported into the mitochondria by steroid acute regulatory protein (StAR). The first step is conversion to pregnenolone by the cholesterol side-chain cleavage enzyme Cyp11A1. An important enzyme at multiple steps in the pathway is Cyp17A1, the 17α-hydroxylase, driving the conversion of pregnenolone to progesterone and other key steroid intermediates. For ovaries and testes, the conversion of dehydroepiandrosterone by 3βHsd, the 3β-Hydroxysteroid dehydrogenase, leads to androstenedione. The androgen androstenedione can be converted to testosterone by 17βHsd, the 17β-Hydroxysteroid dehydrogenase, or alternatively to estrone by Cyp19A1 aromatase. Importantly, testosterone serves as the precursor to 5α-dihydrotestosterone (DHT), the main androgenic hormone through the actions of Srd5α2, also known as 5α-reductase. Aromatase converts testosterone to the main estrogenic hormone oestradiol. Note also that estrone can be converted to oestradiol by 17βHsd. (B) In the steroidogenic cells of the interrenal, Cyp21A2, also known as 21α-hydroxylase, converts progesterone to deoxycorticosterone which in turn is converted by 11β-hydroxylase by Cyp11B1 to the stress hormone corticosterone. Cellular location of the various enzymes determines which steroids and which pathways are active.

receptor antagonist suppressed the expression of courtship behaviour in these newts. In contrast, injections in male Günther's toadlet, *Pseudophryne guentheri*, led to complete inhibition of GnRH-agonist-induced release of sperm (Silla 2010). It is important to note that AVT is a potent stimulator of the stress axis by actions on the corticotrophs to stimulate ACTH (Figure 4.1), and also directly on interrenal cells (Larcher *et al.* 1992). Moreover, it is likely that the multitude of peripheral effects of the nonapeptides on the contraction of smooth muscle, and potentially on blood pressure, lead to negative side effects, limiting their utility for captive breeding. Research targeting the use of analogues that are specific for AVT and MT receptor subtypes (Do-Rego *et al.* 2006) may help advance their use in captive breeding. In this

regard, novel oxytocin analogues have been developed for inducing labour and support of lactation in women (Wiśniewski 2019) and thus could be tested for biological activities in amphibians. Before AVT and MT or their analogues can be safely used for behavioural modifications or for induction of spawning in captive breeding programs more research is recommended.

Dopamine

Considerable evidence points to an inhibitory role for the catecholamine DA in the neuroendocrine control of vertebrate reproduction (Dufour *et al.* 2005; Vu and Trudeau 2016; Nestor *et al.* 2018). Dopamine is a neurotransmitter synthesised from tyrosine through enzymatic conversion by tyrosine hydroxylase and dihydroxyphenylalanine decarboxylase in specific hypothalamic neurons. In many teleost fish, DA acts to inhibit the release of gonadotropins from the pituitary and to attenuate the stimulatory actions of GnRH (Dufour *et al.* 2005). In sheep, DA can inhibit GnRH1 neurons and the release of LH (Nestor *et al.* 2018). Dopamine also inhibits GnRH1-induced release of LH in human males (Leebaw *et al.* 1978). Such mechanisms are suspected to be evolutionarily conserved and active in some amphibian species (Figure 4.1).

Various immunocytochemical studies indicate that frogs exhibit typical hypothalamic catecholaminergic neuronal populations, with the presence of dopaminergic and possibly norepinephrinergic nerve terminals in the median eminence. Thus, the catecholamines would be expected to reach the pituitary via the portal blood, in addition to innervation of the intermediate lobe of the pituitary (Kouki *et al.* 1998; O'Connell *et al.* 2010). A series of early studies demonstrated the existence of inhibitory hypothalamic control, in which LH release and ovulation were induced following electrolytic brain lesions in the nucleus *infundibularis ventralis* of the hibernating grass frog, *Rana temporaria* (Sotowska-Brochocka 1988; Sotowska-Brochocka and Licht 1992). The administration of the DA D2 receptor antagonist metoclopramide (MET) in this species also triggered ovulation, whereas the DA agonist bromocriptine reduced the release of LH (Sotowska-Brochocka *et al.* 1994). Decreased calling rates after injections of the specific dopamine receptor D2 agonist quinpirole were observed in the male green tree frog, *Hyla cinerea* (Creighton and Wilczynski 2014). Treatments with quinpirole have also been shown to delay, or inhibit, oviposition and spawning in the wood frog, *Lithobates sylvaticus* (Vu and Trudeau 2016). The combined injection of a GnRH1 agonist (GnRH-A) and various D2 antagonists results in maximal spawning rates in the leopard frog, *Lithobates pipiens* (see the AMPHIPLEX method; Trudeau *et al.* 2010, 2013) and is effective in numerous amphibians (Vu and Trudeau 2016; Clulow *et al.* 2018). Additionally, MET significantly increased GnRH-A-induced concentrations of sperm in the critically endangered Panamanian golden frog, *Atelopus zeteki* (Della Togna *et al.* 2017). Treatment with MET enhances the stimulatory effect of GnRH-A to increase *lhb* and *fshb* subunit and *gnrhr1* mRNA levels in whole pituitary glands of female *Lithobates pipiens* (Vu *et al.* 2017). Together, these data suggest that DA is involved in controlling reproduction in some amphibians. While definitive evidence requires more experimentation, these data support the proposal that DA inhibits synthesis and secretion of gonadotropins in anurans, as it does in teleosts and some mammals (Dufour *et al.* 2005; Tortonese 2016; Nestor *et al.* 2018).

KEY ANTERIOR PITUITARY HORMONES FOR THE REGULATION OF AMPHIBIAN REPRODUCTION
The gonadotropins: luteinizing hormone (LH) and follicle stimulating hormone (FSH)

The gonadotropins are dimeric proteins composed of the common chorionic gonadotropin subunit α (CGA) and differing LHβ and FSHβ subunits that confer specificity of binding to the LH and FSH receptors, respectively. Both hormones are post-transcriptionally glycosylated; this is important for full biological function and the control of numerous gonadal functions including gonadal steroidogenesis (Figure 4.2A) and ovarian maturation (Figure 4.3). It follows that the specific biological actions of mature LH and FSH are via the respective activation of the G-protein-coupled LH and FSH receptors. Natural ovulation in females and spermiation in males results from a surge-release of LH (Clulow *et al.* 2014; Vu and Trudeau 2016). Final maturation of oocytes and production of testicular androgen in frogs is regulated primarily by LH, whereas FSH promotes early follicular development. Importantly, both LH and FSH have an additional role in female frogs to stimulate the hepatic synthesis and uptake of the egg-yolk protein VTG (a process known as vitellogenesis) directly by actions on the liver, and indirectly through increased ovarian oestradiol (E2) (Vu and Trudeau 2016; Di Fiore *et al.* 2020). Little is known about the neuroendocrine regulation of FSH in Amphibia.

This is a composite view and does not represent one single species. Shown are depictions of a thecal cell of the external thecal layer, a follicular cell, and an oocyte. Upon stimulation, luteinizing hormone (LH) and follicle-stimulating hormone (FSH) are released from the anterior pituitary to the general circulation (see Figure 4.1). Binding of mature bioactive LH to the LH receptor (LHR) on thecal cells stimulates the importation of cholesterol from the cytoplasm into the mitochondrion via the actions of steroid acute regulatory protein (see Figure 4.2A). Activation of the LHR leads to activation of cAMP/protein kinase A (cAMP/PKA)-dependent intracellular signalling pathways, and up-regulation of many enzymes involved in the sequential conversion of steroidal intermediaries.

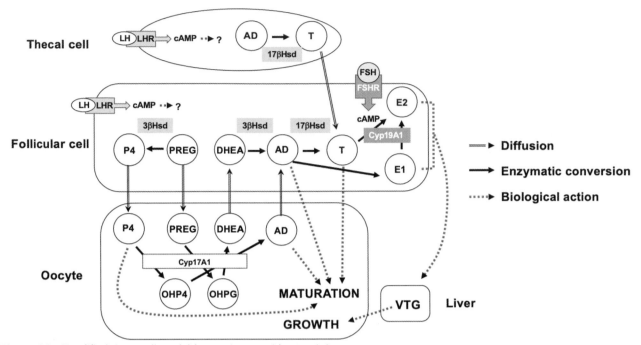

Figure 4.3: Simplified three-cell model for ovarian steroidogenesis in anurans.

For convenience only the androstenedione (AD) to testosterone (T) conversion by 17βHsd within the endoplasmic reticulum is shown. One source of T for follicular cells is by diffusion from thecal cells. Some of the other classic steps in the pathway are assumed to exist in amphibians but have not been fully documented so are not shown. In follicular cells, LHR activation promotes pregnenolone (PREG) and progesterone (P4) production via activation of 3bHsd. These steroids diffuse to the oocyte. In *Xenopus laevis* at least, the exclusive site of expression of Cyp17A1 is in the oocyte, so PREG and P4 are converted to 17α-hydroxypregnenolone (OHPG) and 17α-hydroxyprogesterone (OHP4), respectively. Conversion of OHPG to DHEA and subsequent diffusion to the follicular cells is considered a crucial step for the production of AD and T. In *Xenopus laevis*, AD and T are produced in large amounts just before ovulation and may promote the maturation of oocytes. Although P4 is also capable of promoting maturation, its production before ovulation is quite low; thus, its physiological role in the maturation of oocytes is uncertain. In some species, however, P4 may be important. Signalling of steroids, be it the androgens or P4 for maturation of oocytes, is independent of transcription, so appears to be largely a nongenomic mechanism, and presumably through a membrane-bound classical steroid receptor, or via a

steroid G-protein-coupled receptor. Other mechanisms may also be involved but are beyond the scope of the current discussion. Under the actions of FSH, the FSH receptor (FSHR) is activated, driving the cAMP pathway to activate Cyp19A1 aromatase to enhance the conversion of AD to estrone (E1) and T to oestradiol (E2). These estrogens are released to the blood, and act on oestrogen receptors in the liver to stimulate the synthesis and release of the vitellogenins (VTGs) that will be actively taken up by developing oocytes and promote their growth. Note that females may produce low levels of 5α-dihydrotestosterone (DHT), via 5α-reductase activities in many tissues (not shown). A similar three-cell system is likely present in the testis but with both FSH and LH receptors associated with the Leydig cells and FSH receptors on the Sertoli cells. The androgens reach the Sertoli cell by diffusion through the basal lamina surrounding the seminiferous tubules. While Leydig cells may be the primary location of Cyp17A1, Sertoli and spermatogenic cells also express the enzyme, so steroid metabolite conversion in these cells is likely. Moreover, T can be converted to the principal androgen DHT by 5α-reductase in Leydig and Sertoli cells, depending on the species, and also in many other tissues, including brain and pituitary. Similarly, males can also produce E2, which is important for the regulation of spermatogenesis.

Many details on the cellular localisation of aromatase and the three subtypes of 5α-reductase are lacking for amphibian species and are speculations based on data from other tetrapods. Note that seasonally fluctuating LH and FSH drives seasonal variations in steroid production, thereby regulating steroid-dependent processes such as positive and negative feedback, development of secondary sex characteristics, and sexual behaviours.

Immunolocalisation studies in several ranids indicate that the majority of gonadotrophs (~50–80%) in the pituitary produce both LH and FSH, whereas the remainder produce only one of the gonadotropins. Very little is known about the differential regulation of LH and FSH. In addition to studies of the differential effects of hypothalamic neuropeptides on LH versus FSH, gonadal proteins such as activin and inhibin should be investigated. What is clear is that amphibians are not like laboratory-rodent models. For example, in *Lithobates catesbeianus*, thyrotrophs and to a lesser extent gonadotrophs, are immunoreactive for activin β subunit. Addition of activin B to dispersed anterior pituitary cells *in vitro* stimulated the release not only of FSH but also of LH (Uchiyama *et al.* 2000), not showing the expected FSH-specific responses observed in mammals. Understanding the brain-derived and gonad-derived hormones that differentially regulate the production of LH versus FSH would be very useful for the development of treatments that enhance the production and release of one over the other, thus obtaining differential control over distinct LH-dependent or FSH-dependent gonadal processes in both sexes.

Prolactin (PRL)

Prolactin is synthesised and released from pituitary lactotrophs. While PRL is implicated in a multitude of functions in the various vertebrate classes, ranging from osmoregulation in fishes to lactation and parental care in mammals, data available for amphibians are more limited. Prolactin negatively modulates the pro-metamorphic actions of thyroid hormone in amphibians. Several studies on seasonal variations in the level of PRL in the plasma of adult frogs and newts (Vu and Trudeau 2016; Di Fiore *et al.* 2020) and the presence of PRL receptors in ovaries of the Cayenne caecilian *Typhlonectes compressicauda* (Exbrayat and Morel 2003) suggest direct roles in reproduction. It has been shown *in vitro* that PRL increases the responsiveness of LH cells to GnRH1 from pituitaries of *Lithobates catesbeianus* (Oguchi *et al.* 1997) and induces

release of hepatic VTG in *Pelophylax esculentus* (Carnevali *et al.* 1993). The clearest evidence to date for a role for PRL is in *Cynops pyrrhogaster* (Kikuyama *et al.* 2019). Prolactin from the pituitary acts on the brain to stimulate tail-vibration in males, which is important for courtship. Moreover, the synthesis and release of the female-attracting pheromone, sodefrin, from the abdominal glands of these newts is enhanced by PRL. In these latter examples, the actions of PRL are enhanced by androgens in male newts. Given the numerous suspected roles of PRL in amphibians and the clearly defined roles in birds and mammals (Dobolyi *et al.* 2020), and to some extent also in fish (Whittington and Wilson 2013), it will be important to consider how PRL levels could be controlled and manipulated to optimise reproductive outcomes in conservation breeding programs. It will also be important to understand the relationship between DA-mediated inhibition of both PRL (Nakano *et al.* 2010) and LH (Figure 4.1).

Proopiomelanocortin (POMC) and adrenocorticotropic hormone (ACTH)

Amphibian POMC is a precursor protein 261–264 amino acids long that undergoes extensive tissue-specific processing by various prohormone convertases to numerous biologically active peptides. Important here is that ACTH is produced from the POMC precursor in anterior pituitary corticotrophs in response to stress (Dores *et al.* 1993). The central ~39 amino-acid fragment of POMC represents ACTH. Hypothalamic corticotropin-releasing factor (CRF) and AVT both stimulate corticotrophs (Okada *et al.* 2016) to release ACTH. Binding of ACTH and its shorter processed fragments (e.g. ACTH1–24) to melanocortin type 2 G-protein-coupled receptors (MC2R) in steroidogenic cells in the interrenal (Davis *et al.* 2013, 2019) induces production of CORT (Figure 4.2B), which may, in turn, modulate numerous reproductive processes (Figure 4.1). In the intermediate lobe of the pituitary, ACTH can be further processed to produce a shorter 13-amino-acid N-terminal fragment, MSH, that controls light/dark cutaneous adaptations by actions on pigment cells (Tonon *et al.* 1985).

PRODUCTION AND ROLES OF THE GONADAL STEROIDS

Ovarian and testicular morphology is diverse and has been well described for the major amphibian classes elsewhere (see Ogielska 2009). Under the cell-specific influence of LH and FSH, ovaries and testes both produce steroids from cholesterol, the first critical steps being the transport of cholesterol by steroid acute regulatory (StAR) protein into the mitochondria, and conversion to the universal steroid precursor pregnenolone (Figure 4.2A). Some differences among species are likely to exist regarding the dominant route of conversion (Kime and Hews 1978).

Ovarian sex steroids are essential for follicular growth and the maturation of oocytes (resumption of meiosis from prophase I to metaphase II and ovulation) (Yang *et al.* 2003). Understanding the control of oocytes' maturation is critical for the advancement of captive breeding since it is linked to successful induced breeding. Sex steroids are also critically important in courtship displays in amphibians, such as males' vocalisations, females' phonotaxis, and amplectic clasping that are sequentially exhibited during breeding. These topics have been reviewed previously (Moore *et al.* 2005; Wilczynski and Lynch 2011; Zornik and Kelley 2011; Vu and Trudeau 2016). Here we focus on the roles of sex steroids in the control of ovarian maturation and feedback regulation of the gonadotropins.

Steroids control the maturation of oocytes

Relevant here is the dynamic relationship between thecal cells, heterogeneous follicular cells, and oocytes for the regulation of ovarian steroidogenesis (Figure 4.3). Gonadotropic regulation and interactions between mesenchymal-derived thecal cells and epithelial-derived granulosa follicular cells are at the basis of the classical two-cell model of mammalian ovarian steroidogenesis. Numerous versions of these interactions of thecal-follicular cells exist in the literature, but they do not fully apply to amphibians and are difficult to reconcile with published data (Ahn *et al.* 1999; Yang *et al.* 2003; Rasar and Hammes 2006). This may be due to differences in the various methods of isolating thecal and follicular cells, the contamination of cellular fractions, etc., and likely species' differences. What is clear, however, is that amphibian oocytes play a critical role in the production of the steroids used for their own maturation. Earlier work on three species of *Rana* and the Oriental fire-bellied toad, *Bombina orientalis*, identified a role for thecal cells in the conversion of AD to T (Ahn *et al.* 1999). The source of AD may not be the thecal cells since they do not express Cyp17A1 (Yang *et al.* 2003), and in the ovary of *Xenopus laevis*, oocytes are the exclusive site of Cyp17A1.

Granulosa-like follicular cells may also be a source of AD. As the follicular cells of *Xenopus laevis* do not express CYP17A1, pregnenolone must then enter the surrounding oocytes to be converted to DHEA (Yang *et al.* 2003), which then can be sequentially converted to androgen (Rasar and Hammes 2006). Given the diversity in amphibian ovarian morphology, and potential differences in these intercellular relations, we propose a simplified model that may serve as a framework for future research on diverse species (Figure 4.3).

Gonads express Cyp19A1 aromatase, the enzyme that converts androgens to estrogens, and all three 5α-reductases (Srd5α1, Srd5α2, and Srd5α3). It is mainly Srd5α2 that converts gonadal testosterone to 5α-DHT. The Srd5α1 in the brain is linked to the production of neurosteroids such as allopregnanolone, and Srd5α3 is non-steroidogenic and important for N-glycosylation (Robitaille and Langlois 2020). Cellular localisation of aromatase and reductase enzymes in gonads are not well studied (Rosati *et al.* 2019; Robitaille and Langlois 2020), especially in amphibians. The relative activities of these enzymes determine the predominant steroid produced. For example, in one period of an ovarian cycle the secretion of E2 may predominate over P4 and vice-versa. In Bufonidae, the specialised Bidder's organ is emerging as the main site of the conversion of T to E2 (Scaia *et al.* 2019). It has been compared to a rudimentary ovary because of the presence of previtellogenic follicles. The Bidder's organ is clearly distinguishable from the testes, but in females it is more difficult to differentiate from the ovary (Scaia *et al.* 2019). In mammalian species, the actions of E2 to promote follicular development, expression of gonadotropin receptors in granulosa cells, and intraovarian feedback regulation of steroidogenesis are well described (Rosenfeld *et al.* 2001). While it is likely that similar mechanisms exist in amphibians, studies in a diversity of species remain to be conducted. Oestradiol can regulate spermatogenesis through various mechanisms, including induction of apoptosis, the programmed cell-death of spermatids (Scaia *et al.* 2019). Oestradiol can also inhibit several key steroidogenic enzymes to reduce testicular production of androgen (Fasano *et al.* 1991). Testosterone can be carried in the blood by androgen binding protein, and enzymatically converted to E2 or DHT in the brain, pituitary, and many other organs because aromatase and reductase enzymes are widely distributed (Hu *et al.* 2016; Robitaille and Langlois 2020). The significance of this is that levels of sex steroids circulating in the blood or metabolised and excreted into urine or water are not necessarily reflective of biologically active levels within a target tissue. Careful validations are therefore required.

One of the best-known roles of E2 is in vitellogenesis, the production of egg-yolk proteins. This family of VTGs are cleaved into phosvitin and lipovitellin, and subsequently stored in eggs to provide nutrients for early developing embryos. Vitellogenin is also an immunocompetence factor capable of protecting the host against bacteria and viruses (Li and Zhang 2017). During spawning, the gonadotropins in mature females drive the production of ovarian E2, which is released into the blood and acts on hepatic nuclear oestrogen receptors, inducing the synthesis of VTG proteins (Rasar and Hammes 2006; Di Fiore *et al.* 2020). They are released and actively taken up by developing oocytes, which is essential for early larval development. Details on the maturation of oocytes and on ovulation are well described for anurans such as *Xenopus laevis* (Rasar and Hammes 2006). It has been suggested that oocytes do not express LH receptors (LHR), because the presence of follicular cells and functional gap-junction communication with the oocyte are required for LH to induce the maturation of oocytes (Rasar and Hammes 2006; Toranzo *et al.* 2007; Konduktorova and Luchinskaia 2013). The cellular distribution and differential regulation of LHR and FSHRs have not been adequately investigated in ovaries or testes of amphibians. Obtaining such information would help practitioners of amphibian reproductive technologies to tailor treatment with exogenous gonadotropic hormones for specific outcomes.

Sex steroid feedback control of the gonadotropins

Both negative and positive effects of sex steroids on hypothalamo-pituitary function have been reported *in vivo* with classical gonadectomy and steroid replacement experiments and using *in vitro* tissue-incubation protocols for several amphibian species (Vu and Trudeau 2016). Androgens and estrogens both exert their major actions through the nuclear androgen and oestrogen receptors, which are liganded transcription factors expressed in neurons in the hypothalamus and in gonadotrophs in the pituitary. For example, steroids can act to regulate the production of basal gonadotropin levels by controlling transcription and translation of *lhb*, *fshb*, and *cga* and thus synthesis of biologically active LH and FSH. Steroids may also enhance or suppress pituitary responsiveness to

GnRH or some other releasing factor, and thus regulate the amount of LH or FSH released from gonadotrophs (Trudeau 1997). Important effects of sex steroids at the level of the hypothalamus also exist. Most studies of amphibians to date demonstrate the effectiveness of exogenous steroid treatments in affecting the HPG axis or are correlative studies relating variations in the levels of blood gonadotropins and steroids. These are all suggestive of steroidal feedback and are illustrated in Figure 4.1. However, they do not unequivocally provide evidence for functional feedback loops *in vivo* in amphibians. Earlier work has been reviewed (Vu and Trudeau 2016). Critically missing is information on the physiological and environmental contexts in which negative or positive feedback become activated or disrupted. For example, details on the timing of the switch from a predominant negative feedback to a positive estrogenic drive of the GnRH-LH surge is well described for female-rodent models (Simonneaux *et al.* 2017; Herbison 2020; Silva and Domínguez 2020) but has not been examined in amphibians. Stress associated with transport or translocation could potentially disrupt steroid feedback to alter breeding success of captive amphibians. These data are required to understand why ovulation fails in many induced spawning protocols, and how endocrine disruptors (such as environmental pollutants) may inhibit reproduction in wild populations.

INTERRENAL CORTICOSTERONE AND EFFECTS ON REPRODUCTIVE PROCESSES

Interrenal steroidogenesis is largely controlled by ACTH and activation the MC2R to promote StAR activity, and sequential enzymatic conversion of cholesterol for the production and release of CORT to the general circulation. Corticosterone acts via the glucocorticoid receptor, a liganded transcription factor localised in specific hypothalamic neurons and pituitary cells to exert classical negative effects of feedback on CRF and ACTH to reduce interrenal steroidogenesis (Figure 4.2B). Both inhibitory and permissive actions of CORT at various levels of the amphibian HPG axis (see Figure 4.1) have been reviewed previously (Moore *et al.* 2005). Corticosteroid injections inhibit the release of GnRH1 in a season-dependent manner in the rough-skinned newt, *Taricha granulosa* (Moore and Zoeller 1985). Corticosterone inhibits amplexus within a few minutes (Rose and Moore 1999; Davis *et al.* 2015) by suppressing

AVT-mediated clasping behaviour in this species (Davis *et al.* 2015). Physiologically relevant levels of CORT in the breeding-season reduce *in vitro* production of testicular androgen in the Argentinian toad, *Rhinella arenarum* (Czuchlej *et al.* 2019), providing the only evidence thus far for its direct effects on gonadal function. Narayan *et al.* (2015) tested whether the presence or absence of *Rhinella marina* led to physiological stress and consequences for levels of reproductive hormones and breeding success in the Fijian ground frog, *Platymantis vitiana*, an IUCN-listed endangered species. Exposure to cane toads (an invasive species in Oceania) led to significantly increased urinary CORT, decreased levels of T, E2, and P4 urinary metabolites, and the eventual loss of reproductive output in the native species. In the Indian skipper frog, *Euphlyctis cyanophlyctis*, during the pre-breeding vitellogenic phase, lower doses of ACTH or the synthetic glucocorticoid dexamethasone had no effect on recruitment and growth of oocytes, whereas high doses of both of these impair vitellogenic growth of oocytes and increase follicular atresia (Kupwade and Saidapur 1987).

Short-term increases in CORT and T are facultative for breeding behaviours in some amphibian species (Orchinik *et al.* 1988). This has been demonstrated experimentally in the wild male cane toad *Rhinella marina* under breeding conditions. Narayan *et al.* (2012) exposed adult male toads to capture and handling and reported elevated urinary levels of CORT and T metabolite. Such observations do not support the view that high CORT is necessarily inhibitory to reproduction. The authors suggested that in an explosive breeder like *Rhinella marina*, metabolic actions of CORT would help maintain energetically expensive breeding events (e.g. male–male bouts and male–female mating). During breeding, some amphibians display a significant elevation of levels of CORT with no suppression of reproductive behaviour or physiology (Romero 2002), including *Rhinella arenarum* (Denari and Ceballos 2005) and several Brazilian desert toads (Madelaire and Gomes 2016), among others. Therefore, it is possible that elevated levels of CORT in the plasma may facilitate or modulate several aspects of reproduction. Evidence emerging from research on other vertebrate classes suggests that adequate basal glucocorticoid production helps maintain normal reproduction (Whirledge and Cidlowski 2017). Further discussion on monitoring amphibians' stress physiology is available in Chapter 5.

NON-INVASIVE ASSESSMENT OF REPRODUCTIVE STATUS USING STEROID ANALYSIS

One of the significant challenges for amphibian conservation breeding programs is the early and accurate assessment of reproductive status. Within a breeding season, the best time to apply hormonal therapies must be determined in order to improve success in captive breeding. Typically, body size and the appearance of external secondary sex characters, coupled to the timing of seasonal breeding, is most often used as an indicator of sexual maturity and/or reproductive receptivity. Progress is being made with some species to correlate phenotype with reproductive success (e.g. nuptial pads in male *Silurana tropicalis*; Orton *et al.* 2020). However, external characteristics and morphometrics do not necessarily predict breeding success, as features may appear late in the seasonal recrudescence period, so they may not be the ideal early predictive indicators for the maturation of brood stock. Ultrasonography is a powerful non-invasive tool that may be used for identification of the sex of amphibians (Chapter 3) or for the assessment of ovarian status (Calatayud *et al.* 2018, 2019; Graham *et al.* 2018) (Chapter 6), but the lack of sensitivity of ultrasound probes is a key issue when examining small amphibian species. Moreover, for many endangered species, little is known about their reproductive cycles (Narayan *et al.* 2010; Joshi *et al.* 2018).

It is well established that the production of the classic sex steroids and their metabolites vary in relation to gonadal status in the species studied thus far. Blood levels are traditionally used for evaluations of hormones. Urinary water-borne and faecal steroids are potentially good non-invasive indicators of reproductive physiology (Szymanski *et al.* 2006; Hogan *et al.* 2013; Graham *et al.* 2016; Narayan *et al.* 2019). Earlier studies focused on T and DHT considered the major plasma androgens in different amphibian species (Kime and Hews 1978; Canosa and Ceballos 2002). As there are many possible ways in which amphibians could process steroid hormones, it is difficult to predict the chemical identities of specific conjugated metabolites they excrete. One may anticipate differences among species in the hormone metabolites they produce. Nevertheless, antibodies and enzyme-immunoassay (EIA) protocols that are well established in mammalian and avian species have been applied to amphibians (Munro and Stabenfeldt 1984; Szymanski *et al.* 2006).

To date, only a few studies have assessed cycles in reproductive hormones in anurans using urinary-based T, E1 conjugates (E1C), and P4 EIAs (Germano *et al.* 2009, 2012; Narayan *et al.* 2010 Baugh *et al.* 2018; Joshi *et al.* 2018, 2019, 2020; Baugh and Gray-Gaillard 2021). Narayan *et al.* (2010) reported that urinary P4 quantification can be useful in determining the success of captive breeding programs for cryptic species. In the case of *Platymantis vitiana*, females with high levels of measured urinary P4 and E1C metabolites successfully bred in captivity; however, females with low urinary P4 metabolites were non-vitellogenic (Narayan *et al.* 2010). Although these techniques allow for successful monitoring of the breeding cycle, seasonal variation in hormonal levels poses a challenge for assisted reproduction. Arregui *et al.* (2020) identified that hormonally induced sperm obtained from the natterjack toad, *Epidalea calamita*, during the breeding season was more resilient to cold storage and thus had higher and better motility as compared to sperm obtained during the non-breeding season.

It is crucial to demonstrate that non-invasive measurements of hormones accurately reflect biological events of interest. For example, ovarian cyclicity can be validated by comparing two independent measures of the same reproductive hormone in matched samples (such as faecal versus urinary E2), or by comparing temporal patterns of excretion of hormones (e.g. urinary E2 and P4) with external signs of reproductive status, such as the visual inspection of oocytes through the abdominal skin as an indirect measure of the vitellogenic status of female anurans (Narayan *et al.* 2010). Likewise, administration of hormones that enhance steroid production is useful for demonstrating a cause-and-effect relationship between its exogenous administration and the subsequent excretion of the targeted hormone. Most importantly, a biological validation determines the excretory lag-time between stimulation of the gonads and the appearance of its hormone metabolites in excreta.

Several assays have been physiologically and biologically validated for measurement in excreta, including T, E2, and P4 (Munro *et al.* 1991; Graham *et al.* 2001; Szymanski *et al.* 2006; Kersey *et al.* 2010: see details therein). Urinary T, estrone conjugate (E1C), and P4 EIAs were validated physiologically in *Platymantis vitiana* using challenge by a human chorionic gonadotropin (Narayan *et al.* 2010). The results demonstrated a rise and return to baseline profile for urinary T in male frogs and urinary progesterone and E1C metabolites in female frogs over 2–3 days, thereby reflecting the experimental stimulation of the HPG axis. These data emphasise that care must be taken in the choice of assays used and that commercial kits must be validated for a given species. It is important to carefully design the sampling procedures, storage, and analysis to be able to obtain robust datasets from minimally invasive analysis of hormones. Additional technical limitations have been discussed previously (Narayan *et al.* 2019; Palme 2019). Some key considerations include estimates of extraction and recovery, understanding the specificity and reactivity of the assay, determining the lag time between HPG axis activation, pathways of steroid metabolism, excretion routes, and responsiveness to exogenous hormonal treatments or natural physiological challenges.

CONCLUSIONS

Advancing our understanding of amphibian endocrine systems has enormous potential to inform and direct the development of novel hormonal therapies to promote gamete-release and successful breeding. There is a complex interplay between neuropeptides, catecholamines, pituitary hormones (Figure 4.1), and sex steroids (Figure 4.2) that regulate the diverse reproductive, and often seasonal, modes exhibited by amphibians. Basic research on how and when the numerous neurohormones differentially control LH and FSH is urgently required. It has been known for many years that multiple GnRH forms and multiple GnRH subtypes of receptors exist in amphibians, yet studies on hormonal induction have not taken advantage of this information to promote maturation and release of gametes. Considerable information on the evolution of vertebrate RF-amide peptides exists and, at least for KISS, a stimulatory role within the testes to control production of steroids, proliferation of germ cells, and spermatogenesis has been reported. These RF-amides are also produced by the hypothalamic neurons of amphibians but we do not know much about their roles in the regulation of pituitary function. While definitive evidence requires more experimentation, it appears that DA inhibits the synthesis and secretion of gonadotropins in some anurans, as it does in teleosts and several mammals. There are likely to be other inhibitory neurohormones that await discovery.

A fundamental concept that has emerged from research on anurans is that oocytes play a critical role in the production of steroids. The traditional view is centred on the two-cell model of the interaction of ovarian thecal and follicular cells for the control of steroidogenesis, but oocytes also produce key steroids needed for their own maturation. There is still much research needed to evaluate the exact roles of steroid hormones in readiness to breed, maturation of oocytes, and spermatogenesis. A three-cell model (Figure 4.3) has emerged because oocytes have an important role in controlling their own growth and development by regulating synthesis of the steroids. A similar three-cell system is likely present in the testis where Leydig, Sertoli, and spermatogenic cells interact to regulate steroidogenesis. Critical to the success of future conservation breeding programs will be hormonal therapies that promote final maturation of oocytes and ovulation in the female, which is timed and coordinated with spermatogenesis and spermiation in the male. Data from mammals (Tajima *et al.* 2007) and fish (Mukherjee *et al.* 2017; Zhou *et al.* 2016) indicate that many other hormones and growth factors contribute to gonadal development and steroidogenesis. All await study in amphibians.

Research into steroidal hormones using high-performance liquid chromatography and tandem mass spectroscopy (Miller *et al.* 2013; Jia *et al.* 2016; Havens *et al.* 2020) will be important to elucidate the range of reproductive and stress-hormone metabolites and their exact chemical nature. Phase II metabolism of steroids through sulphation and glucuronidation increases solubility in water for elimination in urine and faeces and the main pathways are well described for humans (Schiffer *et al.* 2019). Little is known for amphibians. Given that current assays developed for the parent steroids typically exhibit low cross-reactivity with their steroidal conjugates, data that could be obtained may not reflect the full range of gonadal steroid activities. The development of rigorous sampling, storage, and protocols for analysis of steroid hormone metabolites would therefore be useful for non-invasive reproductive assessments in amphibian conservation breeding programs.

ACKNOWLEDGEMENTS

The various funding sources (VLT: Environment and Climate Change Canada, Ministère Forêt Faune et Parcs Québec, NSERC-Canada, University of Ottawa Research Chair Program; EN: Australian Academy of Science, Australia-India Fellowship, Griffith University, and University of the South Pacific) for some of our research we cite in this article is acknowledged. We acknowledge, with respect, that our research takes place on traditional lands of numerous Indigenous Peoples around the world.

REFERENCES

Ahn RS, Yoo MS, Kwon HB (1999) Evidence for two-cell model of steroidogenesis in four species of amphibian. *The Journal of Experimental Zoology* **284**, 91–99. doi:10.1002/(SICI)1097-010X(19990615)284:1<91::AID-JEZ12>3.0.CO;2-F

Altmieme Z, Jubouri M, Touma K, Coté G, Fonseca M, Julian T, Mennigen JA (2019) A reproductive role for the nonapeptides vasotocin and isotocin in male zebrafish (*Danio rerio*). *Comparative Biochemistry and Physiology. Part B, Biochemistry & Molecular Biology* **238**, 110333. doi:10.1016/j.cbpb.2019.110333

Álvarez-Viejo M, Cernuda-Cernuda R, DeGrip WJ, Alvarez-López C, García-Fernández JM (2003) Co-localization of mesotocin and opsin immunoreactivity in the hypothalamic preoptic nucleus of *Xenopus laevis*. *Brain Research* **969**, 36–43. doi:10.1016/S0006-8993(03)02273-X

Anjum S, Krishna A, Sridaran R, Tsutsui K (2012) Localization of gonadotropin-releasing hormone (GnRH), gonadotropin-inhibitory Hormone (GnIH), kisspeptin and GnRH receptor and their possible roles in testicular activities from birth to senescence in mice. *Journal of Experimental Zoology. Part A, Ecological Genetics and Physiology* **317**, 630–644. doi:10.1002/jez.1765

Arregui L, Bóveda P, Gosálvez J, Kouba AJ (2020) Effect of seasonality on hormonally induced sperm in *Epidalea calamita* (Amphibia, Anura, Bufonidae) and it's refrigerated and cryopreservated storage. *Aquaculture* **529**, 735677. doi:10.1016/j.aquaculture.2020.735677

Baugh AT, Bastien B, Still MB, Stowell N (2018) Validation of water-borne steroid hormones in a tropical frog (*Physalaemus pustulosus*). *General and Comparative Endocrinology* **261**, 67–80. doi:10.1016/j.ygcen.2018.01.025

Baugh AT, Gray-Gaillard SL (2021) Excreted testosterone and male sexual proceptivity: a hormone validation and proof-of-concept experiment in túngara frogs. *General and Comparative Endocrinology* **300**, 113638. doi:10.1016/j.ygcen.2020.113638

Calatayud NE, Stoops M, Durrant BS (2018) Ovarian control and monitoring in amphibians. *Theriogenology* **109**, 70–81. doi:10.1016/j.theriogenology.2017.12.005

Calatayud NE, Chai N, Gardner NR, Curtis MJ, Stoops MA (2019) Reproductive techniques for ovarian monitoring and control in amphibians. *Journal of Visualized Experiments* **147**, e58675. doi:10.3791/58675

Canosa LF, Ceballos NR (2002) In vitro hCG and human recombinant FSH actions on testicular steroidogenesis in the toad *Bufo arenarum*. *General and Comparative Endocrinology* **126**, 318–324. doi:10.1016/S0016-6480(02)00007-2

Carnevali O, Mosconi G, Yamamoto K, Kobayashi T, Kikuyama S, Polzonetti-Magni AM (1993) In-vitro effects of mammalian and amphibian prolactins on hepatic vitellogenin synthesis in *Rana esculenta*. *The Journal of Endocrinology* **137**, 383–389. doi:10.1677/joe.0.1370383

Chianese R, Ciaramella V, Fasano S, Pierantoni R, Meccariello R (2013) Kisspeptin receptor, GPR54, as a candidate for the regulation of testicular activity in the frog *Rana esculenta*. *Biology of Reproduction* **88**, 1–11. doi:10.1095/biolreprod.112.103515

Chianese R, Ciaramella V, Fasano S, Pierantoni R, Meccariello R (2015) Kisspeptin drives germ cell progression in the anuran amphibian *Pelophylax esculentus*: a study carried out in *ex vivo* testes. *General and Comparative Endocrinology* **211**, 81–91. doi:10.1016/j.ygcen.2014.11.008

Chianese R, Ciaramella V, Fasano S, Pierantoni R, Meccariello R (2017) Kisspeptin regulates steroidogenesis and spermiation in anuran amphibian. *Reproduction (Cambridge, England)* **154**, 403–414. doi:10.1530/REP-17-0030

Clulow J, Trudeau VL, Kouba AJ (2014) Amphibian declines in the twenty-first century: Why we need assisted reproductive technologies. *Advances in Experimental Medicine and Biology* **753**, 275–316. doi:10.1007/978-1-4939-0820-2_12

Clulow J, Pomering M, Herbert D, Upton R, Calatayud N, Clulow S, Mahony MJ, Trudeau VL (2018) Differential success in obtaining gametes between male and female Australian temperate frogs by hormonal induction: A review. *General and Comparative Endocrinology* **265**, 141–148. doi:10.1016/j.ygcen.2018.05.032

Collin F, Chartrel N, Fasolo A, Conlon JM, Vandesande F, Vaudry H (1995) Distribution of two molecular forms of gonadotropin-releasing hormone (GnRH) in the central nervous system of the frog *Rana ridibunda*. *Brain Research* **703**, 111–128. doi:10.1016/0006-8993(95)01074-2

Creighton AE, Wilczynski W (2014) Influence of dopamine D2-type receptors on motor behaviors in the green tree frog, *Hyla cinerea*. *Physiology & Behavior* **127**, 71–80. doi:10.1016/j.physbeh.2014.01.005

Czuchlej SC, Volonteri MC, Regueira E, Ceballos NR (2019) Effect of glucocorticoids on androgen biosynthesis in the testes of the toad *Rhinella arenarum* (Amphibia, Anura). *Journal of Experimental Zoology. Part A, Ecological and Integrative Physiology* **331**, 17–26. doi:10.1002/jez.2232.

d'Anglemont de Tassigny X, Fagg LA, Dixon JPC, Day K, Leitch HG, Hendrick AG, Zahn D, Franceschini I, Caraty A, Carlton MBL, Aparicio SAJR, Colledge WH (2007) Hypogonadotropic hypogonadism in mice lacking a functional Kiss1 gene. *Proceedings of the National Academy of Sciences of the United States of America* **104**, 10714–10719. doi:10.1073/pnas.0704114104

Davis P, Franquemont S, Liang L, Angleson JK, Dores RM (2013) Evolution of the melanocortin-2 receptor in tetrapods: studies on *Xenopus tropicalis* MC2R and *Anolis carolinensis* MC2R. *General and Comparative Endocrinology* **188**, 75–84. doi:10.1016/j.ygcen.2013.04.007

Davis A, Abraham E, McEvoy E, Sonnenfeld S, Lewis C, Hubbard CS, Dolence EK, Rose JD, Coddington E (2015) Corticosterone suppresses vasotocin-enhanced clasping behavior in male rough-skinned newts by novel mechanisms interfering with V1a receptor availability and receptor-mediated endocytosis. *Hormones and Behavior* **69**, 39–49. doi:10.1016/j.yhbeh.2014.12.006

Davis PE, Wilkinson EC, Dores RM (2019) Identifying common features in the activation of melanocortin-2 receptors: Studies on the *Xenopus tropicalis* melanocortin-2 receptor. *International Journal of Molecular Sciences* **20**, 4166. doi:10.3390/ijms20174166

de Roux N, Genin E, Carel JC, Matsuda F, Chaussain JL, Milgrom E (2003) Hypogonadotropic hypogonadism due to loss of function of the KiSS1-derived peptide receptor GPR54. *Proceedings of the National Academy of Sciences of the United States of America* **100**, 10972–10976. doi:10.1073/pnas.1834399100

Decourt C, Robert V, Anger K, Galibert M, Madinier JB, Liu X, Dardente H, Lomet D, Delmas AF, Caraty A *et al.* (2016) A synthetic kisspeptin analog that triggers ovulation and advances puberty. *Scientific Reports* **6**(1), 1–10. doi:10.1038/srep26908

Della Togna G, Trudeau VL, Gratwicke B, Evans M, Augustine L, Chia H, Bronikowski EJ, Murphy JB, Comizzoli P (2017) Effects of hormonal stimulation on the concentration and quality of excreted spermatozoa in the critically endangered Panamanian golden frog (*Atelopus zeteki*). *Theriogenology* **91**, 27–35. doi:10.1016/j.theriogenology.2016.12.033

Denari D, Ceballos NR (2005) 11β-Hydroxysteroid dehydrogenase in the testis of *Bufo arenarum*: Changes in its seasonal activity. *General and Comparative Endocrinology* **143**, 113–120. doi:10.1016/j.ygcen.2005.03.006

Di Fiore MM, Santillo A, Falvo S, Pinelli C (2020) Celebrating 50+ years of research on the reproductive biology and endocrinology of the green frog: an overview. *General and Comparative Endocrinology* **298**, 113578. doi:10.1016/j.ygcen.2020.113578

Dobolyi A, Oláh S, Keller D, Kumari R, Fazekas EA, Csikós V, Renner É, Cservenák M (2020) Secretion and function of pituitary prolactin in evolutionary perspective. *Frontiers in Neuroscience* **14**, 621. doi:10.3389/fnins.2020.00621

DoRego JL, Acharjee S, Seong JY, Galas L, Alexandre D, Bizet P, Burlet A, Kwon HB, Luu-The V, Pelletier G, Vaudry H (2006) Vasotocin and mesotocin stimulate the biosynthesis of neurosteroids in the frog brain. *The Journal of Neuroscience: The Official Journal of the Society for Neuroscience* **26**, 6749–6760. doi:10.1523/JNEUROSCI.4469-05.2006

Dores RM, Sandoval FL, McDonald LK (1993) Proteolytic cleavage of ACTH in corticotropes of sexually mature axolotls (*Ambystoma mexicanum*). *Peptides* **14**, 1029–1035. doi:10.1016/0196-9781(93)90082-R

Dufour S, Weltzien FA, Sebert ME, Le Belle N, Vidal B, Vernier P, Pasqualini C (2005) Dopaminergic inhibition of reproduction in teleost fishes: ecophysiological and evolutionary implications. *Annals of the New York Academy of Sciences* **1040**, 9–21. doi:10.1196/annals.1327.002

Exbrayat JM, Morel G (2003) Visualization of gene expression of prolactin-receptor (PRL-R) by *in situ* hybridization in reproductive organs of *Typhlonectes compressicauda*, a gymnophionan amphibian. *Cell and Tissue Research* **312**, 361–367. doi:10.1007/s00441-003-0730-4

Fasano S, D'Antonio M, Pierantoni R (1991) Sites of action of local estradiol feedback mechanism in the frog (*Rana esculenta*) testis. *General and Comparative Endocrinology* **81**, 492–499. doi:10.1016/0016-6480(91)90177-8

Fasano S, D'Antonio M, Chieffi P, Cobellis G, Pierantoni R (1995) Chicken GnRH-II and salmon GnRH effects on plasma and testicular androgen concentrations in the male frog, *Rana esculenta,* during the annual reproductive cycle. *Comparative Biochemistry and Physiology. Part C, Pharmacology, Toxicology & Endocrinology* **112**, 79–86. doi:10.1016/0742-8413(95)00078-X

Germano JM, Molinia FC, Bishop PJ, Cree A (2009) Urinary hormone analysis assists reproductive monitoring and sex identification of bell frogs (*Litoria raniformis*). *Theriogenology* **72**, 663–671. doi:10.1016/j.theriogenology.2009.04.023

Germano JM, Molinia FC, Bishop PJ, Bell BD, Cree A (2012) Urinary hormone metabolites identify sex and imply unexpected winter breeding in an endangered, subterranean-nesting frog. *General and Comparative Endocrinology* **175**, 464–472. doi:10.1016/j.ygcen.2011.12.003

Graham L, Schwarzenberge F, Möstl E, Galama W, Savage A (2001) A versatile enzyme immunoassay for the determination of progestogens in feces and serum. *Zoo Biology* **20**, 227–236. doi:10.1002/zoo.1022

Graham KM, Kouba AJ, Langhorne CJ, Marcec RM, Willard ST (2016) Biological sex identification in the endangered dusky gopher frog (*Lithobates sevosa*): A comparison of body size measurements, secondary sex characteristics, ultrasound imaging, and urinary hormone analysis methods. *Reproductive Biology and Endocrinology* **14**, 41. doi:10.1186/s12958-016-0174-9

Graham KM, Langhorne CJ, Vance CK, Willard ST, Kouba AJ (2018) Ultrasound imaging improves hormone therapy strategies for induction of ovulation and *in vitro* fertilization in the endangered dusky gopher frog (*Lithobates sevosa*). *Conservation Physiology* **6**, coy020. doi:10.1093/conphys/coy020

Havens SM, Hedman CJ, Hemming JDC, Mieritz MG, Shafer MM, Schauer JJ (2020) Occurrence of estrogens, androgens and progestogens and estrogenic activity in surface water runoff from beef and dairy manure amended crop fields. *The Science of the Total Environment* **710**, 136247. doi:10.1016/j.scitotenv.2019.136247

Herbison AE (2020) A simple model of estrous cycle negative and positive feedback regulation of GnRH secretion. *Frontiers in Neuroendocrinology* **57**, 100837. doi:10.1016/j.yfrne.2020.100837

Hogan LA, Lisle AT, Johnston SD, Goad T, Robertston H (2013) Adult and juvenile sex identification in threatened monomorphic *Geocrinia* frogs using fecal steroid analysis. *Journal of Herpetology* **47**, 112–118. doi:10.1670/11-290

Hu Q, Xiao H, Tian H, Meng Y (2016) Characterization and expression of *cyp19a* gene in the Chinese giant salamander *Andrias davidianus. Comparative Biochemistry and Physiology. Part B, Biochemistry & Molecular Biology* **192**, 21–29. doi:10.1016/j.cbpb.2015.11.005

Irfan S, Ehmcke J, Shahab M, Wistuba J, Schlatt S (2016) Immunocytochemical localization of kisspeptin and kisspeptin receptor in the primate testis. *Journal of Medical Primatology* **45**, 105–111. doi:10.1111/jmp.12212

Jadhao AG, Pinelli C, D'Aniello B, Tsutsui K (2017) Gonadotropin-inhibitory hormone (GnIH) in the amphibian brain and its relationship with the gonadotropin releasing hormone (GnRH) system: an overview. *General and Comparative Endocrinology* **240**, 69–76. doi:10.1016/j.ygcen.2016.09.006

Jia M, Chew WM, Feinstein Y, Skeath P, Sternberg EM (2016) Quantification of cortisol in human eccrine sweat by liquid chromatography–tandem mass spectrometry. *Analyst (London)* **141**, 2053–2060. doi:10.1039/C5AN02387D

Joshi AM, Narayan EJ, Gramapurohit NP (2018) Interrelationship among annual cycles of sex steroids, corticosterone and body condition in *Nyctibatrachus humayuni. General and Comparative Endocrinology* **260**, 151–160. doi:10.1016/j.ygcen.2018.01.013

Joshi AM, Narayan EJ, Gramapurohit NP (2019) Vocalisation and its association with androgens and corticosterone in a night frog (*Nyctibatrachus humayuni*) with unique breeding behaviour. *Ethology* **125**, 774–784. doi:10.1111/eth.12931

Joshi AM, Narayan EJ, Gramapurohit NP (2020) Annual changes in corticosterone and its response to handling, tagging and short-term captivity in *Nyctibatrachus humayuni. The Herpetological Journal* **30**, 118–125. doi:10.33256/hj30.3.118125

Kersey DC, Wildt DE, Brown JL, Huang Y, Snyder RJ, Monfort SL (2010) Parallel and seasonal changes in gonadal and adrenal hormones in male giant pandas (*Ailuropoda melanoleuca*). *Journal of Mammalogy* **91**, 1496–1507. doi:10.1644/09-MAMM-A-404.1

Kikuyama S, Okada R, Hasunuma I, Nakada T (2019) Some aspects of the hypothalamic and pituitary development, metamorphosis, and reproductive behavior as studied in amphibians. *General and Comparative Endocrinology* **284**, 113212. doi:10.1016/j.ygcen.2019.113212

Kime DE, Hews EA (1978) Androgen biosynthesis *in vitro* by testes from amphibia. *General and Comparative Endocrinology* **35**, 280–288. doi:10.1016/0016-6480(78)90073-4

Kline RJ, Holt GJ, Khan IA (2016) Arginine vasotocin V1a2 receptor and GnRH-I co-localize in preoptic neurons of the sex changing grouper, *Epinephelus adscensionis. General and Comparative Endocrinology* **225**, 33–44. doi:10.1016/j.ygcen.2015.07.013

Koda A, Ukena K, Teranishi H, Ohta S, Yamamoto K, Kikuyama S, Tsutsui K (2002) A novel amphibian hypothalamic neuropeptide: isolation, localization, and biological activity. *Endocrinology* **143**, 411–419. doi:10.1210/endo.143.2.8630

Konduktorova VV, Luchinskaia NN (2013) Follicular cells of the amphibian ovary: origin, structure, and functions. *Russian Journal of Developmental Biology* **44**, 232–244. doi:10.1134/S1062360413040024

Kotani M, Detheux M, Vandenbogaerde A, Communi D, Vanderwinden JM, Le Poul E, Brézillon S, Tyldesley R, Suarez-Huerta N, Vandeput F, *et al.* (2001) The metastasis suppressor gene *KiSS-1* encodes kisspeptins, the natural ligands of the orphan G protein-coupled receptor GPR54. *The Journal of Biological Chemistry* **276**, 34631–34636. doi:10.1074/jbc.M104847200

Kouki T, Kawamura K, Kikuyama S (1998) Developmental studies for identification of the inhibitory center of melanotropes in the toad, *Bufo japonicus. Development, Growth & Differentiation* **40**, 651–658. doi:10.1046/j.1440-169X.1998.00395.x

Kupwade VA, Saidapur SK (1987) Effect of dexamethasone and ACTH on oocyte growth and recruitment in the frog *Rana cyanophlyctis* during the prebreeding vitellogenic phase. *General and Comparative Endocrinology* **65**, 394–398. doi:10.1016/0016-6480(87)90124-9

Lapatto R, Pallais JC, Zhang D, Chan YM, Mahan A, Cerrato F, Le Wei W, Hoffman GE, Seminara SB (2007) *Kiss1*–/– mice exhibit more variable hypogonadism than *Gpr54*–/– mice. *Endocrinology* **148**, 4927–4936. doi:10.1210/en.2007-0078

Larcher A, Delarue C, Idres S, Vaudry H (1992) Interactions between vasotocin and other corticotropic factors on the frog adrenal gland. *The Journal of Steroid Biochemistry and Molecular Biology* **41**, 795–798. doi:10.1016/0960-0760(92)90427-K

Lee YR, Tsunekawa K, Mi JM, Haet NU, Hwang JI, Osugi T, Otaki N, Sunakawa Y, Kim K, Vaudry H *et al.* (2009) Molecular evolution of multiple forms of kisspeptins and GPR54 receptors in vertebrates. *Endocrinology* **150**, 2837–2846. doi:10.1210/en.2008-1679

Leebaw WF, Lee LA, Woolf PD (1978) Dopamine affects basal and augmented pituitary hormone secretion. *The Journal of Clinical Endocrinology and Metabolism* **47**, 480–487. doi:10.1210/jcem-47-3-480

Li H, Zhang S (2017) Functions of vitellogenin in eggs. *Results and Problems in Cell Differentiation* **63**, 389–401. doi:10.1007/978-3-319-60855-6_17

Lovejoy DA, Corrigan AZ, Nahorniak CS, Perrin MH, Porter J, Kaiser R, Miller C, Pantoja D, Craig AG, Peter RE (1995) Structural modifications of non-mammalian gonadotropin-releasing hormone (GnRH) isoforms: design of novel GnRH analogues. *Regulatory Peptides* **60**, 99–115. doi:10.1016/0167-0115(95)00116-6

Madelaire CB, Gomes FR (2016) Breeding under unpredictable conditions: annual variation in gonadal maturation, energetic reserves and plasma levels of androgens and corticosterone in anurans from the Brazilian semi-arid. *General and Comparative Endocrinology* **228**, 9–16. doi:10.1016/j.ygcen.2016.01.011

Meccariello R, Fasano S, Pierantoni R (2020) Kisspeptins, new local modulators of male reproduction: A comparative overview. *General and Comparative Endocrinology* **299**, 113618. doi:10.1016/j.ygcen.2020.113618

Mennigen JA, Volkoff H, Chang JP, Trudeau VL (2017) The nonapeptide isotocin in goldfish: evidence for serotonergic regulation and functional roles in the control of food intake and pituitary hormone release. *General and Comparative Endocrinology* **254**, 38–49. doi:10.1016/j.ygcen.2017.09.008

Miller R, Plessow F, Rauh M, Gröschl M, Kirschbaum C (2013) Comparison of salivary cortisol as measured by different immunoassays and tandem mass spectrometry. *Psychoneuroendocrinology* **38**, 50–57. doi:10.1016/j.psyneuen.2012.04.019

Moon JS, Lee YR, Oh DY, Hwang JI, Lee JY, Kim JIl, Vaudry H, Kwon HB, Seong JY (2009) Molecular cloning of the bullfrog kisspeptin receptor GPR54 with high sensitivity to *Xenopus* kisspeptin. *Peptides* **30**, 171–179. doi:10.1016/j.peptides.2008.04.015

Moore FL, Zoeller RT (1979) Endocrine control of amphibian sexual behavior: evidence for a neurohormone-androgen interaction. *Hormones and Behavior* **13**, 207–213. doi:10.1016/0018-506X(79)90038-2

Moore FL, Zoeller RT (1985) Stress-induced inhibition of reproduction: evidence of suppressed secretion of LH-RH in an amphibian. *General and Comparative Endocrinology* **60**, 252–258. doi:10.1016/0016-6480(85)90321-1

Moore FL, Wood RE, Boyd SK (1992) Sex steroids and vasotocin interact in a female amphibian *(Taricha granulosa)* to elicit female-like egg-laying behavior or male-like courtship. *Hormones and Behavior* **26**, 156–166. doi:10.1016/0018-506X(92)90039-X

Moore FL, Boyd SK, Kelley DB (2005) Historical perspective: hormonal regulation of behaviors in amphibians. *Hormones and Behavior* **48**, 373–383. doi:10.1016/j.yhbeh.2005.05.011

Muir AI, Chamberlain L, Elshourbagy NA, Michalovich D, Moore DJ, Calamari A, Szekeres PG, Sarau HM, Chambers JK, Murdock P, *et al.* (2001) AXOR12, a novel human G protein-coupled receptor, activated by the peptide KiSS-1. *The Journal of Biological Chemistry* **276**, 28969–28975. doi:10.1074/jbc.M102743200

Mukherjee D, Majumder S, Roy Moulik S, Pal P, Gupta S, Guha P, Kumar D (2017) Membrane receptor cross talk in gonadotropin-, IGF-I-, and insulin-mediated steroidogenesis in fish ovary: an overview. *General and Comparative Endocrinology* **240**, 10–18. doi:10.1016/j.ygcen.2016.09.002

Munro C, Stabenfeldt G (1984) Development of a microtitre plate enzyme immunoassay for the determination of progesterone. *The Journal of Endocrinology* **101**, 41–49. doi:10.1677/joe.0.1010041

Munro CJ, Stabenfeldt GH, Cragun JR, Addiego LA, Overstreet JW, Lasley BL (1991) Relationship of serum estradiol and progesterone concentrations to the excretion profiles of their major urinary metabolites as measured by enzyme immunoassay and radioimmunoassay. *Clinical Chemistry* **37**, 838–844. doi:10.1093/clinchem/37.6.838

Murthy CK, Turner RJ, Nestor JJ, Jr, Rivier JE, Peter RE (1994) A new gonadotropin-releasing hormone (GnRH) superagonist in goldfish: influence of dialkyl-D-homoarginine at position 6 on gonadotropin-II and growth hormone release. *Regulatory Peptides* **53**, 1–15. doi:10.1016/0167-0115(94)90154-6

Nakano M, Minagawa A, Hasunuma I, Okada R, Tonon MC, Vaudry H, Yamamoto K, Kikuyama S, Machida T, Kobayashi T (2010) D2 Dopamine receptor subtype mediates the inhibitory effect of dopamine on TRH-induced prolactin release from the bullfrog pituitary. *General and Comparative Endocrinology* **168**, 287–292. doi:10.1016/j.ygcen.2010.05.008

Narayan EJ, Molinia FC, Christi KS, Morley CG, Cockrem JF (2010) Annual cycles of urinary reproductive steroid concentrations in wild and captive endangered Fijian ground frogs (*Platymantis vitiana*). *General and Comparative Endocrinology* **166**, 172–179. doi:10.1016/j.ygcen.2009.10.003

Narayan EJ, Molinia FC, Cockrem JF, Hero JM (2012) Changes in urinary testosterone and corticosterone metabolites during short-term confinement with repeated handling in wild male cane toads (*Rhinella marina*). *Australian Journal of Zoology* **59**, 264–269. doi:10.1071/ZO11070

Narayan EJ, Jessop TS, Hero JM (2015) Invasive cane toad triggers chronic physiological stress and decreased reproductive success in an island endemic. *Functional Ecology* **29**, 1435–1444. doi:10.1111/1365-2435.12446

Narayan EJ, Gabor CR, Forsburg ZR, Davis D (2019) Non-invasive methods for measuring and monitoring stress physiology in imperiled amphibians. *Frontiers in Ecology and Evolution* **7**, 431. doi:10.3389/fevo.2019.00431

Nestor CC, Bedenbaugh MN, Hileman SM, Coolen LM, Lehman MN, Goodman RL (2018) Regulation of GnRH pulsatility in ewes. *Reproduction (Cambridge, England)* **156**, R83–R99. doi:10.1530/REP-18-0127

O'Connell LA, Matthews BJ, Ryan MJ, Hofmann HA (2010) Characterization of the dopamine system in the brain of the túngara frog, *Physalaemus pustulosus*. *Brain, Behavior and Evolution* **76**, 211–225. doi:10.1159/000321715

Ogielska M (Ed.) (2009) *Reproduction of Amphibians*. Series: Biological Systems in Vertebrates. Science Publishers, Enfield NH, USA.

Oguchi A, Tanaka S, Aida T, Yamamoto K, Kikuyama S (1997) Enhancement by prolactin of the GnRH-induced release of LH from dispersed anterior pituitary cells of the bullfrog (*Rana catesbeiana*). *General and Comparative Endocrinology* **107**, 128–135. doi:10.1006/gccn.1997.6904

Oh DY, Wang L, Ahn RS, Park J-Y, Seong JY, Kwon HB (2003) Differential G protein coupling preference of mammalian and nonmammalian gonadotropin-releasing hormone receptors. *Molecular and Cellular Endocrinology* **205**, 89–98. doi:10.1016/S0303-7207(03)00204-1

Okada R, Yamamoto K, Hasunuma I, Asahina J, Kikuyama S (2016) Arginine vasotocin is the major adrenocorticotropic hormone-releasing factor in the bullfrog *Rana catesbeiana*. *General and Comparative Endocrinology* **237**, 121–130. doi:10.1016/j.ygcen.2016.08.014

Olivereau M, Vandesande F, Boucique E, Ollevier F, Olivereau JM (1987) Immunocytochemical localization and spatial relation to the adenohypophysis of a somatostatin-like and a corticotropin-releasing factor-like peptide in the brain of four amphibian species. *Cell and Tissue Research* **247**, 317–324. doi:10.1007/BF00218313

Orchinik M, Licht P, Crews D (1988) Plasma steroid concentrations change in response to sexual behavior in *Bufo marinus*. *Hormones and Behavior* **22**, 338–350. doi:10.1016/0018-506X(88)90006-2

Orton F, Svanholm S, Jansson E, Carlsson Y, Eriksson A, Uren Webster T, McMillan T, Leishman M, Verbruggen B, Economou T *et al.* (2020) A laboratory investigation into features of morphology and physiology for their potential to predict reproductive success in male frogs. *PLoS One* **15**, e0241625. doi:10.1371/journal.pone.0241625

Palme R (2019) Non-invasive measurement of glucocorticoids: advances and problems. *Physiology & Behavior* **199**, 229–243. doi:10.1016/j.physbeh.2018.11.021

Pierantoni R, Iela L, d'Istria M, Fasano S, Rastogi RK, Delrio G (1984) Seasonal testosterone profile and testicular responsiveness to pituitary factors and gonadotrophin releasing hormone during two different phases of the sexual cycle of the frog (*Rana esculenta*). *The Journal of Endocrinology* **102**, 387–392. doi:10.1677/joe.0.1020387

Pinto FM, Cejudo-Román A, Ravina CG, Fernández-Sánchez M, Martín-Lozano D, Illanes M, Tena-Sempere M, Candenas ML (2012) Characterization of the kisspeptin system in human spermatozoa. *International Journal of Andrology* **35**, 63–73. doi:10.1111/j.1365-2605.2011.01177.x

Rasar MA, Hammes SR (2006) The physiology of the *Xenopus laevis* ovary. *Methods in Molecular Biology (Clifton, N.J.)* **322**, 17–30. doi:10.1007/978-1-59745-000-3_2

Richerson BA, DeRoos R (1971) Comparative morphology of the neurohypophysis of the three life stages of the central newt (*Notophthalmus viridescens louisianensis*). *General and Comparative Endocrinology* **17**, 256–267. doi:10.1016/0016-6480(71)90134-1

Robitaille J, Langlois VS (2020) Consequences of steroid-5α-reductase deficiency and inhibition in vertebrates. *General and Comparative Endocrinology* **290**, 113400. doi:10.1016/j.ygcen.2020.113400

Romero LM (2002) Seasonal changes in plasma glucocorticoid concentrations in free-living vertebrates. *General and Comparative Endocrinology* **128**, 1–24. doi:10.1016/S0016-6480(02)00064-3

Rosati L, Di Fiore MM, Prisco M, Di Giacomo Russo F, Venditti M, Andreuccetti P, Chieffi Baccari G, Santillo A (2019) Seasonal expression and cellular distribution of star and steroidogenic enzymes in quail testis. *Journal of Experimental Zoology. Part B, Molecular and Developmental Evolution* **332**, 198–209. doi:10.1002/jez.b.22896

Rose JD, Moore FL (1999) A neurobehavioral model for rapid actions of corticosterone on sensorimotor integration. *Steroids* **64**, 92–99. doi:10.1016/S0039-128X(98)00094-4

Rosenfeld CS, Wagner JS, Roberts RM, Lubahn DB (2001) Intraovarian actions of oestrogen. *Reproduction (Cambridge, England)* **122**, 215–226. doi:10.1530/rep.0.1220215

Roseweir AK, Millar RP (2009) The role of kisspeptin in the control of gonadotrophin secretion. *Human Reproduction Update* **15**, 203–212. doi:10.1093/humupd/dmn058

Scaia MF, Volonteri MC, Czuchlej SC, Ceballos NR (2019) Estradiol and reproduction in the South American toad *Rhinella arenarum* (Amphibian, Anura). *General and Comparative Endocrinology* **273**, 20–31. doi:10.1016/j.ygcen.2018.03.018

Schiffer L, Barnard L, Baranowski ES, Gilligan LC, Taylor AE, Arlt W, Shackleton CHL, Storbeck KH (2019) Human steroid biosynthesis, metabolism and excretion are differentially reflected by serum and urine steroid metabolomes: A comprehensive review. *The Journal of Steroid Biochemistry and Molecular Biology* **194**, 105439. doi:10.1016/j.jsbmb.2019.105439

Seminara SB, Messager S, Chatzidaki EE, Thresher RR, Acierno JS, Shagoury JK, Bo-Abbas Y, Kuohung W, Schwinof KM, Hendrick AG *et al.* (2003) The GPR54 gene as a regulator of puberty. *The New England Journal of Medicine* **349**, 1614–1627. doi:10.1056/NEJMoa035322

Sharma A, Thaventhiran T, Minhas S, Dhillo WS, Jayasena CN (2020) Kisspeptin and testicular function – is it necessary? *International Journal of Molecular Sciences* **21**(8), 1–14. doi:10.3390/ijms21082958

Silla AJ (2010) Effects of luteinizing hormone-releasing hormone and arginine-vasotocin on the sperm-release response of Günther's toadlet, *Pseudophryne guentheri*. *Reproductive Biology and Endocrinology* **8**, 139. doi:10.1186/1477-7827-8-139

Silla AJ, Byrne PG (2019) The role of reproductive technologies in amphibian conservation breeding programs. *Annual Review of Animal Biosciences* **7**, 499–519. doi:10.1146/annurev-animal-020518-115056

Silla AJ, Calatayud NE, Trudeau VL (2021) Amphibian reproductive technologies: approaches and welfare considerations. *Conservation Physiology* **9**, coab011. doi:10.1093/conphys/coab011

Silva CC, Domínguez R (2020) Clock control of mammalian reproductive cycles: looking beyond the pre-ovulatory surge of gonadotropins. *Reviews in Endocrine & Metabolic Disorders* **21**, 149–163. doi:10.1007/s11154-019-09525-9

Simonneaux V, Bahougne T, Angelopoulou E (2017) Daily rhythms count for female fertility. *Best Practice & Research. Clinical Endocrinology & Metabolism* **31**, 505–519. doi:10.1016/j.beem.2017.10.012

Smith JT, Dungan HM, Stoll EA, Gottsch ML, Braun RE, Eacker SM, Clifton DK, Steiner RA (2005) Differential regulation of KiSS-1 mRNA expression by sex steroids in the brain of the male mouse. *Endocrinology* **146**, 2976–2984. doi:10.1210/en.2005-0323

Somoza GM, Mechaly AS, Trudeau VL (2020) Kisspeptin and GnRH interactions in the reproductive brain of teleosts. *General and Comparative Endocrinology* **298**, 113568. doi:10.1016/j.ygcen.2020.113568

Sotowska-Brochocka J (1988) The stimulatory and inhibitory role of the hypothalamus in the regulation of ovulation in grass frog, *Rana temporaria*. *General and Comparative Endocrinology* **70**, 83–90. doi:10.1016/0016-6480(88)90096-2

Sotowska-Brochocka J, Licht P (1992) Effect of infundibular lesions on GnRH and LH release in the frog, *Rana temporaria*, during hibernation. *General and Comparative Endocrinology* **85**, 43–54. doi:10.1016/0016-6480(92)90170-O

Sotowska-Brochocka J, Martyńska L, Licht P (1994) Dopaminergic inhibition of gonadotropic release in hibernating frogs, *Rana temporaria*. *General and Comparative Endocrinology* **93**, 192–196. doi:10.1006/gcen.1994.1022

Szymanski DC, Gist DH, Roth TL (2006) Anuran gender identification by fecal steroid analysis. *Zoo Biology* **25**, 35–46. doi:10.1002/zoo.20077

Tajima K, Orisaka M, Mori T, Kotsuji F (2007) Ovarian theca cells in follicular function. *Reproductive Biomedicine Online* **15**, 591–609. doi:10.1016/S1472-6483(10)60392-6

Tena-Sempere M, Felip A, Gómez A, Zanuy S, Carrillo M (2012) Comparative insights of the kisspeptin/kisspeptin receptor system: Lessons from non-mammalian vertebrates. *General and Comparative Endocrinology* **175**, 234–243. doi:10.1016/j.ygcen.2011.11.015

Tonon MC, Leroux P, Jenks BG, Gouteux L, Jegou S, Guy J, Pelletier G, Vaudry H (1985) Le lobe intermédiaire de l'hypophyse des amphibiens: une glande endocrine à sécrétions multiples et sous contrôle pluri-hormonal. [Intermediate lobe of the amphibian pituitary gland: an endocrine gland with multiple secretions and under multi-hormonal control]. *Annales d'Endocrinologie* **46**, 69–87.

Toranzo GS, Oterino J, Zelarayán L, Bonilla F, Bühler MI (2007) Spontaneous and LH-induced maturation in *Bufo arenarum* oocytes: importance of gap junctions. *Zygote (Cambridge, England)* **15**, 65–80. doi:10.1017/S0967199406004023

Tortonese DJ (2016) Intrapituitary mechanisms underlying the control of fertility: Key players in seasonal breeding. *Domestic Animal Endocrinology* **56**, S191–S203. doi:10.1016/j.domaniend.2016.01.002

Toyoda F, Yamamoto K, Ito Y, Tanaka S, Yamashita M, Kikuyama S (2003) Involvement of arginine vasotocin in reproductive events in the male newt *Cynops pyrrhogaster*. *Hormones and Behavior* **44**, 346–353. doi:10.1016/j.yhbeh.2003.06.001

Trudeau VL (1997) Neuroendocrine regulation of gonadotrophin II release and gonadal growth in the goldfish, *Carassius auratus*. *Reviews of Reproduction* **2**, 55–68. doi:10.1530/ror.0.0020055

Trudeau VL, Somoza GM (2020) Multimodal hypothalamo-hypophysial communication in the vertebrates. *General and Comparative Endocrinology* **293**, 113475. doi:10.1016/j.ygcen.2020.113475

Trudeau VL, Somoza GM, Natale GS, Pauli B, Wignall J, Jackman P, Doe K, Schueler FW (2010) Hormonal induction of spawning in 4 species of frogs by coinjection with a gonadotropin-releasing hormone agonist and a dopamine antagonist. *Reproductive Biology and Endocrinology* **8**, 36. doi:10.1186/1477-7827-8-36

Trudeau VL, Schueler FW, Navarro-Martin L, Hamilton CK, Bulaeva E, Bennett A, Fletcher W, Taylor L (2013) Efficient induction of spawning of northern leopard frogs (*Lithobates pipiens*) during and outside the natural breeding season. *Reproductive Biology and Endocrinology* **11**, 14. doi:10.1186/1477-7827-11-14

Uchiyama H, Koda A, Komazaki S, Oyama M, Kikuyama S (2000) Occurrence of immunoreactive activin/inhibin beta(B) in thyrotropes and gonadotropes in the bullfrog pituitary: possible paracrine/autocrine effects of activin B on gonadotropin secretion. *General and Comparative Endocrinology* **118**, 68–76. doi:10.1006/gcen.2000.7456

Ukena K, Koda A, Yamamoto K, Kobayashi T, Iwakoshi-Ukena E, Minakata H, Kikuyama S, Tsutsui K (2003) Novel neuropeptides related to frog growth hormone-releasing peptide: isolation, sequence, and functional analysis. *Endocrinology* **144**, 3879–3884. doi:10.1210/en.2003-0359

Vu M, Trudeau VL (2016) Neuroendocrine control of spawning in amphibians and its practical applications. *General and Comparative Endocrinology* **234**, 28–39. doi:10.1016/j.ygcen.2016.03.024

Vu M, Weiler B, Trudeau VL (2017) Time- and dose-related effects of a gonadotropin-releasing hormone agonist and dopamine antagonist on reproduction in the northern leopard frog (*Lithobates pipiens*). *General and Comparative Endocrinology* **254**, 86–96. doi:10.1016/j.ygcen.2017.09.023

Wang L, Bogerd J, Choi HS, Seong JY, Soh JM, Chun SY, Blomenröhr M, Troskie BE, Millar RP, Wen HY (2001) Three distinct types of GnRH receptor characterized in the bullfrog. *Proceedings of the National Academy of Sciences of the United States of America* **98**, 361–366. doi:10.1073/pnas.98.1.361

Whirledge S, Cidlowski JA (2017) Glucocorticoids and reproduction: traffic control on the road to reproduction. *Trends in Endocrinology and Metabolism* **28**, 399–415. doi:10.1016/j.tem.2017.02.005

Whittington CM, Wilson AB (2013) The role of prolactin in fish reproduction. *General and Comparative Endocrinology* **191**, 123–136. doi:10.1016/j.ygcen.2013.05.027

Wilczynski W, Lynch KS (2011) Female sexual arousal in amphibians. *Hormones and Behavior* **59**, 630–636. doi:10.1016/j.yhbeh.2010.08.015

Wilczynski W, Quispe M, Muñoz MI, Penna M (2017) Arginine vasotocin, the social neuropeptide of amphibians and reptiles. *Frontiers in Endocrinology* **8**, 186. doi:10.3389/fendo.2017.00186

Wiśniewski K (2019) Design of oxytocin analogs. *Methods in Molecular Biology (Clifton, N.J.)* **2001**, 235–271. doi:10.1007/978-1-4939-9504-2_11

Yang WH, Lutz LB, Hammes SR (2003) *Xenopus laevis* ovarian CYP17 is a highly potent enzyme expressed exclusively in oocytes. Evidence that oocytes play a critical role in *Xenopus* ovarian androgen production. *The Journal of Biological Chemistry* **278**, 9552–9559. doi:10.1074/jbc.M212027200

Yoo MS, Kang HM, Choi HS, Kim JW, Troskie BE, Millar RP, Kwon HB (2000) Molecular cloning, distribution and pharmacological characterization of a novel gonadotropin-releasing hormone ([Trp8] GnRH) in frog brain. *Molecular and Cellular Endocrinology* **164**, 197–204. doi:10.1016/S0303-7207(00)00221-5

Zhou R, Yu SM, Ge W (2016) Expression and functional characterization of intrafollicular GH-IGF system in the zebrafish ovary. *General and Comparative Endocrinology* **232**, 32–42. doi:10.1016/j.ygcen.2015.11.018

Zornik E, Kelley DB (2011) A neuroendocrine basis for the hierarchical control of frog courtship vocalizations. *Frontiers in Neuroendocrinology* **32**, 353–366. doi:10.1016/j.yfrne.2010.12.006

5   Non-invasive monitoring of stress physiology during management and breeding of amphibians in captivity

Edward J. Narayan

INTRODUCTION

Loss of biodiversity is one of the most severe anthropogenic environmental issues (Roe *et al.* 2019) of our time, with defaunation driven by natural and anthropogenic stressors such as global climatic change, conversion of habitat, and invasive organisms (Doherty *et al.* 2016; Young *et al.* 2016; Alroy 2017; Thompson *et al.* 2017; O'Connor *et al.* 2020) (also see Chapter 2). These factors interact in synergistic ways, leading to catastrophic declines (Cohen *et al.* 2019; Rollins-Smith 2020). In an attempt to cope with extreme environmental changes, wildlife populations are forced to adapt, thus influencing their survival, distribution, movement, physiology, and reproductive success (Tuomainen and Candolin 2011; Malisch *et al.* 2020). Tools to quantify the physiological responses of wildlife towards extreme environmental change are valuable in conservation studies because they can be used to monitor species' recovery rates and the success of management programs (Bergman *et al.* 2019). Environmental stressors, such as climatic change or alteration of habitat, are manifest at a physiological level and can be measured in real time using biomarkers to predict a population's decline (Ellis *et al.* 2012). Comprehensive understanding of the sensitivity of wildlife towards critical changes in their surroundings is urgently needed for determining species' vulnerability and for setting conservation priorities (Hofmann and Todgham 2010). An important component of species' recovery is the development of conservation breeding programs (McGowan *et al.* 2017; Browne *et al.* 2018) because these provide valuable assurance of genetic stock against perturbation factors such as climatic change that continue to affect wild amphibian populations globally. However, numerous *ex situ* breeding programs have met with many complex challenges such as lack of normal ovarian activity, reduced survival of young, and associated costs of rearing larvae in large numbers (see the review by Smith and Sutherland 2014; https://www.conservation evidence.com/actions/835). These associated challenges could be attributed to environmental conditions (Flach *et al.* 2020), physiological stress linked to translocation (Lèche *et al.* 2016; Batson *et al.* 2017), characteristics of enclosures (Zhang *et al.* 2017), or husbandry practices (Marshall *et al.* 2016). By understanding how acute and chronic stressors influence the physiology of amphibians, such factors can be strictly controlled to increase the overall health and numbers of viable offspring (Taylor 2017; Donaldson 2019).

Conservation physiology has been defined as 'an integrative scientific discipline applying physiological concepts, tools, and knowledge to characterising biological diversity and its ecological implications; understanding and predicting how organisms, populations, and ecosystems respond to environmental change and stressors; and solving conservation problems across the broad range of taxa (i.e. including microbes, plants, and animals)'

(Cooke *et al.* 2013). This expanding discipline provides an organismic view of the environmental challenges and the threats to conservation faced both by wild and captive populations (Wikelski and Cooke 2006; Mingramm *et al.* 2020). The fields of stress and reproductive endocrinology are integral to the discipline of conservation physiology and have made valuable contributions to the conservation of threatened and rare wildlife (Berger *et al.* 1999; Gordon *et al.* 2004; Wikelski and Cooke 2006; Mingramm *et al.* 2020). For example, non-invasive endocrinology provides a direct measure of reproductive hormone cycles and the physiological stress responses of terrestrial and aquatic amphibians in captivity as well as *in situ* populations (Dittami *et al.* 2008; Narayan *et al.* 2012a; Hunt *et al.* 2019). Thus, non-invasive endocrinology can provide crucial insight into a population's eco-physiological relationship with its environment (Narayan *et al.* 2019)

Interest in amphibian conservation physiology has had a recent resurgence with the global realisation that amphibians are rapidly disappearing (Stuart *et al.* 2004; Bickford *et al.* 2018; Browne *et al.* 2019; Bolochio *et al.* 2020; Meza-Parral *et al.* 2020). While there are several proposed causes for the demise of amphibian populations, they are all based on a common mechanism: the inability of the amphibian biology to cope with new, often unnatural, environmental perturbations. However, some species of amphibians are also able to persist with the infectious disease, chytridiomycosis (Brannelly *et al.* 2021). Integration of the assessment of stress hormones paired with a knowledge of the basic biology of amphibians can provide a powerful diagnostic tool for assessments of health and welfare both in captive and wild populations. As demonstrated in other wildlife such as tigers, a mild ecological disturbance, such as eco-tourism (Tyagi *et al.* 2019), can induce high levels of stress hormones (indexed using faecal-based monitoring of stress hormones); this signifies the value of non-invasive monitoring of hormones for management and conservation (Wikelski and Cooke 2006).

Non-invasive endocrinology (monitoring of hormones using biological samples other than blood, e.g. urine and faeces) brings a multifocal approach to solving conservation problems, especially those related to management of captive amphibians (Narayan 2013; Narayan *et al.* 2019). This chapter highlights the usefulness of stress physiology in amphibian biotechnology and guides the development of successful conservation breeding programs.

PHYSIOLOGY OF AMPHIBIAN STRESS

Several comprehensive reviews related to the key theory of stress physiology have been published over the past decade (McEwen and Wingfield 2003; Romero 2004; Valsamakis *et al.* 2019; Kinlein and Karatsoreos 2020). Research on the physiological impacts of stress has developed independently across various fields of biology and as a result the definition of stress often is ambiguous (McEwen and Wingfield 2003; Schulte 2014; Harris 2020). Selye (1950) characterised the stress response as a common set of generalised physiological responses that were experienced by all organisms exposed to a variety of environmental challenges, like changes in ambient temperature or exposure to noise. Stress threatens the biological process of homeostasis, which is the ability of organisms to maintain internal stability within limits that allow survival (McEwen 2010; Collier *et al.* 2017). It is also defined as a self-regulating phenomenon whereby physiological systems such as the stress endocrine system returns to a set-point required for survival. However, in reality, a set-point is not always maintained and this led to the introduction of the terms 'allostasis', 'allostatic load', and 'allostatic overload' (Sterling *et al.* 1988; McEwen 2010).

Allostasis refers to achieving stability through change, and basal stress-hormone levels support allostasis via regulation of the distribution of energy to vital processes (such as metabolism and growth) that help animals cope with predictable environmental and life-history changes. Allostatic load refers to the cumulative result of maintaining allostasis and accounts for the seasonal increases in stress hormones caused by the predictable elevation of energy requirements during more demanding periods, such as winter and during breeding. Allostatic overload occurs when an unpredictable event increases the demand for energy beyond the point at which modulation of energy can successfully deal with it. Such an overload could lead to suppressing effects on other physiological systems such as the reproductive system or immune function. In some cases in captive settings, allostatic overload could lead to social dysfunction (McEwen and Wingfield 2010). It is important that researchers apply stress theory and concepts when studying the stress biology of amphibians because such approaches assist in gauging and understanding physiological responses to environmental change (Romero *et al.* 2009). While an acute stress response is necessary to ensure survival and allow adaptation to changes in the environment, it is only a concern when the response becomes chronic and threatens the animal's wellbeing by exerting deleterious effects on its biological state. This approach has implications for welfare during energetically costly periods, such as during growth and reproduction, or when animals are suffering from parasites or other diseases (Monfort 2003; Kindermann *et al.* 2012; Rollins-Smith 2017; Hammond *et al.* 2020). For this chapter, a stressor (or stress event) will be defined as a noxious stimulant that exposes the amphibian to energetic costs outside of those required to meet predictable daily and seasonal needs.

Corticosterone, secreted from the interrenal cells of the hypothalamo-pituitary interrenal (HPI axis) glands during a physiological stress response in amphibians, aids in regulating metabolic, reproductive, and immune function (Rollins-Smith 2017; Titon *et al.* 2018). The effects of stress on growth, development, and morphology were first demonstrated by Glennemeier and Denver (2002) by exposing tadpoles of the northern leopard frog (*Rana pipiens*) to exogenous corticosterone. The HPI axis is a combination of neural transmitters and hormonal responses, which allow the individual to react to a stressful event in a way that maximises the chance of survival (Hill *et al.* 2008). The first phase of activation of the HPI axis involves the central nervous system (CNS) receiving signals that a stress event has occurred. Within seconds, signals are sent to the hypothalamus where corticotrophin-releasing hormone (CRH) is released. The CRH travels in the hypothalamo-hypophyseal (HH) portal system to the anterior pituitary gland where adrenocorticotropic hormone (ACTH) is secreted. The ACTH is released from the cells of the anterior pituitary into the general circulation of the blood, where it travels to the interrenal cortex (adrenal cortex in higher vertebrates) to stimulate the secretion of glucocorticoids. Glucocorticoids are stress-associated metabolic hormones whose levels rise above a 'baseline' during a stressful activity, leading to gluconeogenesis and the mobilisation of energy (Hau *et al.* 2016; MacDougall-Shackleton *et al.* 2019). Glucocorticoids also have a wide range of other actions including effects on intermediary metabolism, immune function, behaviour, electrolyte balance, growth, and reproduction (Guillette *et al.* 1995). Secretion of glucocorticoids also causes increased heart rate, ventilation, vasoconstriction, and catabolism of fat; and a decrease in digestion and the production of insulin (Hill *et al.* 2008; Demas *et al.* 2011; Sadoul and Vijayan 2016). Glucocorticoids are released when an animal responds to a stressor so that the animal can adjust to the stressor but if the stressor is prolonged, then chronically elevated glucocorticoid levels may have detrimental effects on the individual (Bliley and Woodley 2012). Several studies have recorded the effect of stressors such as restricted intake of food, low pH, and infection on corticosterone levels of amphibians and their larval forms (Burraco and Gomez-Mestre 2016; Gabor *et al.* 2016; Kirschman *et al.* 2018). For example, Kirschman *et al.* (2018) demonstrated the complexity of interactions between glucocorticoids and infection from ranavirus using larvae of the wood frog (*Lithobates sylvaticus*). They showed how acute and chronic exposure to corticosterone influenced infection rates, growth, and survival of larvae. The relationship between amphibians' fitness and their endocrinal response to stress arising from their exposure to environmental trauma and diseases (e.g. exposure to atrazine and chytridiomycosis) is a complex one and not primarily mediated by corticosterone (see the results of Gabor *et al.* 2016). Rather, multidimensional interactions between the HPI axis and various other components of the animal's biology, such as the immune system and other cellular biomarkers, are required (Blaustein *et al.* 2012).

Measuring baseline levels of circulating glucocorticoids over extended periods affords the capacity to infer whether an animal is acutely or chronically stressed or has acclimated to cope with the known stressor (Vitousek *et al.* 2019). This approach can be useful in predicting amphibians' responses to physical stressors related to captive breeding, thereby aiding in the development of management strategies that promote successful breeding programs (Hammond *et al.* 2020). For example, it is important to assess the stress response of amphibian species that were freshly transported from the wild into captivity or translocated between facilities (Narayan and Hero 2011). In another example, Cikanek *et al.* (2014) studied the strategies for housing harlequin frogs (*Atelopus* spp.) in groups, using behavioural and physiological indicators. Researchers applied faecal-based glucocorticoid monitoring to test the stress responses of frogs in grouped housing situations and showed that in the long term male frogs can be housed together. There needs to be further research to evaluate the physiological stress responses of amphibians in conservation breeding programs; to my knowledge little attention has been paid to tools monitoring or evaluating the success of breeding programs.

STRESS IN AMPHIBIAN CONSERVATION BREEDING PROGRAMS

Animals living in their natural environment have the ability to escape from stressful situations, with feedback mechanisms that ascertain when glucocorticoids return to baseline levels after cessation of the stressor. In contrast, animals housed in enclosed settings, such as breeding chambers, cannot escape stressful conditions, thereby potentially leading to elevated glucocorticoid levels over extended periods of time (Narayan *et al.* 2015; Titon *et al.* 2017; Fischer *et al.* 2018).

With the advent of the Anthropocene and its increase in rates of extinction, conservation breeding programs are gaining momentum, although their success is debatable (Madliger *et al.* 2017; Grant *et al.* 2019). Maintaining and breeding amphibians in captivity have proven to be challenging, owing to the animals' complex needs. Only 55% of such programs reviewed by Smith and Sutherland (2014) produced tadpoles, metamorphs, or juveniles. Thus, it is crucial to understand the effects of extrinsic factors (e.g. husbandry practices and housing conditions) and intrinsic factors (e.g. age and sex) on the survival and success of captive breeding programs. Some factors that must be considered in order to minimise stress in *ex situ* programs include examining species' social structures, monitoring their ability to use refugia, identifying overcrowding, and maintaining visual barriers to reduce stress from individuals in adjacent enclosures (Ferrie *et al.* 2014). Issues pertaining to amphibians' life histories, ecological needs, and responses to small shifts in environmental conditions are not fully understood, thus making it difficult to develop generalised husbandry practices (Ferrie *et al.* 2014). Poor knowledge of husbandry may limit our ability to use *ex situ* methods of conservation for the purposes of reintroduction or augmentation of existing populations; many studies have suggested developing practices that are taxon-specific (Michaels *et al.* 2014; Craioveanu *et al.* 2017; Linhoff 2018). Each phase in the captive breeding process has the potential to cause stress (Cikanek *et al.* 2014; Hammond *et al.* 2018; Holmes *et al.* 2018; Petrović *et al.* 2020).

Previously, amphibian tadpoles have been transported using fish-transport vehicles that provide oxygenated water and maintain controlled temperature, as well as plastic bags placed on ice and plastic containers with moist cotton wool (Narayan and Hero 2011; Muths *et al.* 2014). The stress of transport and captivity in endangered wild Fijian ground frogs was assessed using two physiological methods – assessing concentration of urinary corticosterone metabolites and the haematological response to stress (Narayan and Hero 2011). Initial capture and subsequent transport for 6 hours were found to cause a stress response that lasted more than 15 days in captivity. In a similar study, African clawed frogs (*Xenopus laevis*) were transported using water, damp sponges, or damp sphagnum moss. Analysis of water-borne corticosterone found that corticosterone level was high for the first 7 days, only returning to baseline after 35 days, indicating that the process of transport and re-homing elicits a stress response in this species. This study also demonstrated that, irrespective of the transport medium used, the process of transport causes an increase in corticosterone levels (Holmes *et al.* 2018). Contrary to the findings from the previously mentioned studies, Bombay night frogs (*Nyctibatrachus humayuni*) did not elicit a stress response to handling, tagging, or short-term captivity, suggesting interspecific variability in the stress response to capture and handling protocols (Joshi *et al.* 2020). Since transport can be stressful in some species, it is important to ensure that these procedures are tailored towards the species being transported (Teixeira *et al.* 2007). Kenison *et al.* (2016) developed a novel, chiller-based system for transporting hellbender salamanders (*Cryptobranchus alleganiensis*) that was designed to maintain environmental conditions and reduce exposure to stress-inducing stimuli. The chiller maintained cool, relatively constant temperatures that could be set to match appropriate rearing conditions at the proposed destination and thereby avoid shock. Battery-powered air stones maintained dissolved oxygen levels above 90% at all times.

In amphibians, captive conditions and duration of captivity also affect levels of stress. A study by de Assis *et al.* (2015) demonstrated an effect of the intensity of a stressor on hormonal levels. Placing male cururu toads (*Rhinella icterica*) in individual plastic containers (restraint) caused a threefold increase in plasma corticosterone levels. However, placing them in a moistened cloth bag and then in the individual plastic containers (restriction with restriction of movement) led to a ninefold increase in plasma corticosterone levels. This result indicates that more stressful conditions result in stronger activation of the endocrine axis. In response to captivity, *Rhinella schneideri* displayed a sustained threefold increase in plasma corticosterone levels, as opposed to field conditions (Titon *et al.* 2017). However, *Rhinella marina* showed increased plasma corticosterone levels after the first day of captivity, with levels decreasing over time (Narayan and Hero 2014). The difference in the response of these two species to captivity indicates inter-specific differences in attenuation of the stress response.

Space is a major limiting factor in *ex situ* programs, with group housing being the most feasible way to meet the goals for managing populations. However, group housing has resulted in aggressive behaviour during the initial stages. Although quantifying the effects of housing is challenging, Narayan *et al.* (2011a) measured the changes in urinary corticosterone concentrations of cane toads (*Rhinella marina*) under different conditions of housing. Cane toads housed in groups showed higher urinary corticosterone concentrations, which declined after they were moved to individual enclosures. Similar results were reported by Cikanek *et al.* (2014), who evaluated the differences in stress levels under varying housing scenarios in male Harlequin frogs (*Atelopus* spp.), one of the most threatened genera of amphibians in the world (Lötters 2007). Wild-caught males were co-housed both in small and large groups. Ethograms were developed as a behavioural indicator and stress was quantified using faecal analysis. Co-housed males displayed aggressive behaviour and elevated faecal glucocorticoid levels for the first 3 weeks, after which levels subsided to baseline, indicating that it is safe to co-house males long term in either large or small groups.

The lack of appropriate shelters to act as hiding places causes elevated corticosterone levels and is detrimental to numerous species (Bonnet *et al.* 2013; Näslund *et al.* 2013). Crested newt (*Triturus* spp.) larvae housed without shelters displayed higher oxidative stress than did larvae that had access to refugia (Petrović *et al.* 2020). Larval forms of the red-eyed treefrog (*Agalychnis callidryas*) also were found to benefit from the provision of shelter; they displayed reduced stress and larger body size after metamorphosis compared to larval forms without access to shelter (Michaels and Preziosi 2015). Using naturalistic shelters also may facilitate the development of wild-type behaviours (Linhoff 2018), thereby facilitating successful reintroduction.

Crucial factors such as capture, transport, and captivity protocols must be taken into consideration to minimise stress in conservation programs for captive amphibians. The inability of animals to adapt to the conditions of captive housing has the potential to cause chronic stress, leading to harmful physiological changes

(Eisenberg *et al.* 2020). By studying the variation in glucocorticoid hormone levels of amphibians housed in captivity, researchers can distinguish between basal and acute responses to stress and provide useful data for early intervention and management of stress.

METHODS FOR MEASURING STRESS IN AMPHIBIANS

Early studies reported cortisol to be the primary glucocorticoid in fully aquatic amphibians, and corticosterone to be the primary glucocorticoid in terrestrial or semi-terrestrial amphibians (Nandi and Bern 1965; Nandi 1967). This has been reiterated by Hopkins *et al.* (2020) for aquatic hellbender salamanders (*Cryptobranchus alleganiensis*); they concluded that cortisol is the predominant glucocorticoid while corticosterone may be a metabolic by-product of the production of cortisol. Jungreis *et al.* (1970) suggested the possibility of an ontogenetic shift in the steroidogenic pathway from cortisol in the aquatic larvae to corticosterone in the semi-terrestrial or terrestrial adult.

Glucocorticoid levels are most commonly quantified using competitive binding immunoassays, such as radio immunoassays (RIA) or enzyme immunoassays (EIA), both of which are highly sensitive. Less common methods include liquid chromatography or gas chromatography coupled with mass spectrometry (Webb *et al.* 2007; Sheriff *et al.* 2011). Yalow and Berson (1959) achieved a milestone in endocrinology when they established a radioimmunological method to measure insulin in human plasma. RIAs have been in use increasingly since the 1960s and have also been applied to whole-body homogenates of tadpoles (Burraco *et al.* 2015). This technique relies on the use of radioactive isotopes, such as iodine or tritium, which generate a signal to quantify levels of glucocorticoids (Sheriff *et al.* 2011). RIAs are expensive, owing to special laboratory equipment, safety concerns, and additional costs of disposal (Narayan 2013; Burraco *et al.* 2015).

Starting in the late 1960s, efforts were made to replace radioactive labels with alternatives like enzymes (Lequin 2005). EIAs are nearly as sensitive as RIA and have been used in the measurement of the metabolites of amphibians' stress and reproductive hormones (Germano *et al.* 2009; Narayan *et al.* 2010; Kindermann *et al.* 2012). The development of commercial EIA kits in the 1980s enabled conduction of hormonal assays without the use of radioactivity (Burraco *et al.* 2015). For EIAs, enzymes are used to create a colorimetric change, which quantifies glucocorticoid levels (Sheriff *et al.* 2011). Historically, when EIAs were first developed, they were thoroughly validated by the researchers who employed them (Wudy *et al.* 2018). Currently, commercially developed EIA kits are commonly used and include quality-control standards with known cross-reactivities that have been published and are available. Since these commercial kits primarily are meant for use with laboratory animals or humans, yet frequently adapted for research on exotic animals, they need to be validated for use with select wildlife species (Sheriff *et al.* 2011; Wudy *et al.* 2018). For example, validation of analytical methods should include determination of accuracy, precision, specificity, sensitivity, reproducibility, and stability (Wudy *et al.* 2018). These analyses of stress must be validated in order to demonstrate that the measured glucocorticoids and their metabolites are an accurate reflection of the biological events of interest (Palme *et al.* 2005; Narayan 2013). Validations are also being developed for numerous other reasons, including collection, storage, analysis, and interpretation of samples, and for optimising well-established tools for use with different species (Madliger *et al.* 2017). The aim of these validations is to test whether circulating glucocorticoid values are reflected in the concentrations of FCM (Palme 2019).

Physiological validation is conducted by artificially activating the HPI axis, thus increasing the concentration of circulating glucocorticoid. This can be performed by injecting a synthetic exogenous hormone such as adreno-corticotropic hormone, wherein a cause-and-effect relationship is established between the administration of the hormone and subsequent excretion of the targeted hormone (ACTH challenge) (Sheriff *et al.* 2011; Narayan 2013). Biological validation is conducted by exposing an individual to a potential stressor (such as capture, handling, transport, food deprivation, and risk of predation) that also activate the HPI axis (Sheriff *et al.* 2011; Baker *et al.* 2013; Santymire *et al.* 2018). In amphibians, the response to stress and the elevated glucocorticoids have been recorded after conducting an ACTH challenge (Narayan *et al.* 2010; Barsotti *et al.* 2017) and after being handled (Narayan *et al.* 2013; Santymire *et al.* 2018).

Procedures for invasive sampling
Early studies of amphibians used traditional samples of blood plasma via aortic puncture as well as whole-body homogenates to quantify corticosterone (Orchinik *et al.*

1988; Newcomb Homan *et al.* 2003; Burraco *et al.* 2013). These methods are invasive and the act of simply handling the animal can trigger a stress response (Langkilde and Shine 2006). Therefore, sampling must be conducted within 3 min of capture – this is challenging especially when dealing with tadpoles and small juveniles (Romero and Reed 2005; Narayan 2013). Hormones are secreted into the blood in a pulsatile pattern and analysing point samples may not provide a long-term representation of the level of hormone (Romero and Reed 2005). Additionally, since it is rarely ethical to capture and sample the same individual repeatedly, long-term trends in glucocorticoids cannot be studied easily using samples of blood plasma (Kersey and Dehnhard 2014).

Glucocorticoids also can be measured using whole-body homogenates, which is a more integrated measure, wherein animals are sacrificed and ground up in a homogeniser (Burraco *et al.* 2015). This procedure requires euthanising individuals, which is counterproductive to conservation (Narayan 2019). Although blood sampling and whole-body homogenates provide accurate results and are still useful in monitoring amphibians' hormones, they can be viewed as invasive and ethically questionable. In particular, worldwide efforts to refine, replace, and reduce procedures that might impact animals' welfare are occurring. With the advent of non-invasive sampling procedures, these invasive methods are rapidly being replaced.

Procedures for non-invasive sampling
As opposed to point measures such as blood plasma, non-invasive biological samples like faeces, urine, pond water, saliva, and dermal swabs have been used successfully to provide earlier indicators of chronic stress (Kindermann *et al.* 2012; Dickens and Romero 2013; Gabor *et al.* 2013; Hogan *et al.* 2013; Narayan 2013; Hammond *et al.* 2018; Santymire *et al.* 2018). Owing to their non-invasive nature, these methods can be used repeatedly to measure the cumulative effect of stress over time (Touma *et al.* 2003).

Glucocorticoids circulating in the plasma primarily are metabolised in the liver and excreted in the bile or urine as conjugates (Taylor 1971). In the intestine, bacterial enzymes may further metabolise the steroids secreted via the bile, while some metabolites may also be reabsorbed. Glucocorticoid metabolites appear in the faeces after a time-lag that corresponds to the time of passage through the gut and are species-specific (Palme *et al.* 1996; Palme

2019). This time lag ensures an accurate measure of the individual's response to the environment, since non-invasive samples are a reflection of the glucocorticoids that occurred over the previous 6–24 h and not from the stress of capture and handling (Santymire *et al.* 2018). Non-invasive measures of hormones could be affected by issues such as the time of sample collection, age of sample, and techniques for storage (Millspaugh and Washburn 2004). Factors including seasonality, duration in captivity, reproductive status, sex, and body condition also can confuse the interpretation of data (Marra *et al.* 1995; Whittier *et al.* 1997; Touma *et al.* 2003). Although monitoring of faecal hormones is a reliable tool, it is not always practical. Collection of faecal samples is challenging in small amphibians, such as Pickersgill's reed frog (*Hyperolius pickersgilli*), which is 22–30 mm in length (Scheun *et al.* 2019). Amphibians housed in groups pose another concern as samples cannot be traced to particular individuals (Nagel *et al.* 2019). Passage of food through the guts in many amphibians is slow (< 33 h), making the collection of repeated samples time-consuming (Jiang and Claussen 1993; Scheun *et al.* 2019).

Recently, urine samples are being used more frequently than faecal samples for hormone analysis. Urine can be collected from several species by rubbing the underbelly to promote urination. Alternatively, urine can be collected by inserting a small tube or catheter into the cloaca to stimulate urination (Kouba *et al.* 2012; Browne *et al.* 2019; Silla *et al.* 2019) (see Chapter 7). In cane toads (*Rhinella marina)* and Fijian ground frogs (*Platymantis vitiana*), urination generally occurs between 30 s and 1 min of initial handling (Narayan and Hero 2011; Narayan *et al.* 2011b; Kouba *et al.* 2012). The volume of urine expelled can vary both within and between species, depending on factors such as size and the state of hydration, with 50 µL to 3 mL typically collected from Fijian ground frogs (Narayan and Hero 2011). With this non-invasive method, baseline and short-term corticosterone stress response can be assessed in amphibians over various periods (Narayan *et al.* 2011a). The magnitude of the stress response of urinary corticosterone in cane toads was affected by the duration and intensity of handling, as well as by the temperature of acclimation under laboratory conditions (Narayan *et al.* 2012b; Narayan *et al.* 2012c).

Another method of analysing stress is by analysis of water-borne hormones, which can be used for threatened and endangered species, regardless of their size. It is advantageous as it does not require handling the animals and allows for repeated measurements for the same individuals (Gabor *et al.* 2013). The basic principle of this method is to extract steroid hormones that are passively diffused into the water via urine, faeces, skin, and gills. Individuals are placed in a standard volume of water for 30–60 min and the level of glucocorticoids is measured as the rate of release (per hour). Hormones are then extracted from these aqueous samples by solid-phase extraction columns and analysed using EIA kits (Gabor *et al.* 2013, 2016; Narayan 2019). Unlike blood samples, samples in water do not need to be immediately processed or frozen once collected. They can be transported with ice packs in a cooler until returning to a laboratory where they can be stored at –20°C (Forsburg *et al.* 2019). A potential disadvantage of this method is that individuals have to be confined to a beaker, or similar restrictive container, for at least 1 h, which may cause stress (Middlemis Maher *et al.* 2013).

The most recently developed method for monitoring levels of glucocorticoids employs the use of dermal secretions (Santymire *et al.* 2018). Since skin is recognised as an endocrine gland containing hormone receptors and producing hormones, it was hypothesised that it could be possible to measure hormonal activity from the skin (Zouboulis 2004, 2009). This involves swabbing the animal dorsally along the mid-back with a cotton-tipped swab. The novel study conducted by Santymire *et al.* (2018) involved a minimal capture time of < 5 min and included amphibians representing a range of habitats and types of skin. Species that were used included green treefrogs (*Hyla cinerea*) representative of semi-terrestrial/ arboreal species, American toads (*Anaxyrus americanus*) representative of terrestrial species with a high density of granular glands, northern leopard frogs (*Lithobates pipiens*) representative of semi-aquatic species, axolotls (*Ambystoma mexicanum*) and a mudpuppy (*Necturus maculosus*) representative of fully aquatic species, and adult, red-spotted newts (*Notophthalamus viridescens*) representative of aquatic species with a high density of granular glands. Manual restraint and an acute ACTH challenge was used to test the feasibility of this method. Further supporting the use of dermal secretions, Scheun *et al.* (2019) were the first to validate this method in the bullfrog (*Pyxicephalus edulis*). Biological validation was performed using handling and restraint as the stressors, whereas physiological validation was performed using an ACTH as a challenge. Dermal swabs from the dorsal region were preferred over swabs from the ventral region, as they resulted in limited contamination (Scheun *et al.* 2019). As is the case of sampling urine, individuals have to be picked up in order to maintain consistent pressure by the swab during collection, which otherwise could induce repeated stress (Santymire *et al.* 2018). Although this technique needs to be further validated, it presents a viable alternative technique for non-invasive monitoring of hormones. Table 5.1 summarises the main methods used for monitoring non-invasive stress hormones in amphibians.

Table 5.1:   Summary of research in amphibians using minimally invasive hormone-monitoring methods.

Species	Study site	Sample type/ analytical method used	Reagents	Hormone	Stressor	Key result	Reference
American bullfrog American toad Blue-spotted salamander Green frog Green treefrog Grey treefrog Northern leopard frog Red-spotted newt Spotted salamander Spring peeper Tiger salamander Western chorus frog Cricket frog Axolotl Mudpuppy	Laboratory/ field	Dermal skin secretions EIA	Cortisol and corticosterone kits (CJM006 and R4866)	Cortisol	Trapping or netting in the wild or sampling in laboratory	Successfully detected cortisol from all species collected in the field.	Santymire *et al.* (2018)
Cane toad (*Rhinella Marina*)	Captive	Urinary EIA	CJM006, R156/7	CORT, T	Manual restraint for 5, 15 or 30 min	CORT much higher after 15 or 30 min restraint than after 5 min restraint. T low after 5, 15 or 30 min restraint.	Narayan *et al.* (2012a)
Cane toad (*Rhinella marina*)	Captive	Urinary RIA	MP Biomedicals, USA RIA Kit	CORT	ACTH challenge Short-term capture Captivity	CORT increased 1–2 days after ACTH. CORT elevated 2 h after capture handling. No sex difference.	Narayan *et al.* (2011a)
Cane toad (*Rhinella Marina*)	Free-living	Urinary EIA	CJM006, R156/7	CORT, T	Toe-clipping	CORT increased 6 h after clipping and remained elevated at 72 h. T decreased 6 h after clipping and remained low at 72 h.	Narayan *et al.* (2011b)
Common midwife toad (*Alytes obstetricans*) Mallorcan midwife toad (*Alytes muletensis*)	Laboratory	Aquatic media EIA	EIA kit No. 500651, Cayman Chemical	CORT	Chytrid fungus	CORT release rates were higher in infected populations of two species of tadpoles than in an uninfected population for both species.	Gabor *et al.* (2013)

Species	Study site	Sample type/ analytical method used	Reagents	Hormone	Stressor	Key result	Reference
Common toad (*Bufo bufo*)	Free-living	Urinary EIA	EIA kit No. 500651, Cayman Chemical	CORT	Habitat availability and fragmentation	CORT significantly altered by habitat availability and fragmentation.	Janin *et al.* (2011)
Common toad (*Bufo bufo*)	Outdoor enclosure	Saliva EIA	EIA kit No. 500651, Cayman Chemical	CORT	Matrix resistance (substrate choice experiments)	Adult toads had higher CORT on ploughed soil than on forest litter or meadow substrates.	Janin *et al.* (2012)
Fijian ground frog (*Platymantis vitiana*)	Captive	Urinary EIA	CJM006	CORT	Cane toad (*Rhinella marina*)	CORT elevated at 6 h exposure to sight of a cane toad.	E. Narayan, unpublished data
Fijian ground frog (*Platymantis vitiana*)	Captive	Urinary EIA	CJM006	CORT	Transportation Captivity	CORT higher 6 h after transportation and after 5 and 15 days in captivity. Returned to baseline after 25 days of captive enrichment.	Narayan *et al.* (2011a)
Fijian ground frog (*Platymantis vitiana*)	Free-living Captive	Urinary EIA	CJM006	CORT	ACTH challenge Capture handling	CORT increased in frog urine at 6 h, 1–2 days after ACTH. CORT elevated 3–4 h after capture handling. No seasonal pattern in CORT.	Narayan *et al.* (2010)
Fijian tree frog (*Platymantis vitiensis*)	Free-living	Urinary EIA	CJM006	CORT	Capture handling	CORT elevated 4–5 h after capture handling.	Narayan *et al.* (2013)
Great barred frog (*Mixophyes fasciolatus*)	Free-living	Urinary EIA	CJM006	CORT	Natural altitudinal gradients	Frogs living at high elevation (660–790 m) had much higher baseline CORT than frogs living lowland (60 m).	Narayan, unpublished data
Stony creek frog (*Litoria wilcoxii*	Free-living	Urinary EIA	CJM006	CORT	Chytrid fungus	CORT levels high for chytrid fungus positive frogs.	Kindermann *et al.* (2012)

Abbreviations: EIA – enzyme-immunoassay; RIA – radio-immunoassay; CORT – corticosterone; T – testosterone; CJM006 – corticosterone antibody; R4866 – cortisol antibody; R156/7 – testosterone antibody.

CONCLUSIONS

Amphibian conservation has received increased attention since the publication of the Amphibian Conservation Action Plan in 2007 (Gascon 2007). Since then, 77 species have been involved in breeding programs for conservation (Harding *et al.* 2016) with various levels of sustainability. Although significant gains have been made in establishing successful captive breeding, continued efforts are needed to conserve amphibians in their natural habitat. Various *ex situ* strategies have been used to sustainably reproduce amphibians for recovery programs, but there is a lack of standardisation of protocols and of manuals for taxon-specific husbandry (Linhoff 2018; Della Togna *et al.* 2020). Stress, as discussed in this chapter, is one of the key factors that can influence the progress and success of breeding programs for conservation. With the advent of improved capability for diagnostics, hormones can be rapidly quantified in amphibian bodily fluids other than blood, thereby providing highly sensitive and robust quantitative data to assess the functioning of the stress endocrine system of amphibians in relation to management, such as male–female pairing for initiation of breeding, social groupings, and long-term tracking of stress as an index of health and welfare. Through the applications of non-invasive technology for monitoring hormones, researchers can establish baseline datasets crucial for bolstering other aspects of amphibian husbandry and breeding-intervention, such as nutrition, reproductive technology, and maintenance of genetic brood stocks. Thus, non-invasive monitoring of glucocorticoids has a strong future in conservation breeding programs.

REFERENCES

Alroy J (2017) Effects of habitat disturbance on tropical forest biodiversity. *Proceedings of the National Academy of Sciences of the United States of America* **114**, 6056–6061. doi:10.1073/pnas.1611855114

Baker MR, Gobush KS, Vynne CH (2013) Review of factors influencing stress hormones in fish and wildlife. *Journal for Nature Conservation* **21**, 309–318. doi:10.1016/j.jnc.2013.03.003

Barsotti AMG, de Assis VR, Titon SCM, Junior BT, da Silva Ferreira ZF, Gomes FR (2017) ACTH modulation on corticosterone, melatonin, testosterone and innate immune response in the tree frog *Hypsiboas faber*. *Comparative Biochemistry and Physiology. Part A, Molecular & Integrative Physiology* **204**, 177–184. doi:10.1016/j.cbpa.2016.12.002

Batson WG, Gordon IJ, Fletcher DB, Portas T, Manning AD (2017) The effect of pre-release captivity on the stress physiology of a reintroduced population of wild eastern bettongs. *Journal of Zoology* **303**, 311–319. doi:10.1111/jzo.12494

Berger J, Testa JW, Roffe T, Monfort SL (1999) Conservation endocrinology: a noninvasive tool to understand relationships between carnivore colonization and ecological carrying capacity. *Conservation Biology* **13**, 980–989. doi:10.1046/j.1523-1739.1999.98521.x

Bergman JN, Bennett JR, Binley AD, Cooke SJ, Fyson V, Hlina BL, Reid CH, Vala MA, Madliger CL (2019) Scaling from individual physiological measures to population-level demographic change: case studies and future directions for conservation management. *Biological Conservation* **238**, 108242. doi:10.1016/j.biocon.2019.108242

Bickford DP, Alford R, Crump ML, Whitfield S, Karraker N, Donnelly MA (2018) Impacts of climate change on amphibian biodiversity. In *Encyclopedia of the Anthropocene*. (Eds DA Dellasala and MI Goldstein) pp. 113–121. Elsevier, Oxford, UK. doi.org/10.1016/B978-0-12-809665-9.10022-9

Blaustein AR, Gervasi SS, Johnson PT, Hoverman JT, Belden LK, Bradley PW, Xie GY (2012) Ecophysiology meets conservation: understanding the role of disease in amphibian population declines. *Philosophical Transactions of the Royal Society of London. Series B, Biological Sciences* **367**, 1688–1707. doi:10.1098/rstb.2012.0011

Bliley JM, Woodley SK (2012) The effects of repeated handling and corticosterone treatment on behavior in an amphibian (Ocoee salamander: *Desmognathus ocoee*). *Physiology & Behavior* **105**, 1132–1139. doi:10.1016/j.physbeh.2011.12.009

Bolochio BE, Lescano JN, Cordier JM, Loyola R, Nori J (2020) A functional perspective for global amphibian conservation. *Biological Conservation* **245**, 108572. doi:10.1016/j.biocon.2020.108572

Bonnet X, Fizesan A, Michel CL (2013) Shelter availability, stress level and digestive performance in the aspic viper. *The Journal of Experimental Biology* **216**, 815–822.

Brannelly LA, McCallum HI, Grogan LF, Briggs CJ, Ribas MP, Hollanders M, Sasso T, Familiar López M, Newell DA, Kilpatrick AM (2021) Mechanisms underlying host persistence following amphibian disease emergence determine appropriate management strategies. *Ecology Letters* **24**, 130–148. doi:10.1111/ele.13621

Browne R, Janzen P, Bagaturov M, van Houte D (2018) Amphibian keeper conservation breeding programs. *Journal of Zoological Research* **2**, 29–46.

Browne RK, Silla AJ, Upton R, Della-Togna G, Marcec-Greaves R, Shishova NV, Uteshev VK, Proaño B, Pérez OD, Mansour N (2019) Sperm collection and storage for the sustainable management of amphibian biodiversity. *Theriogenology* **133**, 187–200. doi:10.1016/j.theriogenology.2019.03.035

Burraco P, Gomez-Mestre I (2016) Physiological stress responses in amphibian larvae to multiple stressors reveal marked anthropogenic effects even below lethal levels. *Physiological and Biochemical Zoology* **89**, 462–472. doi:10.1086/688737

Burraco P, Duarte LJ, Gomez-Mestre I (2013) Predator-induced physiological responses in tadpoles challenged with herbicide pollution. *Current Zoology* **59**, 475–484. doi:10.1093/czoolo/59.4.475

Burraco P, Arribas R, Kulkarni SS, Buchholz DR, Gomez-Mestre I (2015) Comparing techniques for measuring corticosterone in tadpoles. *Current Zoology* **61**, 835–845. doi:10.1093/czoolo/61.5.835

Cikanek SJ, Nockold S, Brown JL, Carpenter JW, Estrada A, Guerrel J, Hope K, Ibáñez R, Putman SB, Gratwicke B (2014) Evaluating group housing strategies for the *ex-situ* conservation of harlequin frogs (*Atelopus* spp.) using behavioral and physiological indicators. *PLoS One* **9**, e90218. doi:10.1371/journal.pone.0090218

Cohen JM, Civitello DJ, Venesky MD, McMahon TA, Rohr JR (2019) An interaction between climate change and infectious disease drove widespread amphibian declines. *Global Change Biology* **25**, 927–937. doi:10.1111/gcb.14489

Collier RJ, Renquist BJ, Xiao Y (2017) A 100-year review: stress physiology including heat stress. *Journal of Dairy Science* **100**, 10367–10380. doi:10.3168/jds.2017-13676

Cooke SJ, Sack L, Franklin CE, Farrell AP, Beardall J, Wikelski M, Chown SL (2013) What is conservation physiology? Perspectives on an increasingly integrated and essential science. *Conservation Physiology* **1**(1), cot001.

Craioveanu O, Craioveanu C, Cosma I, Ghira I, Mireşan V (2017) Shelter use assessment and shelter enrichment in captive bred common toads (*Bufo bufo*, Linnaeus 1758). *North-Western Journal of Zoology* **13**, 341–346.

de Assis VR, Titon SCM, Barsotti AMG, Titon B, Jr, Gomes FR (2015) Effects of acute restraint stress, prolonged captivity stress and transdermal corticosterone application on immunocompetence and plasma levels of corticosterone on the cururu toad (*Rhinella icterica*). *PLoS One* **10**, e0121005. doi:10.1371/journal.pone.0121005

Della Togna G, Howell LG, Clulow J, Langhorne CJ, Marcec-Greaves R, Calatayud NE (2020) Evaluating amphibian biobanking and reproduction for captive breeding programs according to the Amphibian Conservation Action Plan objectives. *Theriogenology* **150**, 412–431. doi:10.1016/j.theriogenology.2020.02.024

Demas GE, Zysling DA, Beechler BR, Muehlenbein MP, French SS (2011) Beyond phytohaemagglutinin: assessing vertebrate immune function across ecological contexts. *Journal of Animal Ecology* **80**, 710–730. doi:10.1111/j.1365-2656.2011.01813.x

Dickens MJ, Romero LM (2013) A consensus endocrine profile for chronically stressed wild animals does not exist. *General and Comparative Endocrinology* **191**, 177–189. doi:10.1016/j.ygcen.2013.06.014

Dittami J, Katina S, Möstl E, Eriksson J, Machatschke IH, Hohmann G (2008) Urinary androgens and cortisol metabolites in field-sampled bonobos (*Pan paniscus*). *General and Comparative Endocrinology* **155**, 552–557. doi:10.1016/j.ygcen.2007.08.009

Doherty TS, Glen AS, Nimmo DG, Ritchie EG, Dickman CR (2016) Invasive predators and global biodiversity loss. *Proceedings of the National Academy of Sciences of the United States of America* **113**, 11261–11265. doi:10.1073/pnas.1602480113

Donaldson C (2019) Analyzing factors influencing reproductive success of the mountain chicken: Nordens Ark Captive Breeding Program. Student thesis. University of Skövde, Sweden.

Eisenberg T, Fawzy A, Kaim U, Nesseler A, Riße K, Völker I, Hechinger S, Schauerte N, Geiger C, Knauf-Witzens T (2020) Chronic wasting associated with *Chlamydia pneumoniae* in three *ex situ* breeding facilities for tropical frogs. *Antonie van Leeuwenhoek* **113**, 2139–2154. doi:10.1007/s10482-020-01483-6

Ellis RD, McWhorter TJ, Maron M (2012) Integrating landscape ecology and conservation physiology. *Landscape Ecology* **27**, 1–12. doi:10.1007/s10980-011-9671-6

Ferrie GM, Alford VC, Atkinson J, Baitchman E, Barber D, Blaner WS, Crawshaw G, Daneault A, Dierenfeld E, Finke M (2014) Nutrition and health in amphibian husbandry. *Zoo Biology* **33**, 485–501. doi:10.1002/zoo.21180

Fischer CP, Wright-Lichter J, Romero LM (2018) Chronic stress and the introduction to captivity: how wild house sparrows (*Passer domesticus*) adjust to laboratory conditions. *General and Comparative Endocrinology* **259**, 85–92. doi:10.1016/j.ygcen.2017.11.007

Flach EJ, Feltrer Y, Gower DJ, Jayson S, Michaels CJ, Pocknell A, Rivers S, Perkins M, Rendle ME, Stidworthy MF (2020) Postmortem findings in eight species of captive caecilian (Amphibia: Gymnophiona) over a ten-year period. *Journal of Zoo and Wildlife Medicine* **50**, 879–890. doi:10.1638/2019-0047

Forsburg ZR, Goff CB, Perkins HR, Robicheaux JA, Almond GF, Gabor CR (2019) Validation of water-borne cortisol and corticosterone in tadpoles: Recovery rate from an acute stressor, repeatability, and evaluating rearing methods. *General and Comparative Endocrinology* **281**, 145–152. doi:10.1016/j.ygcen.2019.06.007

Gabor CR, Bosch J, Fries JN, Davis DR (2013) A non-invasive water-borne hormone assay for amphibians. *Amphibia-Reptilia* **34**, 151–162. doi:10.1163/15685381-00002877

Gabor CR, Zabierek KC, Kim DS, da Barbiano LA, Mondelli MJ, Bendik NF, Davis DR (2016) A non-invasive water-borne assay of stress hormones in aquatic salamanders. *Copeia* **104**, 172–181. doi:10.1643/OT-14-207

Gascon C (2007) Amphibian conservation action plan. In *Proceedings IUCN/SSC Amphibian Conservation Summit 2005*. (Eds C Gascon, JP Collins, RD Moore, DR Church, JE McKay and JR Mendelson III) pp. 35–39. World Conservation Union (IUCN), Gland, Switzerland.

Germano J, Molinia F, Bishop P, Cree A (2009) Urinary hormone analysis assists reproductive monitoring and sex identification of bell frogs (*Litoria raniformis*). *Theriogenology* **72**, 663–671. doi:10.1016/j.theriogenology.2009.04.023

Glennemeier KA, Denver RJ (2002) Small changes in whole-body corticosterone content affect larval *Rana pipiens* fitness components. *General and Comparative Endocrinology* **127**, 16–25. doi:10.1016/S0016-6480(02)00015-1

Gordon M, Bartol SM, Bartol S (2004) *Experimental Approaches to Aonservation Biology* University of California Press, California, US.

Grant EHC, Muths E, Schmidt BR, Petrovan SO (2019) *Amphibian Conservation in the Anthropocene*. Elsevier, San Diego CA, USA.

Guillette LJ, Cree A, Rooney AA (1995) Biology of stress: interactions with reproduction, immunology and intermediary metabolism. In *Health and Welfare of Captive Reptiles*. (Eds C Warwick, FL Frye and JB Murphy) pp. 32–81. Springer, Dordrecht, The Netherlands.

Hammond TT, Au ZA, Hartman AC, Richards-Zawacki CL (2018) Assay validation and interspecific comparison of salivary glucocorticoids in three amphibian species. *Conservation Physiology* **6**, coy055. doi:10.1093/conphys/coy055

Hammond TT, Curtis MJ, Jacobs LE, Tobler MW, Swaisgood RR, Shier DM (2020) Behavior and detection method influence detection probability of a translocated, endangered amphibian. *Animal Conservation* **24**(3), 401–411.

Harding G, Griffiths RA, Pavajeau L (2016) Developments in amphibian captive breeding and reintroduction programs. *Conservation Biology* **30**, 340–349. doi:10.1111/cobi.12612

Harris BN (2020) Stress hypothesis overload: 131 hypotheses exploring the role of stress in tradeoffs, transitions, and health. *General and Comparative Endocrinology* **288**, 113355. doi:10.1016/j.ygcen.2019.113355

Hau M, Casagrande S, Ouyang JQ, Baugh AT (2016) Glucocorticoid-mediated phenotypes in vertebrates: multilevel variation and evolution. *Advances in the Study of Behavior* **48**, 41–115. doi:10.1016/bs.asb.2016.01.002

Hill MN, Carrier EJ, Ho WSV, Shi L, Patel S, Gorzalka BB, Hillard CJ (2008) Prolonged glucocorticoid treatment decreases cannabinoid CB1 receptor density in the hippocampus. *Hippocampus* **18**, 221–226. doi:10.1002/hipo.20386

Hofmann GE, Todgham AE (2010) Living in the now: physiological mechanisms to tolerate a rapidly changing environment. *Annual Review of Physiology* **72**, 127–145. doi:10.1146/annurev-physiol-021909-135900

Hogan LA, Lisle AT, Johnston SD, Goad T, Robertston H (2013) Adult and juvenile sex identification in threatened monomorphic *Geocrinia* frogs using fecal steroid analysis. *Journal of Herpetology* **47**, 112–118. doi:10.1670/11-290

Holmes AM, Emmans CJ, Coleman R, Smith TE, Hosie CA (2018) Effects of transportation, transport medium and re-housing on *Xenopus laevis* (Daudin). *General and Comparative Endocrinology* **266**, 21–28. doi:10.1016/j.ygcen.2018.03.015

Hopkins WA, DuRant SE, Beck ML, Ray WK, Helm RF, Romero LM (2020) Cortisol is the predominant glucocorticoid in the giant paedomorphic hellbender salamander (*Cryptobranchus alleganiensis*). *General and Comparative Endocrinology* **285**, 113267. doi:10.1016/j.ygcen.2019.113267

Hunt KE, Innis C, Merigo C, Burgess EA, Norton T, Davis D, Kennedy AE, Buck CL (2019) Ameliorating transport-related stress in endangered Kemp's ridley sea turtles (*Lepidochelys kempii*) with a recovery period in saltwater pools. *Conservation Physiology* **7**, coy065. doi:10.1093/conphys/coy065

Janin A, Jean-Paul L, Pierre J (2011) Beyond occurrence: body condition and stress hormone as integrative indicators of habitat availability and fragmentation in the common toad. *Biological Conservation* **144**(3), 1008–1016.

Janin A, Léna JP, Joly P (2012) Habitat fragmentation affects movement behavior of migrating juvenile common toads. *Behavioral Ecology and Sociobiology* **66**(9), 1351–1356.

Jiang S, Claussen DL (1993) The effects of temperature on food passage time through the digestive tract in Notophthalmus viridescens. *Journal of Herpetology* **27**(4), 414–419. doi:10.2307/1564829

Joshi AM, Narayan EJ, Gramapurohit NP (2020) Annual changes in corticosterone and its response to handling, tagging and short-term captivity in *Nyctibatrachus humayuni*. *The Herpetological Journal* **30**(3), 118–125. doi:10.33256/hj30.3.118125

Jungreis AM, Huibregtse WH, Ungar F (1970) Corticosteroid identification and corticosterone concentration in serum of *Rana pipiens* during dehydration in winter and summer. *Comparative Biochemistry and Physiology* **34**, 683–689. doi:10.1016/0010-406X(70)90293-8

Kenison EK, Olson ZH, Williams RN (2016) A novel transport system for hellbender salamanders (*Cryptobranchus alleganiensis*). *Herpetological Conservation and Biology* **11**, 355–361.

Kersey DC, Dehnhard M (2014) The use of noninvasive and minimally invasive methods in endocrinology for threatened mammalian species conservation. *General and Comparative Endocrinology* **203**, 296–306. doi:10.1016/j.ygcen.2014.04.022

Kindermann C, Narayan EJ, Hero J-M (2012) Urinary corticosterone metabolites and chytridiomycosis disease prevalence in a free-living population of male Stony Creek frogs (*Litoria wilcoxii*). *Comparative Biochemistry and Physiology. Part A, Molecular & Integrative Physiology* **162**, 171–176. doi:10.1016/j.cbpa.2012.02.018

Kinlein SA, Karatsoreos IN (2020) The hypothalamic-pituitary-adrenal axis as a substrate for stress resilience: interactions with the circadian clock. *Frontiers in Neuroendocrinology* **56**, 100819. doi:10.1016/j.yfrne.2019.100819

Kirschman LJ, Crespi EJ, Warne RW (2018) Critical disease windows shaped by stress exposure alter allocation trade-offs between development and immunity. *Journal of Animal Ecology* **87**, 235–246. doi:10.1111/1365-2656.12778

Kouba A, Vance C, Calatayud N, Rowlison T, Langhorne C, Willard S (2012) Assisted reproductive technologies (ART) for amphibians. In *Amphibian Husbandry Resource Guide*. 2nd edn. (Eds VA Poole and S Grow) pp. 60–118. Association of Zoos and Aquariums, Silver Spring MA, USA.

Langkilde T, Shine R (2006) How much stress do researchers inflict on their study animals? A case study using a scincid lizard, *Eulamprus heatwolei*. *The Journal of Experimental Biology* **209**, 1035–1043. doi:10.1242/jeb.02112

Lèche A, Cortez MV, Della Costa NS, Navarro JL, Marin RH, Martella MB (2016) Stress response assessment during translocation of captive-bred Greater Rheas into the wild. *Journal of Ornithology* **157**, 599–607. doi:10.1007/s10336-015-1305-3

Lequin RM (2005) Enzyme immunoassay (EIA)/enzyme-linked immunosorbent assay (ELISA). *Clinical Chemistry* **51**, 2415–2418. doi:10.1373/clinchem.2005.051532

Linhoff LJ (2018) Linking husbandry and behavior to enhance amphibian reintroduction success. PhD thesis. Florida International University, USA.

Lötters S (2007) The fate of the harlequin toads – help through a synchronous multi-disciplinary approach and the IUCN 'Amphibian Conservation Action Plan'? *Zoosystematics and Evolution* **83**, 69–73. doi:10.1002/mmnz.200600028

MacDougall-Shackleton SA, Bonier F, Romero LM, Moore IT (2019) Glucocorticoids and 'stress' are not synonymous. *Integrative Organismal Biology* **1**, obz017. doi:10.1093/iob/obz017

Madliger CL, Franklin CE, Hultine KR, Van Kleunen M, Lennox RJ, Love OP, Rummer JL, Cooke SJ (2017) Conservation physiology and the quest for a 'good' Anthropocene. *Conservation Physiology* **5**, 1–10. doi:10.1093/conphys/cox003

Malisch JL, Garland T, Jr, Claggett L, Stevenson L, Kohl EA, John-Alder HB (2020) Living on the edge: glucocorticoid physiology in desert iguanas (*Dipsosaurus dorsalis*) is predicted by distance from an anthropogenic disturbance, body condition, and population density. *General and Comparative Endocrinology* **294**, 113468. doi:10.1016/j.ygcen.2020.113468

Marra PP, Lampe KT, Tedford BL (1995) Plasma corticosterone levels in two species of Zonotrichia sparrows under captive and free-living conditions. *The Wilson Bulletin* **107**(2), 296–305.

Marshall AR, Deere NJ, Little HA, Snipp R, Goulder J, Mayer-Clarke S (2016) Husbandry and enclosure influences on penguin behavior and conservation breeding. *Zoo Biology* **35**, 385–397. doi:10.1002/zoo.21313

McEwen B (2010) Stress: homeostasis, rheostasis, allostasis and allostatic load. In *Stress Science: Neuroendocrinology.* (Ed. G Fink) pp. 10–14. Academic Press, Cambridge MA, USA.

McEwen BS, Wingfield JC (2003) The concept of allostasis in biology and biomedicine. *Hormones and Behavior* **43**, 2–15. doi:10.1016/S0018-506X(02)00024-7

McEwen BS, Wingfield JC (2010) What's in a name? Integrating homeostasis, allostasis and stress. *Hormones and Behavior* **57**, 105. doi:10.1016/j.yhbeh.2009.09.011

McGowan PJ, Traylor-Holzer K, Leus K (2017) IUCN guidelines for determining when and how *ex situ* management should be used in species conservation. *Conservation Letters* **10**, 361–366. doi:10.1111/conl.12285

Meza-Parral Y, García-Robledo C, Pineda E, Escobar F, Donnelly MA (2020) Standardized ethograms and a device for assessing amphibian thermal responses in a warming world. *Journal of Thermal Biology* **89**, 102565. doi:10.1016/j.jtherbio.2020.102565

Michaels CJ, Gini BF, Preziosi RF (2014) The importance of natural history and species-specific approaches in amphibian *ex-situ* conservation. *The Herpetological Journal* **24**, 135–145.

Michaels CJ, Preziosi RF (2015) Fitness effects of shelter provision for captive amphibian tadpoles. *The Herpetological Journal* **25**, 21–26.

Middlemis Maher J, Werner EE, Denver RJ (2013) Stress hormones mediate predator-induced phenotypic plasticity in amphibian tadpoles. *Proceedings. Biological Sciences* **280**, 20123075. doi:10.1098/rspb.2012.3075

Millspaugh JJ, Washburn BE (2004) Use of fecal glucocorticoid metabolite measures in conservation biology research: considerations for application and interpretation. *General and Comparative Endocrinology* **138**, 189–199. doi:10.1016/j.ygcen.2004.07.002

Mingramm FM, Keeley T, Whitworth DJ, Dunlop RA (2020) Blubber cortisol levels in humpback whales (Megaptera novaeangliae): a measure of physiological stress without effects from sampling. *General and Comparative Endocrinology* **291**, 113436. doi:10.1016/j.ygcen.2020.113436

Monfort S (2003) 10 non-invasive endocrine measures of reproduction and stress in wild populations. In *Reproductive Science and Integrated Conservation.* (Eds WV Holt, AR Pickard, JC Rodger and DE Wildt) pp. 147–165. Cambridge University Press, Cambridge, UK.

Muths E, Bailey LL, Watry MK (2014) Animal reintroductions: an innovative assessment of survival. *Biological Conservation* **172**, 200–208. doi:10.1016/j.biocon.2014.02.034

Nagel AH, Beshel M, DeChant CJ, Huskisson SM, Campbell MK, Stoops MA (2019) Non-invasive methods to measure inter-renal function in aquatic salamanders – correlating fecal corticosterone to the environmental and physiologic conditions of captive *Necturus*. *Conservation Physiology* **7**, coz074. doi:10.1093/conphys/coz074

Nandi J (1967) Comparative endocrinology of steroid hormones in vertebrates. *American Zoologist* **7**, 115–133. doi:10.1093/icb/7.1.115

Nandi J, Bern HA (1965) Chromatography of corticosteroids from teleost fishes. *General and Comparative Endocrinology* **5**, 1–15. doi:10.1016/0016-6480(65)90063-8

Narayan E, Hero J-M (2011) Urinary corticosterone responses and haematological stress indicators in the endangered Fijian ground frog (*Platymantis vitiana*) during transportation and captivity. *Australian Journal of Zoology* **59**, 79–85. doi:10.1071/ZO11030

Narayan EJ (2013) Non-invasive reproductive and stress endocrinology in amphibian conservation physiology. *Conservation Physiology* **1**, cot011. doi:10.1093/conphys/cot011

Narayan EJ (2019) Introductory chapter: applications of stress endocrinology in wildlife conservation and livestock science. In *Comparative Endocrinology of Animals.* (Ed. E Narayan) pp. 1–8. BoD – Books on Demand.

Narayan EJ, Hero J-M (2014) Repeated thermal stressor causes chronic elevation of baseline corticosterone and suppresses the physiological endocrine sensitivity to acute stressor in the cane toad (*Rhinella marina*). *Journal of Thermal Biology* **41**, 72–76. doi:10.1016/j.jtherbio.2014.02.011

Narayan E, Molinia F, Christi K, Morley C, Cockrem J (2010) Urinary corticosterone metabolite responses to capture, and annual patterns of urinary corticosterone in wild and captive endangered Fijian ground frogs (*Platymantis vitiana*). *Australian Journal of Zoology* **58**, 189–197. doi:10.1071/ZO10010

Narayan EJ, Cockrem JF, Hero J-M (2011a) Urinary corticosterone metabolite responses to capture and captivity in the cane toad (*Rhinella marina*). *General and Comparative Endocrinology* **173**, 371–377. doi:10.1016/j.ygcen.2011.06.015

Narayan EJ, Molinia FC, Kindermann C, Cockrem JF, Hero J-M (2011b) Urinary corticosterone responses to capture and toe-clipping in the cane toad (*Rhinella marina*) indicate that toe-clipping is a stressor for amphibians. *General and Comparative Endocrinology* **174**, 238–245. doi:10.1016/j.ygcen.2011.09.004

Narayan E, Hero J-M, Evans N, Nicolson V, Mucci A (2012a) Non-invasive evaluation of physiological stress hormone responses in a captive population of the greater bilby *Macrotis lagotis*. *Endangered Species Research* **18**, 279–289. doi:10.3354/esr00454

Narayan EJ, Cockrem JF, Hero J-M (2012b) Effects of temperature on urinary corticosterone metabolite responses to short-term capture and handling stress in the cane toad (*Rhinella marina*). *General and Comparative Endocrinology* **178**, 301–305. doi:10.1016/j.ygcen.2012.06.014

Narayan EJ, Hero J-M, Cockrem JF (2012c) Inverse urinary corticosterone and testosterone metabolite responses to different durations of restraint in the cane toad (*Rhinella marina*). *General and Comparative Endocrinology* **179**, 345–349. doi:10.1016/j.ygcen.2012.09.017

Narayan EJ, Cockrem JF, Hero J-M (2013) Are baseline and short-term corticosterone stress responses in free-living amphibians repeatable? *Comparative Biochemistry and Physiology. Part A, Molecular & Integrative Physiology* **164**, 21–28. doi:10.1016/j.cbpa.2012.10.001

Narayan EJ, Jessop TS, Hero JM (2015) Invasive cane toad triggers chronic physiological stress and decreased reproductive success in an island endemic. *Functional Ecology* **29**, 1435–1444. doi:10.1111/1365-2435.12446

Narayan EJ, Forsburg ZR, Davis DR, Gabor CR (2019) Non-invasive methods for measuring and monitoring stress physiology in imperiled amphibians. *Frontiers in Ecology and Evolution* **7**, 431. doi:10.3389/fevo.2019.00431

Näslund J, Rosengren M, Del Villar D, Gansel L, Norrgård JR, Persson L, Winkowski JJ, Kvingedal E (2013) Hatchery tank enrichment affects cortisol levels and shelter-seeking in Atlantic salmon (*Salmo salar*). *Canadian Journal of Fisheries and Aquatic Sciences* **70**, 585–590. doi:10.1139/cjfas-2012-0302

Newcomb Homan R, Regosin JV, Rodrigues DM, Reed JM, Windmiller BS, Romero LM (2003) Impacts of varying habitat quality on the physiological stress of spotted salamanders (*Ambystoma maculatum*). *Animal Conservation* **6**, 11–18. doi:10.1017/S1367943003003032

O'Connor B, Bojinski S, Röösli C, Schaepman ME (2020) Monitoring global changes in biodiversity and climate essential as ecological crisis intensifies. *Ecological Informatics* **55**, 101033. doi:10.1016/j.ecoinf.2019.101033

Orchinik M, Licht P, Crews D (1988) Plasma steroid concentrations change in response to sexual behavior in Bufo marinus. *Hormones and Behavior* **22**, 338–350. doi:10.1016/0018-506X(88)90006-2

Palme R (2019) Non-invasive measurement of glucocorticoids: advances and problems. *Physiology & Behavior* **199**, 229–243. doi:10.1016/j.physbeh.2018.11.021

Palme R, Fischer P, Schildorfer H, Ismail M (1996) Excretion of infused 14C-steroid hormones via faeces and urine in domestic livestock. *Animal Reproduction Science* **43**, 43–63. doi:10.1016/0378-4320(95)01458-6

Palme R, Rettenbacher S, Touma C, El-Bahr S, Möstl E (2005) Stress hormones in mammals and birds: comparative aspects regarding metabolism, excretion, and noninvasive measurement in fecal samples. *Annals of the New York Academy of Sciences* **1040**, 162–171. doi:10.1196/annals.1327.021

Petrović TG, Vučić TZ, Nikolić SZ, Gavrić JP, Despotović SG, Gavrilović BR, Radovanović TB, Faggio C, Prokić MD (2020) The effect of shelter on oxidative stress and aggressive behavior in crested newt larvae (*Triturus* spp.). *Animals (Basel)* **10**, 603. doi:10.3390/ani10040603

Roe D, Seddon N, Elliott J (2019) Biodiversity loss is a development issue: a rapid review of evidence. *International Institute for Environment and Development* **798**, 678–683.

Rollins-Smith LA (2017) Amphibian immunity – stress, disease, and climate change. *Developmental and Comparative Immunology* **66**, 111–119. doi:10.1016/j.dci.2016.07.002

Rollins-Smith LA (2020) Global amphibian declines, disease, and the ongoing battle between *Batrachochytrium* fungi and the immune system. *Herpetologica* **76**, 178–188. doi:10.1655/0018-0831-76.2.178

Romero LM (2004) Physiological stress in ecology: lessons from biomedical research. *Trends in Ecology & Evolution* **19**, 249–255. doi:10.1016/j.tree.2004.03.008

Romero LM, Reed JM (2005) Collecting baseline corticosterone samples in the field: is under 3 min good enough? *Comparative Biochemistry and Physiology. Part A, Molecular & Integrative Physiology* **140**, 73–79. doi:10.1016/j.cbpb.2004.11.004

Romero LM, Dickens MJ, Cyr NE (2009) The reactive scope model – a new model integrating homeostasis, allostasis, and stress. *Hormones and Behavior* **55**, 375–389. doi:10.1016/j.yhbeh.2008.12.009

Sadoul, B, Vijayan, MM (2016) Stress and growth. In *Fish physiology*. Volume 35. (Eds CB Schreck, L Tort, AP Farrell and CJ Brauner) pp. 167–205. Academic Press, San Diego CA, USA.

Santymire R, Manjerovic M, Sacerdote-Velat A (2018) A novel method for the measurement of glucocorticoids in dermal secretions of amphibians. *Conservation Physiology* **6**, coy008. doi:10.1093/conphys/coy008

Scheun J, Greeff D, Medger K, Ganswindt A (2019) Validating the use of dermal secretion as a matrix for monitoring glucocorticoid concentrations in African amphibian species. *Conservation Physiology* **7**, coz022. doi:10.1093/conphys/coz022

Schulte PM (2014) What is environmental stress? Insights from fish living in a variable environment. *The Journal of Experimental Biology* **217**, 23–34. doi:10.1242/jeb.089722

Selye H (1950) Stress and the general adaptation syndrome. *British Medical Journal* **1**, 1383. doi:10.1136/bmj.1.4667.1383

Sheriff MJ, Dantzer B, Delehanty B, Palme R, Boonstra R (2011) Measuring stress in wildlife: techniques for quantifying glucocorticoids. *Oecologia* **166**, 869–887. doi:10.1007/s00442-011-1943-y

Silla AJ, McFadden MS, Byrne PG (2019) Hormone-induced sperm-release in the critically endangered Booroolong frog (*Litoria booroolongensis*): effects of gonadotropin-releasing hormone and human chorionic gonadotropin. *Conservation Physiology* **7**, coy080. doi:10.1093/conphys/coy080

Smith RK, Sutherland WJ (Eds) (2014) *Amphibian Conservation: Global Evidence for the Effects of Interventions*. Pelagic Publishing, Exeter, UK.

Sterling P, Eyer J, Fisher S, Reason J (1988) Allostasis: a new paradigm to explain arousal pathology. In *Handbook of Life Stress, Cognition and Health*. (Eds S Fisher and J Reason) pp. 629–649. John Wiley and Sons, New York NY, USA.

Stuart SN, Chanson JS, Cox NA, Young BE, Rodrigues AS, Fischman DL, Waller RW (2004) Status and trends of amphibian declines and extinctions worldwide. *Science* **306**, 1783–1786. doi:10.1126/science.1103538

Taylor EN (2017) Amphibian and reptile adaptations to the environment: interplay between physiology and behavior. *Copeia* **105**, 171–173.

Taylor W (1971) The excretion of steroid hormone metabolites in bile and feces. *Vitamins and Hormones* **29**, 201–285. doi:10.1016/S0083-6729(08)60050-3

Teixeira CP, De Azevedo CS, Mendl M, Cipreste CF, Young RJ (2007) Revisiting translocation and reintroduction programmes: the importance of considering stress. *Animal Behaviour* **73**, 1–13. doi:10.1016/j.anbehav.2006. 06.002

Thompson PL, Rayfield B, Gonzalez A (2017) Loss of habitat and connectivity erodes species diversity, ecosystem functioning, and stability in metacommunity networks. *Ecography* **40**, 98–108. doi:10.1111/ecog.02558

Titon SCM, Assis VR, Titon Junior B, Cassettari Bd O, Fernandes PACM, Gomes FR (2017) Captivity effects on immune response and steroid plasma levels of a Brazilian toad (*Rhinella schneideri*). *Journal of Experimental Zoology. Part A, Ecological and Integrative Physiology* **327**, 127–138. doi:10.1002/jez.2078

Titon SCM, Junior BT, Assis VR, Kinker GS, Fernandes PACM, Gomes FR (2018) Interplay among steroids, body condition and immunity in response to long-term captivity in toads. *Scientific Reports* **8**(1), 1–13. doi:10.1038/s41598-018-35495-0

Touma C, Sachser N, Möstl E, Palme R (2003) Effects of sex and time of day on metabolism and excretion of corticosterone in urine and feces of mice. *General and Comparative Endocrinology* **130**, 267–278. doi:10.1016/S0016-6480(02)00620-2

Tuomainen U, Candolin U (2011) Behavioural responses to human-induced environmental change. *Biological Reviews of the Cambridge Philosophical Society* **86**, 640–657. doi:10.1111/ j.1469-185X.2010.00164.x

Tyagi A, Kumar V, Kittur S, Reddy M, Naidenko S, Ganswindt A, Umapathy G (2019) Physiological stress responses of tigers due to anthropogenic disturbance especially tourism in two central Indian tiger reserves. *Conservation Physiology* **7**, coz045. doi:10.1093/conphys/coz045

Valsamakis G, Chrousos G, Mastorakos G (2019) Stress, female reproduction and pregnancy. *Psychoneuroendocrinology* **100**, 48–57. doi:10.1016/j.psyneuen.2018.09.031

Vitousek MN, Johnson MA, Downs CJ, Miller ET, Martin LB, Francis CD, Donald JW, Fuxjager MJ, Goymann W, Hau M (2019) Macroevolutionary patterning in glucocorticoids suggests different selective pressures shape baseline and stress-induced levels. *American Naturalist* **193**, 866–880. doi:10.1086/703112

Webb MA, Allert JA, Kappenman KM, Marcos J, Feist GW, Schreck CB, Shackleton CH (2007) Identification of plasma glucocorticoids in pallid sturgeon in response to stress. *General and Comparative Endocrinology* **154**, 98–104. doi:10.1016/j.ygcen.2007.06.002

Whittier JM, Corrie F, Limpus C (1997) Plasma steroid profiles in nesting loggerhead turtles (*Caretta caretta*) in Queensland, Australia: relationship to nesting episode and season. *General and Comparative Endocrinology* **106**, 39–47. doi:10.1006/ gcen.1996.6848

Wikelski M, Cooke SJ (2006) Conservation physiology. *Trends in Ecology & Evolution* **21**, 38–46. doi:10.1016/j.tree.2005.10.018

Wudy S, Schuler G, Sánchez-Guijo A, Hartmann M (2018) The art of measuring steroids: principles and practice of current hormonal steroid analysis. *The Journal of Steroid Biochemistry and Molecular Biology* **179**, 88 103. doi:10.1016/j.jsbmb.2017. 09.003

Yalow RS, Berson SA (1959) Assay of plasma insulin in human subjects by immunological methods. *Nature* **184**, 1648–1649. doi:10.1038/1841648b0

Young HS, McCauley DJ, Galetti M, Dirzo R (2016) Patterns, causes, and consequences of Anthropocene defaunation. *Annual Review of Ecology, Evolution, and Systematics* **47**, 333–358. doi:10.1146/annurev-ecolsys-112414-054142

Zhang F, Yu J, Wu S, Li S, Zou C, Wang Q, Sun R (2017) Keeping and breeding the rescued Sunda pangolins (*Manis javanica*) in captivity. *Zoo Biology* **36**, 387–396. doi:10.1002/zoo.21388

Zhouboulis CC (2004) The human skin as a hormone target and an endocrine gland. *Hormones (Athens, Greece)* **3**, 9–26. doi:10.14310/horm.2002.11109

Zhouboulis CC (2009) The skin as an endocrine organ. *Dermato-Endocrinology* **1**, 250–252. doi:10.4161/derm.1.5.9499

6 Ultrasonography to assess female reproductive status and inform hormonally induced ovulation

Katherine M. Graham and Carrie K. Kouba

INTRODUCTION

In amphibian conservation breeding programs the reproductive failure of a species is often the result of the limited number or poor quality of eggs produced by females (Kouba *et al.* 2009). To counter this, assisted reproductive technologies can be employed to stimulate the development of oocytes, ovulation, and oviposition, as well as to synchronise the collection of eggs and sperm to facilitate *ex situ* breeding efforts. Although there is a great deal of understanding on how to stimulate, collect, and store sperm from male amphibians of numerous genera and species (Chapter 7), the parallel knowledge for managing egg production by females is limited due to the wide variety of amphibian reproductive strategies (Lofts 1984; Duellman and Trueb 1994), as well as the lengthy and complex dynamics of the maturation and release of eggs. To date, most assisted reproduction focuses on applying exogenous hormones to obtain eggs on demand from externally fertilising species (Kouba *et al.* 2009; 2012a). Obtaining mature and viable eggs is key to successful natural breeding or *in vitro* fertilisation and subsequent production of offspring, which are needed to meet the recovery goals of amphibian conservation breeding programs, including sustaining *ex situ* breeding colonies, managing genetic diversity (Chapter 8), or linking *in situ* and *ex situ* populations (Chapter 11).

Challenges of assisted reproduction for female amphibians

The obstacles faced by amphibian breeding programs regarding the reproduction of females span the entire suite of events from initiation of oogenesis to oviposition of viable eggs. Categorically, in externally fertilising amphibians the primary challenges to reproduction resulting in failed breeding are insufficient development and maturation of oocytes; failure of mature oocytes to be ovulated into the oviduct; retention of eggs rather than oviposition; or, if oviposition occurs, inviability of eggs (Kouba *et al.* 2009; Clulow *et al.* 2018; Bronson *et al.* 2021). The first three challenges stem from the complex nature of gametogenesis in females, which is influenced by poorly understood species-specific environmental and behavioural cues that trigger neuroendocrine control of oogenesis and oviposition. Moreover, the progression of gamete development may occur over weeks, months, or even years depending on the natural history of a species, thereby confounding interpretation of the reproductive cycles of females at single points in time (Hartel *et al.* 2007; Muths *et al.* 2010; Bronson *et al.* 2021). In a limited number of species with translucent skin, eggs can be visualised in the abdomen (Reyer and Bättig 2004; Calatayud *et al.* 2018; Bronson *et al.* 2021; Silla and Byrne 2021). However, for most species, practitioners that breed amphibians are blind to the reproductive status of females (e.g. stage of oogenesis, maturity of eggs, timeframe of oviposition), which can lead to ineffective hormonal treatments for inducing ovulation and/or missed opportunities for breeding. Finally, if eggs are laid, fertilisation may fail if eggs are immature, over-ripened, damaged, undergoing atresia, have insufficient egg jelly, or have been prematurely auto-activated (Michael *et al.* 2004; Toro and Michael 2004; Silla 2011; Kouba *et al.* 2012a). Asynchronicity in the availability of sperm within the window of release of viable eggs is also commonly observed for externally fertilising species, arising from differential responses of male and female amphibians to exogenous hormonal treatments (Kouba *et al.* 2012a; Silla and Byrne 2019). Internally fertilising species present the same confounding issues in terms of assessing the stage of egg development before attempts at assisted reproduction, but eggs may be even more vulnerable to damage and premature activation by mechanical manipulation, or premature expression without ample constituents in the egg jelly needed for fertilisation (Takahashi *et al.* 2006).

Ultimately, breeding outcomes could be improved by monitoring the natural reproductive state of a female through routine, non-invasive analysis of steroid hormones or ultrasonography. Although analysis of faecal and urinary steroid hormones has been successfully used for sexing amphibians, and elevations in reproductive hormones have been linked with mating and egg laying in some species, such analysis requires time-consuming processing of samples and validation of biochemical assays (Szymanski *et al.* 2006; Germano *et al.* 2009; Narayan *et al.* 2010). Alternatively, ultrasonography can be conducted in real time, and without invasive sampling, giving practical information about a female's overall reproductive status that can be used to coordinate breeding (Calatayud *et al.* 2018; Graham *et al.* 2018).

REPRODUCTIVE CYCLES OF FEMALE AMPHIBIANS

Challenges to assisted reproduction mentioned above (production and maturation of oocytes, ovulation) generally reflect the sequential stages of gametogenesis in which reproductive failure coincides with arrested development at key points of the reproductive cycle. In conservation breeding programs highly important environmental reproductive cues may be missing or insufficient,

development and maturation of eggs may not follow the natural patterns of a species, or the timing and developmental stage of eggs across females within a population may be broadly distributed rather than synchronous. As such, applications of reproductive technologies, specifically hormonal regimens, need to be coordinated with the reproductive state of females in order to be effective, while simultaneously minimising possible negative physiological responses from stress (Chapter 5), or overstimulating the female with improper types or doses of hormones that may lead to down-regulation of hormone receptors, hyper-stimulation syndrome, hyper-recruitment of immature oocytes, severe osmotic regulatory imbalances, damage to organs, or in rare instances, fatality (Michael *et al.* 2004; Green *et al.* 2007; Crawshaw 2008; Rowlison 2013; Sato and Tokmakov 2020). Ultrasonography can be aligned with the stages of oogenesis (described below), giving direct insight into the maturation of oocytes, thus informing the timing and application of hormonal therapies for assisted breeding.

Oogenesis is a molecular and cellular process

The process of oogenesis in amphibians is similar to that of most vertebrates and is coordinated by endocrine control of molecular and cellular events (Uribe 2011; Chapter 4). During development, oocytes progress through primary and secondary phases of growth before the final maturation needed for ovulation, at which point the oocytes formally become eggs. In amphibians, the development of oocytes is described in seven stages, with three stages in each of the primary and secondary phases of growth, followed by a final stage of maturation and ovulation (Sharon *et al.* 1997; Sretarugsa *et al.* 2001; Uribe 2011; Sato and Tokmakov 2020). Uribe (2009, 2011) can be referenced for an in-depth review of amphibian oogenesis and its endocrine control, with detailed information on key processes and comparative sizes of oocytes at each stage of development, based on the model anurans *Hyla eximia* and *Osteopilus septentrionalis*, and the model urodeles *Ambystoma dumerilii* and *Ambystoma mexicanum*.

Growth and maturation of oocytes

Primary growth of the oocyte (stages 1–3) is categorised as pre-vitellogenic growth and includes significant increases in transcriptional and translational processes to support biochemical activity leading to the development of cellular organelles, cytoplasmic machinery, cytoskeletal structure, and reorganisation of membranes (Gall *et al.* 2004). In stage 1, initial growth of oocytes from ~10 to 30 µm occurs along with initiation of folliculogenesis. Changes in chromosome structure occur in stage 2, along with completion of folliculogenesis and growth of the oocyte up to 100 µm. In the third stage, the oocyte grows to nearly 300 µm and accommodates substantial numbers of mitochondria, ribosomes, oil droplets, and cortical granules (Grey *et al.* 1974; Oterino *et al.* 2006; Ogielska and Bartmanska 2009).

The secondary growth of oocytes is characterised by the formation of yolk, or vitellogenesis, and is again described in three stages (stages 4–6). Steroid hormones produced from the follicular cells are essential in activating the liver to release vitellogenin. In stage 4, yolk platelets accumulate on the periphery of the oocyte as it increases in diameter to nearly ~700 µm and integrates with the follicular cells. Further growth up to 1.0 mm and aggregation of melanin pigment that defines the animal hemisphere occurs in stage 5. In the aforementioned model species, significant accumulation of yolk increases the diameter of the full-grown oocyte to nearly ~1.8 mm in stage 6, and the large yolk platelets migrate to the vegetal hemisphere, while glycogen, oil droplets, and melanin continue to accumulate in the animal hemisphere (Bohun and Breward 2009; Uribe 2009, 2011). Prior to ovulation, final maturation of the oocyte occurs (stage 7), meiosis resumes, and structural changes take place in preparation for fertilisation, although further volumetric growth is minimal (Uribe 2011).

Ovulation and oviposition

When ovulation occurs, eggs are directed into the oviduct where deposition of the vitelline membrane occurs before the eggs enter the *pars convoluta*. In the tubule, secretory cells deposit a thick coat of jelly composed of oviducal mucins and secretions that are essential for communication between sperm and eggs during fertilisation (Fernandez and Ramos 2003, Takahashi *et al.* 2006; Uribe 2009). Stimulated by progesterone from the ovary, the jelly coat is layered evenly onto the eggs as they are rotated through ciliary activity of the oviducal lining (Fernandez and Ramos 2003; Ogielska and Bartmanska 2009). At the terminal end of the *pars convoluta*, eggs are deposited into the ovisac, which opens into the cloaca.

Oviposition, or spawning, occurs when eggs are released from the cloaca to the external environment. For externally fertilising species, eggs may be pushed through the cloaca by amplexing males as they apply pressure to the abdomen while simultaneously depositing sperm directly onto the emerging eggs. For oviparous amphibians with internal fertilisation, sperm that has been previously taken up and stored by the female in the spermatheca directly fertilises the eggs immediately before they pass through the cloaca (Sever *et al.* 1996; Uribe 2009), although internal embryonic development can occur in the ovisac of some viviparous species (Wake 2015).

Resorption and retention of eggs

If ovulation does not occur, mature oocytes can undergo atresia or be reabsorbed. The follicular cells transform into phagocytic cells and begin to digest the oocyte (Gilbert 2000; Exbrayat 2009). Atresia and resorption of oocytes are widely recognised but not well studied, and are commonly found in malnourished individuals that obtain nutrients from the resorbed eggs (Ogielska and Bartmanska 2009).

Retained eggs, or 'egg binding', in female amphibians is a serious health problem in captive breeding programs and occurs when oocytes are ovulated, but not oviposited. Instead, eggs become caught in the tubular fold of the oviduct during deposition of the vitelline membrane and jelly coat, such that an egg adhering to the oviducal lining forms a cyst-like occlusion, which then blocks the egg mass from advancing (Wright and Whitaker 2001; Uribe 2011). When bound, eggs deteriorate within the female, become necrotic and potentially lead to a condition known as egg sepsis (Kouba *et al.* 2009; Eustace *et al.* 2018). Females of some anuran species, such as the Panamanian golden frog (*Atelopus zeteki*), Wyoming toad (*Anaxyrus baxteri*), and Puerto Rican crested toad (*Peltophryne lemur*) have died due to egg binding (Wright and Whitaker 2001; Crawshaw 2008; Bronson *et al.* 2021). For instance, gravid female Panamanian golden frogs exhibit up to ~90% morbidity and mortality during the breeding season when they fail to undergo oviposition, despite prolonged amplexus with males (Eustace *et al.* 2018; Bronson *et al.* 2021). The physiological stress of retaining a large mass of mature eggs within the coelomic cavity possibly renders the females immune-compromised and susceptible to opportunistic bacterial and fungal infections. Egg binding and egg sepsis are not commonly recognised in caudates; however, with fewer captive breeding programs for caudate species available this condition may not have been notably observed (Marcec 2016).

Egg size, clutch size, and the regeneration of clutches

Proliferation of the next generation of oogonia begins following ovulation of the previously developed clutch of eggs, thus renewing the store of oocytes that will advance to maturity by the beginning of the next breeding cycle (Gilbert 2000; Exbrayat 2009). The size and number of eggs deposited in a clutch is influenced by individual fitness and nutrition, seasonal reproductive patterns, and species' evolution and life history. Within a species, older animals tend to lay clutches of larger eggs that are more evenly distributed in size compared to those of younger animals (Kaplan 1985; Bernardo 1996; Crespi and Lessig 2004). Generally, egg size and clutch size tend to be inversely related since the distribution of limited vitellogenic resources needs to be concentrated in a few oocytes or spread across many. Phylogenetic analysis of amphibian life history and reproductive traits has also demonstrated a correlation between clutch size and fecundity (Kupfer *et al.* 2016), and variation in egg size may offer certain advantages depending on environmental conditions (Kaplan 1985; Semlitsch and Gibbons 1990; Dziminski and Alford 2005). Behaviour and mate choice may also influence clutch size (Reyer *et al.* 1999).

The frequency of egg-clutch production also varies across amphibian species with different life histories. For instance, tropical species, or populations of temperate species residing along warm springs, may undergo several cycles of oogenesis in a year or even year-round (Lofts 1984; Norris and Lopez 2005; Rorabaugh and Sredl 2014). Most temperate species experience oogenesis annually, while biennial cycles of oogenesis have been observed in some alpine anurans and urodeles (Bruce 1975; Whittier and Crews 1987; Norris 2007; Muths *et al.* 2010, 2013). Sufficient progression through the stages of oocyte development and maturation must occur for successful expression of viable eggs, and consideration of the species' natural reproductive pattern is likely to influence the timing and number of eggs that can be obtained using reproductive technologies.

HORMONAL TREATMENTS FOR FEMALE AMPHIBIANS

Exogenous hormonal therapies are often used by amphibian conservation breeding programs to stimulate oogenesis, egg maturation, ovulation, and spawning in females in order to overcome reproductive dysfunction (Griffiths and Pavajeau 2008; Kouba and Vance 2009; Clulow *et al.* 2014; Calatayud *et al.* 2018; Silla and Byrne 2019). A majority of studies using exogenous hormonal treatments have been conducted on anurans, and although there is some information available on the use of hormones in urodeles, these primarily focus on common laboratory species, for example *Ambystoma mexicanum* (Trottier and Armstrong 1975; Mansour *et al.* 2011). Unfortunately, even less information is available for gymnophionans. The ovoviviparous and viviparous reproductive strategies of many salamanders and caecilians can bring additional challenges when applying reproductive technologies.

For many amphibians, the reproductive cycle is heavily influenced by environmental cues (e.g. light, humidity, barometric pressure, temperature) and behavioural cues (Delgado and Vivien-Roels 1989; Norris 2011). Missing cues, along with factors such as nutritional deficits and stress from handling or housing in captivity (Narayan and Hero 2011; Titon *et al.* 2017, 2018), may contribute to reduced reproductive output. Some breeding programs have attempted to manipulate the design of habitats and simulate environmental signals to promote natural reproduction (Stoops *et al.* 2014; Santana *et al.* 2015; Behr and Rodder 2018), but cues may be unknown or difficult to replicate (Kouba *et al.* 2012a). Chapter 3 reviews housing and environmental considerations leading to improved reproduction.

Reproductive physiology and endocrinology of amphibians

Successful use of exogenous hormones requires a strong understanding of amphibian endocrinology, gametogenesis, and reproductive physiology (reviewed in detail in Chapter 4). Briefly, abiotic and biotic factors stimulate the hypothalamic-pituitary-gonadal axis, which is the primary driver of gametogenesis, courtship, and spawning. Gonadotropin releasing hormone (GnRH), a neuropeptide secreted from the hypothalamus, acts on the pituitary gland to promote secretion of the gonadotropins luteinizing hormone (LH) and follicle stimulating hormone (FSH) (Daniels and Licht 1980; Stamper and Licht 1990). In turn, LH and FSH moderate early follicular development, final maturation of oocytes, and stimulate steroidogenesis in the gonads (Polzonetti-Magni *et al.* 1998; Morrill *et al.* 2006). A surge in LH is critical to triggering ovulation of mature oocytes into the oviduct. The sex steroids, including estrogens, progestogens, and androgens, are associated with the development and maturation of oocytes, vitellogenesis, and courtship and breeding behaviours (Polzonetti-Magni *et al.* 2004; Deng, *et al.* 2009; Norris 2011). Recent evidence supports the role of secondary hormones and compounds (e.g. arginine vasotocin, prolactin, dopamine, kisspeptins, endocannabinoids) as mediators of the hypothalamic-pituitary-gonadal axis (Rowlison 2013; Trudeau 2015; Vu and Trudeau 2016).

Developing hormonal therapies for female amphibians

It is widely stated that the effectiveness of exogenous hormonal treatments for amphibian-assisted reproduction is highly variable across species, even among those that are closely related. Although a valid argument, making true comparisons of the efficacy of treatments can be challenging, as published studies vary widely in the types, dosages, routes, and timing of the application of hormones (both relative to breeding season and interval of successive hormonal treatments) (Table 6.1). Also of note, most studies do not account for females being in various reproductive phases. In other words, the response rate to a particular hormonal regimen assumes that all females in the study are reproductively synchronised, and thus the efficacy is interpreted as a function of the type of hormone and its dosage and timing without also considering the female's condition. Additionally, many studies contain only small numbers of individuals per treatment, or the same individuals may be cycled through treatments, meaning that the observed response may be the result of the compound sequence of hormones rather than the independent regimen being tested. It has also been proposed that responses to hormones may be tied to phylogenetic relationships (Silla and Roberts 2012; Clulow *et al.* 2018) or to sexual selection (Silla and Byrne 2019).

Determining the optimal dosage is a critical component of hormonal therapy. Dosages of hormone that are too low may be insufficient to stimulate the release of gametes; whereas excessively high or repeated dosing could result in desensitisation or down-regulation of receptors (Gore 2002), or reduced viability and fertilisation of gametes (Silla *et al.* 2018; Bronson *et al.* 2021). In several species, hormone dose–response curves have been constructed to study spermiation in males (Obringer *et al.* 2000; Kouba *et al.* 2012b; Silla *et al.* 2019), however, dose–response data are limited for females (Vu *et al.* 2017; Silla *et al.* 2018; Bronson *et al.* 2021). Nonetheless, effective dose–responses to any hormone will depend on the

female's stage of oogenesis, which dictates the availability of mature eggs for oviposition.

Additionally, the frequency interval and seasonal timing of hormonal treatments may impact responses. Multiple doses of hormones administered over time (e.g. days or weeks) may promote development and maturation of eggs, prior to a final ovulatory treatment; or hormones aligned with a species' natural breeding season may be more effective than those given 'out of season' (Clulow *et al.* 2014; Guy *et al.* 2020). For species in which reproduction is highly linked to season, obtaining only a single clutch of eggs per year may be possible, but for other species, administering exogenous hormones may allow for multiple annual spawnings (Guy *et al.* 2020). However, an important consideration is that reproduction is energetically expensive and care must be taken not to overstimulate animals beyond a safe capacity (Green *et al.* 2007).

The route of hormone administration is also an important consideration, both from practical standpoints (e.g. size of individual; sensitivity to handling) and efficacy of treatment. The efficacy of a hormone depends on its stability *in vivo* and the migration pathway to the targeted tissues (e.g. pituitary gland or gonads) from the point of administration. Exogenous hormones typically are delivered through intraperitoneal, subcutaneous (particularly in the area of the dorsal lymph sac), or intramuscular injections (Table 6.1). Intranasal administration is a recent novel, non-invasive method of delivering hormones that has successfully induced release of eggs and sperm in caudates (*Ambystomatid tigrinum*; C. Kouba, unpublished data) and release of sperm in anurans (Julien *et al.* 2019). Intranasal delivery has been demonstrated as a more direct route of hormone uptake that acts on the pituitary gland, compared to routes such as intraperitoneal injection that depend on diffusion of the hormone from distant sites (Julien 2021). Transdermal (via topical absorption or uptake in a water bath) and oral (via feeding hormone solutions directly or in an insect vehicle) delivery routes also have been successful. Effectiveness of hormone-delivery methods in male amphibians is extensively covered in Chapter 7.

Use of exogenous hormones to induce ovulation and spawning

Exogenous hormones have been used successfully to induce ovulation and spawning in a diversity of

Table 6.1: Comparison of hormonal treatments, spawning response, and fertilisation rates in studies of assisted reproduction for female amphibians.

Species (common name)	Route	Dosage of hormone	Spawning rate^ [fertilisation rate]	Reference
Acris crepitans (Northern cricket frog)	WB	O: 0.17µg GnRHa+0.42 µg MET / µL	84% (32/38)^ [27%]	Snyder et al. (2012)
Ambystoma mexicanum (Mexican axolotl)	IM	O: 200–300 IU hCG	100% (10/10) [50 ± 6%]	Mansour et al. (2011)
	IM	O: 1/2 Ovopel® pellet (5 µg GnRHa + MET)	100% (10/10) [50 ± 6%]	Mansour et al. (2011)
Ambystoma tigrinum (Tiger salamander)	SC	O: 4 IU/g hCG + 0.1 µg/g GnRHa	100% (9/9) [–]	Marcec (2016)
	IN	O: 20 µg GnRHa	100% (3/3) [–]	Julien (2021)
Anaxyrus baxteri (Wyoming toad)	IP	O: 500 IU hCG + 4 µg GnRHa	0% (0/10) [–]	Browne et al. (2006a)
	IP	P: 100 IU hCG + 0.8 µg GnRHa; O: 500 IU hCG + 4 µg GnRHa	70% (7/10) [6.4 ± 3.6%]	Browne et al. (2006a)
	IP	P: 500 IU hCG + 4 µg GnRHa; P: 100 IU hCG + 0.8 µg GnRHa; O: 500 IU hCG + 4 µg GnRHa	88% (8/9) [12.7 ± 3.4%]	Browne et al. (2006a)
Anaxyrus boreas boreas (Boreal toad)	IP	P: 2x 3.7 IU/g hCG; O: 13.5 IU/g hCG + 0.4 µg/g CnRHa	77% (17/22) [19%]	Calatayud et al. (2015)
Anaxyrus fowleri (Fowler's toad)	IP	O: 500 IU hCG + 4 µg GnRHa	29% (2/7) [0%]	Browne et al. (2006b)
	IP	O: 20 µg GnRHa	0% (0/7) [–]	Browne et al. (2006b)
	IP	O: 20 µg GnRHa + 0.25 mg PIM	14% (1/7) [34%]	Browne et al. (2006b)
	IP	O: 20 µg GnRHa + 0.25 mg PIM + 5 mg P4	58% (4/7) [35 ± 11%]	Browne et al. (2006b)
	IP	O: 60 µg GnRHa + 5 mg P4	71% (5/7) [73 ± 5%]	Browne et al. (2006b)
	IP	O: 60 µg GnRHa + 0.25 mg PIM + 5 mg P4 + 500 IU hCG	85% (6/7) [20 ± 8%]	Browne et al. (2006b)
	IP	P: 5 mg P4; O: 5 mg P4	0% (0/5) [–]	Browne et al. (2006b)
	IP	P: 2x 100 IU hCG; O: 500 IU hCG + 15 µg GnRHa	100% (7/7) [52 ± 10%]	Guy et al. (2020)

(Continued)

Table 6.1: *Continued*

Species (common name)	Route	Dosage of hormone	Spawning rate^ [fertilisation rate]	Reference
Atelopus zeteki (Panamanian golden frog)	IP	O: 1–2× 0.5, 1, 2, 4 µg GnRHa	46% (23/50)^ [–]	Bronson *et al.* (2021)
	IP	O: 1–2× 0.5–4 µg GnRHa + 150 µg MET	44% (15/34)^ [–]	Bronson *et al.* (2021)
Eleutherodactylus coqui (Common coqui)	SC	O: 3–33 µg mammalian GnRH	13% (2/15) [–]	Michael *et al.* (2004)
	SC	O: 3–33 µg avian GnRH	8% (1/13) [–]	Michael *et al.* (2004)
	SC	O: 3–33 µg fish GnRH	13% (2/16) [–]	Michael *et al.* (2004)
	SC	O: 5–20 µg synthetic GnRHa	52% (12/23) [10 ± 3%]	Michael *et al.* (2004)
	SC	O: 25–200 µg hCG	15% (2/13) [–]	Michael *et al.* (2004)
Heleioporus eyrei (Moaning frog)	SC	P: 0.04 µg/g GnRHa O: 2 µg/g GnRHa	100% (6/6) [24 ± 15%]	Silla and Byrne (2021)
Hoplobatrachus occipitalis (African tiger frog)	IM	O: 8 µL/g salmon GnRHa + DOM	100% (4/4)^ [93 ± 3%]	Godome *et al.* (2021)
Limnodynastes tasmaniensis (Spotted grass frog)	IP	O: 3× PIT + 100 IU hCG	69% (37/54) [–]	Clulow *et al.* (2018)
	IP	O: 2× 0.9–1.2 µg/g GnRHa + 10 µg/g PIM	44% (4/9) [–]	Clulow *et al.* (2018)
Lithobates chiricahuensis (Chiricahua leopard frog)	IP	P: 2× 100 IU hCG O: 2× 10 IU/g hCG + 15 µg GnRHa	83% (5/6) [33%]+	C. Kouba, unpublished data
Lithobates pipiens (Northern leopard frog)	IP	O: 0.4 µg/g GnRHa + 10 µg/g PIM	50% (6/12)^ [–]	Trudeau *et al.* (2010)
	IP	O: 0.4 µg/g GnRHa + 5 µg/g MET	44% (4/9)^ [–]	Trudeau *et al.* (2010)
	IP	O: 0.4 µg/g GnRHa + 10 µg/g MET	82% (18/22)^ [–]	Trudeau *et al.* (2010)
	IP	O: 0.4 µg/g GnRHa + 10 µg/g MET	42% (5/12)^ [–]	Trudeau *et al.* (2010)
	IP	O: 0.4 µg/g GnRHa	87% (13/15)^ [90 ± 4%]	Vu *et al.* (2017)
Lithobates sevosus (Mississippi gopher frog)	IP	O: 0.4 µg/g GnRHa	45% (5/11) [57 ± 10%]	Graham *et al.* (2018)
	IP	O: 0.4 µg/g GnRHa + 10 µg/g MET	73% (8/11) [55 ± 7%]	Graham *et al.* (2018)

Species	Route	Hormone treatment	Result	Reference
L. sevosus continued	IP	O: 0.4 µg/g GnRHa + 13.5 IU/g hCG	64% (7/11) [29 ± 8%]	Graham *et al.* (2018)
	IP	P: 2× 3.7 IU/g hCG O: 0.4 µg/g GnRHa + 13.5 IU/g hCG	73% (8/11) [59 ± 6%]	Graham *et al.* (2018)
Mixophyes fasciolatus (Barred frog)	SC	O: 900–1400 IU hCG	27% (10/37) [–]	Clulow *et al.* (2012)
	SC	O: 1200–1500 IU hCG + 6× PIT	29% (2/7) [–]	Clulow *et al.* (2012)
	SC	P: 2× 50 + 25 IU PMSG O: 2× 100 IU hCG	47% (32/68) [–]	Clulow *et al.* (2012)
Peltophryne lemur (Puerto Rican crested toad)	IP	O: 20 µg GnRHa	100% (3/3) [20%]	Burger *et al.* (2021)
Pseudophryne bibronii (Brown toadlet)	SC	P: 0.04 µg/g GnRHa O: 2 µg/g GnRHa	95% (20/21) [85 ± 4%]	Silla and Byrne (2021)
Pseudophryne coriacea (Red-backed toadlet)	SC	P: 0.04 µg/g GnRHa O: 2 µg/g GnRHa	95% (18/19) [67 ± 8%]	Silla and Byrne (2021)
Pseudophryne corroboree (Southern corroboree frog)	SC	P: 1 µg/g GnRHa O: 5 µg/g GnRHa	30% (5/17) [55 ± 13%]	Byrne and Silla (2010)
Pseudophryne pengilleyi (Northern corroboree frog)	SC	O: 0.5 µg/g GnRHa	100% (9/9)^ [61 ± 16%]	Silla *et al.* (2018)
	SC	O: 1 µg/g GnRHa	89% (8/9)^ [16 ± 11%]	Silla *et al.* (2018)
	SC	O: 2 µg/g GnRHa	100% (9/9)^ [22 ± 11%]	Silla *et al.* (2018)
	T	O: 25 µg/g GnRHa	77% (10/13)^ [48 ± 13%]	Silla *et al.* (2018)
	T	O: 50 µg/g GnRHa	62% (8/13)^ [31 ± 10%]	Silla *et al.* (2018)
Pseudophryne guentheri (Günther's toadlet)	SC	O: 2 µg/g GnRHa	25% (2/8) [0%]	Silla (2011)
	SC	P: 0.04 µg/g GnRHa O: 2 µg/g GnRHa	100% (8/8) [97 ± 1%]	Silla (2011)
	SC	P: 2× 0.04 µg/g GnRHa O: 2 µg/g GnRHa	100% (8/8) [0%]	Silla (2011)
	SC	P: 0.04 µg/g GnRHa O: 2 µg/g GnRHa	100% (38/38) [89 ± 3%]	Silla and Byrne (2021)
Rana temporaria (Common European frog)	IP	O: 50 µg GnRHa	100% (5/5) [95%]	Uteshev *et al.* (2018)
Xenopus laevis (African clawed frog)	SC	P: 100 IU PMSG O: 500 IU hCG	62% (18/29) [–]	Ogawa *et al.* (2011)

(Continued)

Table 6.1: *Continued*

Species (common name)	Route	Dosage of hormone	Spawning rate^ [fertilisation rate]	Reference
X. laevis continued	WB	O: 10–40 µM P4	53% (10/19) [–]	Ogawa et al. (2011)
	WB	O: 4 µM P4 + 4 µM E	18% (1/11) [–]	Ogawa et al. (2011)
	WB	O: 10 µM P4 + 4 µM E	18% (2/11) [–]	Ogawa et al. (2011)
	WB	O: 20 µM P4 + 10 µM E	83% (20/24) [–]	Ogawa et al. (2011)
	SC	P: 50 IU PMSG O: 500–550 IU hCG	83% (10/12) [–]	Wlizla et al. (2017)
	SC	P: 50 IU PMSG O: 150–300 µg ovine LH	89% (25/28) [–]	Wlizla et al. (2017)
	SC	P: 50 IU PMSG O: 200–250 µg bovine LH	83% (10/12) [–]	Wlizla et al. (2017)
	SC	P: 50 IU PMSG O: 200–250 µg human LH	100% (12/12) [–]	Wlizla et al. (2017)
	SC	P: 50 IU PMSG O: 200–250 µg porcine LH	0% (0/12) [–]	Wlizla et al. (2017)
Xenopus tropicalis (Western clawed frog)	SC	P: 10 IU hCG O: 100–150 IU hCG	75% (9/12) [–]	Wlizla et al. (2017)
	SC	P: 10 IU hCG O: 100–150 IU rhCG	67% (8/12) [–]	Wlizla et al. (2017)
	SC	P: 10 IU hCG O: 25–100 µg ovine LH	87% (27/31) [–]	Wlizla et al. (2017)
	SC	P: 15 IU PMSG O: 100–150 IU hCG	67% (8/12) [–]	Wlizla et al. (2017)
	SC	P: 15 IU PMSG O: 100–150 IU rhCG	64% (9/14) [–]	Wlizla et al. (2017)
	SC	P: 15 IU PMSG O: 25–100 µg ovine LH	81% (29/36) [–]	Wlizla et al. (2017)

Abbreviations: Route: IM – intramuscular; IN – intranasal; IP – intraperitoneal; SC – subcutaneous; T – topical; WB – water bath. **Hormone:** DOM – domperidone; E – oestradiol; GnRHa – gonadotropin – releasing hormone agonist; hCG – human chorionic gonadotropin; rhCG – recombinant human chorionic gonadotropin; LH – luteinizing hormone; MET – metoclopramide; PIM – Pimozide™; PMSG™, PMSG – pregnant mare serum gonadotropin; P4 – progesterone; PIT – pituitary extracts. **Dosage:** P – priming (anovulatory) dose; O – ovulatory dose. Doses listed as X/g indicate concentration given on the basis of per gram of bodyweight. ^ indicates female paired with male during hormonal treatment. Fertilisation rate includes eggs that were determined to be fertilised, or counted at two-cell stage or later; + indicates fertilisation conducted using cryopreserved sperm; – – not available.

amphibian species. Different types of hormones administered to amphibians target different tissues in the hypothalamic-pituitary-gonadal axis. Table 6.1 provides an overview of types of hormones, doses, and routes of administration, along with the subsequent rates of oviposition and fertilisation for selected species.

Gonadotropin releasing hormone (GnRH)

Gonadotropin releasing hormone (GnRH) works by stimulating the anterior pituitary to secrete endogenous LH and FSH, which subsequently exert downstream effects on the reproductive system. Several endogenous subforms of GnRH exist, with mammalian (mGnRH) and chicken (cGnRH-II) found most commonly in the amphibian brain, and mGnRH thought to be primarily responsible for stimulating LH in amphibians (King and Millar 1986; Licht et al. 1994). However, commercially available synthetic GnRH analogues (GnRHa) are primarily administered to amphibians for assisted reproduction. A study testing various GnRH subforms in the common coqui (Eleutherodactylus coqui), found that a synthetic GnRH analogue was more effective at inducing ovulation (52%) compared to mGnRH (13%) or cGnRH (8%) (Michael et al. 2004). A single dose of synthetic GnRHa stimulates ovulatory responses (rates ranging from 40–100%) in several species, including Northern leopard frogs (Lithobates pipiens) (Vu et al. 2017), Panamanian golden frogs (Bronson et al. 2021), and Puerto Rican crested toads (Burger et al. 2021), among others (Table 6.1).

Gonadotropins

Purified amphibian LH and FSH are rarely used for hormonal therapy due to the difficulty of obtaining them in sufficient quantities, and use of amphibian pituitary homogenates is discouraged because of the risk of disease and the need to sacrifice donor animals (Clulow et al. 2014). Instead, gonadotropins derived from mammalian sources, including pregnant mare serum gonadotropin (PMSG) and human chorionic gonadotropin (hCG), commonly are used. PMSG and hCG are thought to stimulate LH-like effects on the ovaries; however, relatively high concentrations are needed to achieve a response in amphibians, probably due to low affinity of amphibian LH receptors to the mammalian-derived compounds (Seong et al. 2003). Gonadotropins have been used with moderate success in anurans, including the barred frog (Mixophyes fasciolatus; Clulow et al. 2012),

spotted grass frog (Limnodynastes tasmaniensis; Clulow et al. 2018), African clawed frog (Xenopus laevis; Ogawa et al. 2011; Wlizla et al. 2017), and Western clawed frog (Xenopus tropicalis; Wlizla et al. 2017). A 100% ovulation rate (10/10) was achieved in the Mexican axolotl (Ambystoma mexicanum) using hCG at doses of 200–300 IU, thereby demonstrating that this hormone is successful in some salamander species as well (Mansour et al. 2011) (Table 6.1).

Steroid hormones

In females, endogenous estrogens, progestogens, and androgens synthesised by ovarian follicles are involved in vitellogenesis and final maturation of oocytes (Polzonetti-Magni et al. 2004; Deng et al. 2009; Norris 2011); therefore, administering exogenous reproductive steroid hormones may advance the maturation of oocytes. Ogawa et al. (2011) demonstrated a 50% ovulation response in female African clawed frogs following doses of 10–40 μM progesterone alone, and addition of 10 μM oestradiol further increased ovulation rates to 83%. In Fowler's toads (Anaxyrus fowleri), progesterone paired with GnRHa increased ovulatory rates, but could not stimulate ovulation independently (Browne et al. 2006a). Importantly, steroid hormones exert both positive and negative feedback on the reproductive system and the type of feedback may partly be related to the concentration of hormone, reproductive season, or stage of the life cycle (McCreery and Licht 1983; Stamper and Licht 1994; Tsai et al. 2005), making it important to consider dosage and timing when applying steroid hormones for assisted reproduction.

Dopamine antagonists

A growing number of researchers are testing dopamine antagonists (DAs), including pimozide, domperidone, and metoclopramide, alone or in combination with GnRHa. Based on findings in some species of fish (Dufour et al. 2005; Popesku et al. 2011), it is thought that dopamine may inhibit release of LH from the amphibian pituitary (Sotowska-Brochocka et al. 1994) and dopamine antagonists may block this inhibitory effect, resulting in an improved ovulatory response to exogenous GnRHa (Sotowska-Brochocka et al. 1994). Trudeau et al. (2010, 2013) achieved ovulation from female northern leopard frogs by administering metoclopramide in conjunction with GnRHa, although the spawning rate was similar to treatment with GnRHa alone (Vu et al. 2017). Nonetheless, GnRHa paired with DAs has induced ovulation in

Fowler's toads (Browne *et al.* 2006a), several species of Argentinian frogs (Trudeau *et al.* 2013), Mississippi gopher frogs (*Lithobates sevosus*; Graham *et al.* 2018), and Mexican axolotls (Mansour *et al.* 2011). As with some fish (Zohar and Mylonas 2001), success of DAs for amphibians can be variable and rates of ovulation did not significantly increase in American bullfrogs (*Lithobates catesbeianus*) treated with GnRHa + domperidone (do Nascimento *et al.* 2015), or Panamanian golden frogs treated with GnRHa + metoclopramide (Bronson *et al.* 2021) versus GnRHa alone. Clulow *et al.* (2018) also reported that supplementation with DAs had no effect in inducing spawning of eastern dwarf tree frogs (*Litoria fallax*), when zero of eight females oviposited after treatment with GnRHa and metoclopramide.

Hormone 'cocktails'

As noted in some examples above, using a mixture of hormones (a 'cocktail') has produced higher ovulation/spawning rates than does administration of a single type of hormone (Table 6.1). Cocktails often consist of GnRHa paired with hCG, DA, or steroid hormones. For instance, Browne *et al.* (2006a) found administering GnRHa to Fowler's toads was unsuccessful at inducing ovulation until it was paired with hCG, pimozide, and/or progesterone. Graham *et al.* (2018) observed GnRHa alone could stimulate ovulation in Mississippi gopher frogs (52% ovulation rate); however, rates of ovulation up to 73% were achieved when GnRHa was paired with either hCG or metoclopramide.

Hormonal priming and pulsing

For some tested species, response rates increase when low doses of hormones ('priming doses') are administered (typically 1–7 days) before the ovulatory dose (e.g. Browne *et al.* 2006a; Clulow *et al.* 2012; Graham *et al.* 2018; Table 6.1). Priming doses are generally given at 20–25% of the concentration of the ovulatory dose and may assist with the final maturation of eggs, resulting in increased rates of ovulation/spawning or improved quality of eggs (Kouba and Vance 2009). In many cases, hCG is used as the priming hormone, followed by a cocktail of hCG and GnRHa as the ovulatory dose (Calatayud *et al.* 2015; Marcec 2016; Graham *et al.* 2018; Guy *et al.* 2020). Priming doses of GnRHa have also been successfully applied (Silla 2011; Bronson *et al.* 2021; Silla and Byrne 2021). Some studies do not present data comparing ovulation rates in animals administered priming doses versus those

receiving only an ovulatory dose, so the necessity of priming for successful induction of ovulation cannot be ascertained in all cases.

Administering two priming doses may be more effective than a single prime for some species. For instance, female Wyoming toads receiving two priming doses of hCG + GnRHa produced on average twice as many eggs as toads receiving one prime (Browne *et al.* 2006b). However, Bronson *et al.* (2021) showed that ovulatory rates of the Panamanian golden frog did not significantly increase with up to four sequential pulses of GnRHa (with or without the DA metoclopramide), despite females being characterised as gravid (stages 6–7 on a species-defined scale of gravidity) and paired with a male for 2 weeks during the breeding season. Thus, the optimal number of priming doses may be species-specific. There is also risk of inducing hyper-stimulation syndrome with priming doses when a female is already gravid, which causes recruitment of new oocytes before the expulsion of the existing clutch. This phenomenon results in the female's body cavity filling with immature oocytes that crowd the organs and can lead to death and has been observed in the Puerto Rican crested toad and Fowler's toad (C. Kouba, unpublished data) and *X. laevis* (Green *et al.* 2007). It is also possible that eggs may become 'over-ripened' by administering hormones at too high a concentration or frequency, thereby compromising the quality of oocytes. Silla (2011) observed this phenomenon, finding that Günther's toadlets that had received one priming dose had a 100% ovulation rate and 97% fertilisation rate, whereas two priming doses resulted in 100% ovulation, but the eggs failed to fertilise. Table 6.1 includes ovulatory rates for listed hormonal treatments along with fertilisation rates of the ovulated eggs.

Alternative compounds

A developing area of interest is the exploration of alternative neuropeptides and compounds, including kisspeptins, endocannabinoids (ECBs), and prolactin, for assisted reproduction. Much of this research is based on extrapolation from birds, mammals, and teleost fish, with limited studies conducted on amphibians (reviewed by Vu and Trudeau 2016). Kisspeptin genes and receptors have been identified in some amphibians, although their influence on amphibian reproduction remains unclear and possibly variable (Trudeau 2015). Endocannabinoids may interfere with secretion of GnRH, and ECB receptors have been detected in multiple tissues, including the gonadotrophs and gonads of several amphibians (Cesa *et al.* 2002; Pagotto *et al.* 2006; Cottone *et al.* 2008). Behavioural support was shown by Soderstrom *et al.* (2000), when use of levonantradol, a cannabinoid receptor agonist, resulted in decreased clasping behaviour in male rough-skinned newts (*Taricha granulosa*). *In vitro* studies suggested that prolactin may enhance release of LH from gonadotropic cells (Oguchi *et al.* 1997), and prolactin has been linked to courtship behaviours in some urodeles (Toyoda *et al.* 1996; Iwata *et al.* 2000). Overall, much experimental work remains to be completed on this topic and may lead to novel combinations of hormones useful for targeted reproductive responses.

Hormonal treatments to induce spawning in male–female pairs

Although many studies utilise hormonal therapy to collect eggs from isolated females for *in vitro* fertilisation, others administer synchronised hormonal treatments to male–female pairs to induce breeding behaviours, as well as natural spawning (Trudeau *et al.* 2010; Kouba *et al.* 2012a; Snyder *et al.* 2012; Silla *et al.* 2018). Behavioural responses, including phonotaxis, amplexus, and other courtship behaviours, are commonly observed following administration of hormones (Vu and Trudeau 2016). Behaviours stimulated by exogenous hormones in paired animals may, in turn, promote further endogenous release of hormones; thus, for some species, treatments (including optimal doses and/or priming) may differ between conspecific females depending on the presence or absence of a male. Table 6.1 denotes studies with male–female pairs versus individually treated females, as a male's presence could affect spawning rate. Importantly, when relying on male–female pairing, it is necessary to determine the optimal interval to administer hormones such that peak release of gametes is synchronised between the sexes (Kouba *et al.* 2012a).

Collection of eggs

Following administration of hormones, unpaired females may spontaneously oviposit, or eggs may be manually collected. For species that are particularly sensitive to handling, spontaneous oviposition is encouraged, but it may be necessary to manually express ovulated eggs by gentle abdominal massage (also referred to as 'stripping') or using a small probe to open the cloaca to assist the deposition of eggs. Manual collection of eggs allows the benefit of controlling the timing of gamete expression to conduct *in vitro* fertilisation under controlled conditions, such as pairing specific sperm and eggs for optimal genetic crosses (see Chapter 8). Manual collection of eggs must be done carefully to ensure that females are not injured (by possible damage to internal organs or cloacal prolapse) during the process (Chai 2017).

ULTRASONOGRAPHY AS AN EMERGING TOOL FOR REPRODUCTIVE ASSESSMENT IN AMPHIBIANS

Ultrasonography is emerging as an important technology in the field of assisted reproduction in amphibians. As previously discussed, a major challenge in applying hormonal therapies to female amphibians is the ability to diagnose and monitor the reproductive state of individual animals, as this may impact timing, patterns, and effectiveness of treatments. Incorporating ultrasonography into breeding programs has the potential to improve assisted reproduction by enabling characterisation of reproductive readiness, seasonal reproductive cycles, and response to therapies.

Fundamentals of ultrasonography in amphibians

Ultrasonography uses high frequency sound waves to produce images of internal organs. A transducer, or probe, produces sound waves, which are then reflected back, thereby creating an image (sonogram) that appears in gradations of black to white, representing the reflectivity of the waves from a tissue relative to the surrounding tissues (echogenicity) (Mannion 2008). Echogenicity can be described as hyperechoic (white), hypoechoic (grey), or anechoic (black), and depends on the density of tissue, and the level and type of fluid. Typically, regions containing large amounts of fluid are anechoic, and lend to the identification of physiological landmarks such as fluid-filled cavities. In amphibians, the signature characteristic of a gravid female is the distinct anechoic and hyperechoic patterns present throughout the lower body cavity that are indicative of developing follicles/ova (Stetter 2001).

Ultrasonic waves pass easily through the skin of amphibians, making them well suited for non-invasive imaging (Schildger and Triet 2001). In most cases, amphibians do not require anaesthesia and the transducer can be held directly against the skin for rapid assessment. Alternatively, smaller or more sensitive species can be placed in a plastic container containing shallow water while the transducer is held directly against the side of the abdomen; alternatively, conducting gel can be applied to the external surface of the container for indirect imaging (Schildger and Triet 2001; Stetter 2001). Altogether ultrasonography is a low-stress, rapid technique for collecting important information on the reproduction and health of female amphibians.

Identifying relevant structures and patterns on a sonogram requires practice and initial consultation with a veterinarian or experienced practitioner is recommended to train new users. Plate 5 shows a series of sonograms taken using a transducer oriented transversely across the abdomen of female Mississippi gopher frogs exhibiting different degrees ('grades') of oocyte development. Each image shows a cross-section of the female from ventrum (top of image) to dorsum (bottom of image). The skin appears as a thin, hyperechoic line, and the ovaries containing numerous developing follicles/ova can be identified. Each ovum consists of a hypoechoic or anechoic portion and a hyperechoic region representing the fluid and yolk portions, respectively (Stetter 2001). Sonograms vary significantly among species, based on both clutch size and size of individual oocytes. For instance, anuran species that lay hundreds to thousands of eggs will present with numerous small hypoechoic and hyperechoic features, whereas anurans and urodeles that produce only a small number of eggs at a time have fewer and larger hypoechoic and hyperechoic regions visible. Figure 6.1 demonstrates variations in sonographic images across some amphibian genera. Moreover, variation in imaging can be further amplified when using different instrumentation or changing the position of the transducer.

Tools for assessing the reproductive status of females

Ultrasonography largely has been used for medical evaluation in amphibians (Wright and Whitaker 2001) but is increasingly being applied to monitor development of oocytes in females (Johnson *et al.* 2002; Kouba and Vance 2009; Krause *et al.* 2013; Marcec 2016; Holtze *et al.* 2017; Graham *et al.* 2018). Other tools for reproductive assessment include laparoscopy and ovariectomy, but these require invasive surgery (Gentz 2007; Chai 2015). Non-invasive or minimally invasive approaches such as behavioural assessment, visual or tactile inspection, gravidity scaling, candling, and indexing of bodyweight commonly are used but have limited reliability and many do not translate across species or genera (Reyer and Bättig 2004; Lynch

and Wilczynski 2005; Grant *et al.* 2009; Roth *et al.* 2010; Bronson and Vance 2019; Bronson *et al.* 2021). More sophisticated techniques such as magnetic resonance imaging (MRI) or total body electrical conductance (TOBEC) require extremely specialised equipment with very limited accessibility and high cost (Reyer and Bättig 2004; Ruiz-Fernández *et al.* 2020). In a few amphibian species, sampling methodologies and immunoassays have been developed to analyse urinary or faecal steroid hormones to determine sex but have limited applicability in measuring variations in the reproductive status of females (Lynch and Wilczynski 2005; Szymanski *et al.* 2006; Germano *et al.* 2009; Graham *et al.* 2016). Compared to most of these techniques, ultrasound is non-invasive, low-stress, and applicable to nearly all species of amphibians, and allows for rapid characterisation of the development of oocytes in real-time, as well as repeated, longitudinal assessments. Moreover, ultrasound is widely available, easy to use, portable for field work, and relatively inexpensive after initial investment (≤ £18 000/US$25 000/A$32 000).

Many early studies of reproduction utilising ultrasound only evaluated females at a single point in time (Johnson *et al.* 2002; Kouba *et al.* 2009). However, if imaging is conducted in a standardised manner, longitudinal mapping of the ovarian cycle could be used to track the growth and maturation of oocytes and the occurrence of ovulation; this mapping is not only valuable for assessing the reproductive stage of a female at a given time, but can also be applied to monitor responses to hormonal therapies.

Ultrasound grading scales

A useful practice for assessing reproductive readiness and development of oocytes in female amphibians is to build a 'grading scale' that aids in a consistent, comparative characterisation of ovarian status based on the echogenicity patterns observed on the sonogram (Plate 5). Studies have used different numbering schemes, typically a 0–3, 1–3, or 1–5 grading scale (e.g. Marcec 2016; Calatayud *et al.* 2018; Graham *et al.* 2018, Bronson and Vance 2019), in which 0 or 1 represents low development of oocytes and 3 or 5 (depending on the scale) represents highly developed oocytes. Researchers utilising a 1–3 grading scale occasionally include half-grades (i.e. grades 1.5 or 2.5 out of 3), ultimately resulting in five separate categories, some grading scales may also incorporate a category for retained eggs or regressing oocytes (Calatayud *et al.* 2018). When applying ultrasonography to a new species, the

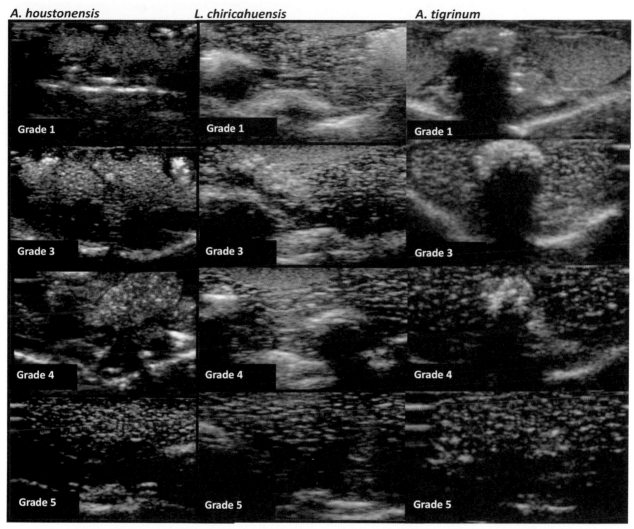

A. houstonensis   **L. chiricahuensis**   **A. tigrinum**

Figure 6.1: Sonograms of female amphibians demonstrating differences between genera based on egg size and clutch size. Left column: Houston toad (*Anaxyrus houstonensis*) 2000–5000 eggs per clutch. Centre column: Chiricahua leopard frog (*Lithobates chiricahuensis*) 100–400 eggs per clutch. Right column: tiger salamander (*Ambystoma tigrinum*) up to 1000 eggs per clutch with eggs ~1–2 mm in diameter. Ultrasound grades 1, 3, 4, and 5 are shown for each species. Images by Allison Julien, Devin Chen, Li-Dunn Chen, and Shaina Lampert, Mississippi State University, Starkville MS, USA.

grading scale may require modification by tailoring the descriptions to species-specific characteristics and clutch sizes. Despite some differences, there is relative consistency in the developmental patterns of oocytes as visualised on sonograms, thereby making the general principles of the grading scale transferable across species.

Ultrasound grades correspond to stages of oogenesis

The characteristics of each progressive ultrasound grade correspond to development of the oocytes throughout oogenesis and maturation. Low grades reflect immature oocytes in early stages of growth, whereas higher grades are representative of vitellogenic growth, maturation, or ovulated eggs. Considering that previtellogenic oocytes in the first two stages of primary growth are < 100 μm in diameter, macroscopic imaging by ultrasound is unlikely to distinguish the early phases of oogenesis and such females would be graded as 0 or 1. As females progress through the lower to mid-grades, there is increased distinction of anechoic/hyperechoic areas on the sonogram, representing continued growth of oocytes and the accompanying biochemical changes. In higher grades (e.g. 3–5), further distinction of anechoic/hyperechoic patterns potentially reflects increases in the size of the oocytes due to the accumulation of yolk during secondary growth, as well as changes in the biochemical makeup of oocytes as they undergo final development. As such, ultrasound grades 3 and 4 (out of 5) likely correspond to oocytes in mid to late stages of phase-2 growth, including vitellogenesis, distinction of pigmentation between the animal and vegetal poles, and growth of the oocytes to their maximal size. The highest ultrasound grade should correspond to fully mature and/or ovulated eggs with an increasing coating of jelly as they move through the oviduct, and present as bright, symmetric hyperechoic points against an anechoic background.

A small number of studies using ultrasound have dissected the ovaries of females at different reproductive states as validation of ultrasound grading scales (Calatayud *et al.* 2018; Graham *et al.* 2018). In a validation study using Mississippi gopher frogs, Graham *et al.* (2018) found that primarily small, unpigmented oocytes predominated in the ovaries at low ultrasound grades (1–2), but at higher grades (3–4) increases in pigmentation and size of oocytes were observed, aligning with late vitellogenic growth stages of the oocytes (Plate 5). At grade 5, fully grown and mature oocytes with distinct animal and vegetal poles were present in the highest proportion, and some ovulated eggs were present in the oviduct. Such validation studies are limited, due to the need to sacrifice individuals; however, available data support that assigned ultrasound grades are representative of physical changes to the oocytes.

Using ultrasonography to inform the application of exogenous hormones

Ultrasonography can identify females with mature oocytes that are most likely to respond to hormonal therapy, allowing practitioners to target the best candidates for assisted reproduction. Ultrasonography has successfully identified high-grade females that oviposit in response to hormonal treatment in American toads (*Anaxyrus americanus*; Kouba and Vance 2009), Mississippi gopher frogs (Graham *et al.* 2018), Puerto Rican crested toads (Burger *et al.* 2021), Houston toads (*Anaxyrus houstonensis*; C. Kouba, unpublished data), Chiricahua leopard frogs (C. Kouba, unpublished data), Asian yellow-spotted climbing toads (*Rentapia hosii*; K. Graham, unpublished data), and tiger salamanders (*Ambystoma tigrinum*; Marcec 2016).

For seasonally breeding species, ultrasound may be particularly valuable for determining whether females have sufficiently developed oocytes before treatment. For example, Calatayud *et al.* (2018) observed that, during a 3-month brumation period, the development of oocytes was variable between individual females of mountain yellow-legged frogs (*Rana muscosa*). Individuals exhibiting higher ultrasound grades upon emergence from brumation were more likely to mate and oviposit, compared to those with lower ultrasound grades. Although many amphibian species naturally breed on an annual or even biennial basis, animals housed in captive breeding colonies can, in some cases, be hormonally stimulated to spawn more than once per year, or outside of the normal breeding season (e.g. Trudeau *et al.* 2013; Guy *et al.* 2020). In these instances, ultrasound may be useful to assess whether sufficient development of oocytes has occurred to produce additional clutches of eggs, thereby potentially increasing the reproductive output of breeding programs. However, because of the high energy demands of reproduction, it is critical that animals are given appropriate time to recover between treatments (Guy *et al.* 2020).

Beyond identifying females with well-developed oocytes, imaging can be used to assess effectiveness of priming and ovulatory doses following administration of hormones. In a study with Mississippi gopher frogs, only females of ultrasound grades 3–4 (indicative of later stages of oocyte development) advanced to grade 5 and successfully oviposited following administration of hormones, whereas females initially categorised as low grade (grades 1–2) did not advance their grade after treatment and failed to spawn (Graham *et al.* 2018). Importantly, this observation suggests that during early stages of growth, the oocytes may not respond to stimulation by hormones, and is supported by Kwon *et al.* (1991), who found small follicles from *Rana nigromaculata* secreted little to no steroid hormones in response to *in vitro* exposure to gonadotropins. In tiger salamanders, Marcec (2016) similarly found females at lower ultrasound grades spawned less frequently following hormonal therapy compared to females of higher grades.

As a future direction, ultrasonography could prove useful in applying hormonal induction protocols tailored to a female's stage of oocyte development. For example, females of high ultrasound grades may only require a single dose to stimulate ovulation, whereas multiple doses (priming) over an extended time may be more effective for females at lower grades. Marcec (2016) administered tiger salamanders specific treatments based on ultrasound grade, with females of high grades (well-developed oocytes), receiving a single dose, whereas females of lower grades received a series of doses over an extended time in an effort to promote the development and maturation of oocytes. Although subsequent data from a broader experiment showed that successful ovulation was more closely related to initial ultrasound grade, rather than to the course of treatment (Marcec 2016), additional research may demonstrate feasibility of individualised hormonal regimens informed by ultrasound.

Diagnosis and treatment of egg binding

Verifying that a female has deposited a full clutch of eggs and is not egg-bound is of critical importance to maintaining the health of breeding amphibians (Bronson and Vance 2019). Retained eggs are problematic as they may impact future development of oocytes, with females retaining older eggs until the following breeding season; alternatively, calcification and reabsorption may occur, as noted in Alabama waterdogs (*Necturus alabamensis*) (Stoops *et al.* 2014). Retention of eggs also can lead to sepsis and mortality as observed in Panamanian golden frogs (Eustace *et al.* 2018; Bronson *et al.* 2021). Egg binding is most commonly diagnosed using abdominal palpation or ultrasound in fish (George *et al.* 2017) and reptiles (DeNardo 2006; Soroori *et al.* 2019), followed by removal of the septic eggs and treatment with antibiotics (George *et al.* 2017). Unfortunately, the retention of eggs in amphibians is rarely detected before the onset of sepsis and is most often determined following rectal prolapse or upon necropsy. Retained eggs may be removed surgically if detected early (Wright and Whitaker 2001), but often females have already succumbed to advanced stages of sepsis and it is too late. Alternatively, ultrasound can detect retained eggs early on, thereby providing the opportunity to perform surgery or administer hormonal therapy to treat egg binding before necrosis and sepsis sets in (Bronson *et al.* 2021). Plate 6 demonstrates detection of bound and necrotic eggs in a Chiricahua leopard frog using ultrasonography, followed by surgical removal of the eggs and subsequent recovery of the female. The practice of routine ultrasonography throughout, and immediately following, the breeding season would be beneficial in mitigating risks to the health of females from retained eggs.

Other applications for ultrasonography in assisted reproduction

Beyond monitoring the development of oocytes, sonography has been used successfully to discriminate between males and females in monomorphic or weakly dimorphic species (Li *et al.* 2010; Graham *et al.* 2016; Holtze *et al.* 2017; Ruiz-Fernández *et al.* 2020) and applied to improve synchronicity of gamete collection for artificial fertilisation (Kouba *et al.* 2012a). When females are treated with exogenous hormones, there can be a delay of hours to days before oviposition, whereas sperm can be collected from males within a few hours. Thus, ultrasonography may assist in predicting the timeframe of oviposition and, in turn, inform optimal timing for collection of sperm from males. Finally, portable ultrasound units taken into field settings could prove valuable for studying *in-situ* amphibian populations, including estimating reproductive rates or assisting with the collection of gametes to increase genetic diversity through linking *in situ* and *ex situ* populations (Burger *et al.* 2021).

CONCLUSIONS

Acquisition of viable eggs from female amphibians is critical to the success of conservation breeding programs, but the process of development, maturation, and ovulation of eggs includes many stages during which reproductive failure can occur. Exogenous hormonal therapies continue to be an essential strategy to overcome reproductive dysfunction in females. Refining existing protocols, as well as potentially exploring novel types of hormone and endocrine pathways, is important for increasing the success of hormonal therapies. When applying hormones, practitioners must consider ovarian dynamics, including the stages of oocyte development and maturation, and seasonal reproductive cycles, in order for treatments to be effective. In this regard, incorporating regular use of ultrasonography may improve the success of assisted reproductive efforts. Relative ease of use and wide availability of equipment makes ultrasonography a practical tool that should be implemented in amphibian breeding programs when possible. When ultrasonography is applied to new species, patterns are likely to emerge that can better predict the females that are most likely to respond to hormonal therapies. Standardisation of grading scales is an important step for increasing the comparability and transferability of the technology between species. Although not widely studied in a controlled manner, an exciting future application may be using ultrasonography to direct the timing of hormonal therapy for enhanced ovulation rates and improved productivity within breeding programs.

ACKNOWLEDGEMENTS

This work was supported by grants from the Institute of Library and Museum Services (IMLS) grant #MG-30–17–0052–17, the Association of Zoos and Aquariums and Disney Conservation Grant Funds #CGF-16–1396 and #CGF-19–1618, and the USDA-ARS Biophotonic Initiative #58–6402–3-018.

REFERENCES

Behr N, Rodder D (2018) Captive management, reproduction, and comparative larval development of Klappenbach's Red-bellied Frog, *Melanophryniscus klappenbachi* Prigioni and Langone. *Amphibian & Reptile Conservation* **12**, 18–26.

Bernardo J (1996) The particular maternal effect of propagule size, especially egg size: patterns, models, quality of evidence and interpretations. *American Zoologist* **36**, 216–236. doi:10.1093/icb/36.2.216

Bohun CS, Breward C (2009) Yolk dynamics in amphibian embryos. *Mathematics-in-Industry Case Studies Journal* **1**, 99–119.

Bronson E, Guy EL, Murphy KJ, Barrett K, Kouba AJ, Poole VA, Kouba CK (2021) Influence of oviposition-inducing hormone on spawning and mortality in the endangered Panamanian golden frog (*Atelopus zeteki*). *BMC Zoology* **6**. doi:10.1186/s40850-021-00076-8.

Bronson E, Vance CK (2019) Anuran reproduction. In *Fowler's Zoo Wild Animal Medicine, Current Therapy*. Volume 9. (Eds RE Miller, N Lamberski and P Calle) pp. 371–379. Elsevier Inc., St Louis MO, USA.

Browne RK, Li H, Seratt J, Kouba AJ (2006*a*) Progesterone improves the number and quality of hormone induced Fowler toad (*Bufo fowleri*) oocytes. *Reproductive Biology and Endocrinology* **4**, 3. doi:10.1186/1477-7827-4-3

Browne RK, Seratt J, Vance CK, Kouba AJ (2006*b*) Hormonal priming, induction of ovulation and *in-vitro* fertilization of the endangered Wyoming toad (*Bufo baxteri*). *Reproductive Biology and Endocrinology* **4**, 34. doi:10.1186/1477-7827-4-34

Bruce RC (1975) Reproductive biology of the mud salamander *Pseudotriton montanus*, in western South Carolina. *Copeia* **1975**, 129–137. doi:10.2307/1442416

Burger I, Julien AR, Kouba AJ, Counsell KR, Barber D, Pacheco C, Kouba CK (2021) Linking *in situ* and *ex situ* populations of the endangered Puerto Rican crested toad (*Peltophryne lemur*) through genome banking. *Conservation Science and Practice* **3**(11), e525. doi:10.1111/csp2.525.

Byrne PG, Silla AJ (2010) Hormonal induction of gamete release, and *in-vitro* fertilisation, in the critically endangered Southern Corroboree Frog, *Pseudophryne corroboree*. *Reproductive Biology and Endocrinology* **8**, 144. doi:10.1186/1477-7827-8-144

Calatayud NE, Langhorne CJ, Mullen AC, Williams CL, Smith T, Bullock L, Kouba AJ, Willard ST (2015) A hormone priming regimen and hibernation affect oviposition in the boreal toad (*Anaxyrus boreas boreas*). *Theriogenology* **84**, 600–607. doi:10.1016/j.theriogenology.2015.04.017

Calatayud NE, Stoops M, Durrant BS (2018) Ovarian control and monitoring in amphibians. *Theriogenology* **109**, 70–81. doi:10.1016/j.theriogenology.2017.12.005

Cesa R, Guastalla A, Cottone E, Mackie K, Beltramo M, Franzoni MF (2002) Relationships between CB1 cannabinoid receptors and pituitary endocrine cells in *Xenopus laevis*: an immuno-histochemical study. *General and Comparative Endocrinology* **125**, 17–24. doi:10.1006/gcen.2001.7720

Chai N (2015) Endoscopy in amphibians. *The Veterinary Clinics of North America. Exotic Animal Practice* **18**(3), 479–491. doi:10.1016/j.cvex.2015.04.006

Chai N (2017) Reproductive medicine in amphibians. *The Veterinary Clinics of North America. Exotic Animal Practice* **20**, 307–325. doi:10.1016/j.cvex.2016.11.001

Clulow J, Clulow S, Guo J, French AJ, Mahony MJ, Archer M (2012) Optimisation of an oviposition protocol employing human chorionic and pregnant mare serum gonadotropins in the Barred Frog *Mixophyes fasciolatus* (Myobatrachidae). *Reproductive Biology and Endocrinology* **10**, 60. doi:10.1186/1477-7827-10-60.

Clulow J, Trudeau VL, Kouba AJ (2014) Amphibian declines in the twenty-first century: why we need assisted reproductive technologies. *Advances in Experimental Medicine and Biology* **753**, 275–316. doi:10.1007/978-1-4939-0820-2_12

Clulow J, Pomering M, Herbert D, Upton R, Calatayud N, Clulow S, Mahoney MJ, Trudeau VL (2018) Differential success in obtaining gametes between male and female Australian temperate frogs by hormonal induction: a review. *General and Comparative Endocrinology* **265**, 141–148. doi:10.1016/j.ygcen.2018.05.032

Cottone E, Guastalla A, Mackie K, Franzoni MF (2008) Endo-cannabinoids affect the reproductive functions in teleosts and amphibians. *Molecular and Cellular Endocrinology* **286**, S41–S45. doi:10.1016/j.mce.2008.01.025

Crawshaw G (2008) Veterinary participation in Puerto Rican crested toad program. In *Zoo and Wild Animal Medicine Current Therapy*. 6th edn. (Eds ME Fowler and RE Miller) pp. 126–136. Elsevier Inc., St Louis MO, USA.

Crespi EJ, Lessig H (2004) Mothers influence offspring body size through post-oviposition maternal effects in the red-backed salamander, *Plethodon cinereus*. *Oecologia* **138**, 306–311. doi:10.1007/s00442-003-1410-5

Daniels EL, Licht P (1980) Effects of gonadotropin-releasing hormone on the levels of plasma gonadotropins (FSH and LH) in the bullfrog, *Rana catesbeiana*. *General and Comparative Endocrinology* **42**, 455–463. doi:10.1016/0016-6480(80)90211-7

Delgado MJ, Vivien-Roels B (1989) Effect of environmental temperature and photoperiod on the melatonin levels in the pineal, lateral eye, and plasma of the frog *Rana perezi*: importance of ocular melatonin. *General and Comparative Endocrinology* **75**, 46–53. doi:10.1016/0016-6480(89)90006-3

DeNardo DF (2006) Dystocias. In *Reptile Medicine and Surgery*. 2nd edn. (Ed. DR Mader) pp. 787–792. Saunders, St Louis MO, USA.

Deng J, Carbajal L, Evaul K, Rasar M, Jamnongjit M, Hammes SR (2009) Nongenomic steroid-triggered oocyte maturation: Of mice and frogs. *Steroids* **74**, 595–601. doi:10.1016/j.steroids.2008.11.010

do Nascimento NF, Silva RC, Valentin FN, Paes Md CF, De Stéfani MV, Nakaghi LSO (2015) Efficacy of buserelin acetate combined with a dopamine antagonist for spawning induction in the bullfrog (*Lithobates catesbeianus*). *Aquaculture Research* **46**, 3093–3096. doi:10.1111/are.12461

Duellman DE, Trueb L (1994) *Biology of Amphibians*. Johns Hopkins University Press, Baltimore MD, USA.

Dufour S, Weltzien FA, Sebert ME, Le Belle N, Vidal B, Vernier P, Pasqualini C (2005) Dopaminergic inhibition of reproduction in teleost fishes: ecophysiological and evolutionary implications. *Annals of the New York Academy of Sciences* **1040**, 9–21. doi:10.1196/annals.1327.002

Dziminski MA, Alford RA (2005) Patterns and fitness consequences of intra clutch variation in egg provisioning in tropical Australian frogs. *Oecologia* **146**, 98–109. doi:10.1007/s00442-005-0177-2

Eustace R, Wack A, Mangus L, Bronson E (2018) Causes of mortality in captive Panamanian golden frogs (*Atelopus zeteki*) at the Maryland Zoo in Baltimore, 2001–2013. *Journal of Zoo and Wildlife Medicine* **49**, 324–334. doi:10.1638/2016-0250.1

Exbrayat JM (2009) Oogenesis and female reproductive system in Amphibia-Gymnophiona. In *Reproduction of Amphibians*. Series: Biological Systems in Vertebrates. (Ed. M Ogielska) pp 305–342. Science Publishers, Enfield NH, USA.

Fernandez SN, Ramos I (2003) Endocrinology of reproduction. In *Reproductive Biology and Phylogeny of Anura*. Volume 2. (Ed. GM Jamieson) Science Publishers, Enfield NH, USA.

Gall JG, Wu Z, Murphy C, Gao H (2004) Structure in the amphibian germinal vesicle. *Experimental Cell Research* **296**, 28–34. doi:10.1016/j.yexcr.2004.03.017

Gentz EJ (2007) Medicine and surgery in amphibians. *Institute for Laboratory Animal Research Journal* **48**, 255–259.

George RH, Steeil J, Baine K (2017) Diagnosis and treatment of common reproductive problems in elasmobranchs. In *The Elasmobranch Husbandry Manual II: Recent Advances in the Care of Sharks, Rays and their Relatives*. (Eds M Smith, D Warmolts, D Thoney, R Hueter, M Murray and J Ezcurra) pp. 363–374. Ohio State University Printing Services, Columbus OH, USA.

Germano JM, Molinia FC, Bishop PJ, Cree A (2009) Urinary hormone analysis assists reproductive monitoring and sex identification of bell frogs (*Litoria raniformis*). *Theriogenology* **72**, 663–671. doi:10.1016/j.theriogenology.2009.04.023

Gilbert SF (2000) *Developmental Biology*. 6th edn. Sinauer Associates, Sunderland MA, USA.

Godome T, Sintondji S, Azon MTC, Tossavi CE, Ouattara NI, Fiogbe ED (2021) Artificial Reproduction and Embryogeny of the Tiger Frog *Hoplobatrachus occipitalis* (Günther 1858). *Proceedings of the Zoological Society* **74**, 43–51. doi:10.1007/s12595-020-00341-7

Gore AC (2002) *GnRH: The Master Molecule of Reproduction*. Kluwer Academic Publishers, Springer, Boston MA, USA.

Graham KM, Kouba AJ, Langhorne CJ, Marcec RM, Willard ST (2016) Biological sex identification in the endangered dusky gopher frog (*Lithobates sevosa*): a comparison of body size measurements, secondary sex characteristics, ultrasound imaging, and urinary hormone analysis methods. *Reproductive Biology and Endocrinology* **14**, 41. doi:10.1186/s12958-016-0174-9.

Graham KM, Langhorne CJ, Vance CK, Willard ST, Kouba AJ (2018) Ultrasound imaging improves hormone therapy strategies for induction of ovulation and *in vitro* fertilization in the endangered dusky gopher frog (*Lithobates sevosa*). *Conservation Physiology* **6**, coy020. doi:10.1093/conphys/coy020

Grant RA, Chadwick EA, Halliday T (2009) The lunar cycle: a cue for amphibian reproductive phenology? *Animal Behaviour* **78**, 349–357. doi:10.1016/j.anbehav.2009.05.007

Green SL, Parker J, Davis C, Bouley DM (2007) Ovarian hyperstimulation syndrome in gonadotropin treated laboratory South African clawed frogs (*Xenopus laevis*). *Journal of the American Association for Laboratory Animal Science; JAALAS* **46**, 64–67.

Grey RD, Wolf DP, Hendrick JL (1974) Formation and structure of the fertilization envelope in *Xenopus laevis*. *Developmental Biology* **54**, 52–60. doi:10.1016/0012-1606(76)90285-2

Griffiths R, Pavajeau L (2008) Captive breeding, reintroduction, and the conservation of amphibians. *Conservation Biology* **22**, 852–861. doi:10.1111/j.1523-1739.2008.00967.x

Guy EL, Martin MW, Kouba AJ, Cole JA, Kouba CK (2020) Evaluation of different temporal periods between hormone-induced ovulation attempts in the female Fowler's toad *Anaxyrus fowleri*. *Conservation Physiology* **8**, coz113. doi:10.1093/conphys/coz113

Hartel T, Sas-Kovacs I, Pernetta AP, Geltsch IC (2007) The reproductive dynamics of temperate amphibians: a review. *North-Western Journal of Zoology* **3**, 127–145.

Holtze S, Lukač M, Cizelj I, Mutschmann F, Szentiks CA, Jelić D, Hermes R, Göritz F, Braude S, Hildebrandt TB (2017) Monitoring health and reproductive status of olms (*Proteus anguinus*) by ultrasound. *PLoS One* **12**, e0182209. doi:10.1371/journal.pone.0182209

Iwata T, Toyoda F, Yamamoto K, Kikuyama S (2000) Hormonal control of urodele reproductive behavior. *Comparative Biochemistry and Physiology. Part B, Biochemistry & Molecular Biology* **126**, 221–229. doi:10.1016/S0305-0491(00)00200-5

Johnson CJ, Vance CK, Roth TL, Kouba AJ (2002) Oviposition and ultrasound monitoring of American toads (*Bufo americanus*) treated with exogenous hormones. In *Proceedings of the American Association of Zoo Veterinarians*. 5–10 October, Milwaukee, Wisconsin. (Ed. CK Baer) pp. 299–301.

Julien AR (2021) The nose glows: investigating amphibian neuroendocrine pathways with quantum dots. PhD thesis. Mississippi State University, USA.

Julien AR, Kouba AJ, Kabelik D, Feugang JM, Willard ST, Kouba CK (2019) Nasal administration of gonadotropin releasing hormone (GnRH) elicits sperm production in Fowler's toads (*Anaxyrus fowleri*). *BMC Zoology* **4**, 3. doi:10.1186/s40850-019-0040-2

Kaplan RH (1985) Maternal influences on offspring development in the California newt, *Taricha torosa*. *Copeia* **1985**, 1028–1035. doi:10.2307/1445258

King JA, Millar RP (1986) Identification of His 5, Trp 7, Tyr 8 GnRH (chicken GnRH II) in amphibian brain. *Peptides* **7**, 827–834. doi:10.1016/0196-9781(86)90102-6

Kouba AJ, Vance CK (2009) Applied reproductive technologies and genetic resource banking for amphibian conservation. *Reproduction, Fertility and Development* **21**(6), 719–737. doi:10.1071/RD09038

Kouba AJ, Vance CK, Willis EL (2009) Artificial fertilization for amphibian conservation: current knowledge and future considerations. *Theriogenology* **71**, 214–227. doi:10.1016/j.theriogenology.2008.09.055

Kouba AJ, Vance CK, Calatayud N, Rowlison T, Langhorne C, Willard ST (2012a) Assisted reproduction technologies (ART) for amphibians. In *Amphibian Husbandry Resource Guide*. 2nd edn. (Eds VA Poole and S Grow) pp. 60–118. Amphibian Taxon Advisory Group, American Association of Zoos and Aquariums, Silver Springs MD, USA.

Kouba AJ, DelBarco-Trillo J, Vance CK, Milam C, Carr M (2012b) A comparison of human chorionic gonadotropin and luteinizing hormone releasing hormone on the induction of spermiation and amplexus in the American toad (*Anaxyrus americanus*). *Reproductive Biology and Endocrinology* **10**, 59–71. doi:10.1186/1477-7827-10-59

Krause ET, von Engelhardt N, Steinfartz S, Trosien R, Caspers BA (2013) Ultrasonography as a minimally invasive method to assess pregnancy in the fire salamanders (*Salamandra salamandra*). *Salamandra (Frankfurt)* **49**, 211–214.

Kupfer A, Maxwell E, Reinhard S, Kuehnel S (2016) The evolution of parental investment in caecilian amphibians: a comparative approach. *Biological Journal of the Linnean Society. Linnean Society of London* **119**, 4–14. doi:10.1111/bij.12805

Kwon HB, Choi HH, Ahn RS, Yoon YD (1991) Steroid production by amphibian (*Rana nigromaculata*) ovarian follicles at different developmental stages. *The Journal of Experimental Zoology* **260**, 66–73. doi:10.1002/jez.1402600109

Li PQ, Zhu BC, Wang YF, Xiang XJ (2010) Sex identification of Chinese giant salamander (*Andrias davidianus*) by Doppler B-ultrasound method. *Journal of Biology* **27**, 94–96.

Licht P, Tsai PS, Sotowska-Brochocka J (1994) The nature and distribution of gonadotropin-releasing hormones in brains and plasma of ranid frogs. *General and Comparative Endocrinology* **94**, 186–198. doi:10.1006/gcen.1994.1075

Lofts B (1984) Amphibians. In *Marshall's Physiology of Reproduction*. Volume 1. (Ed. GE Lamming) pp. 127–205. Churchill Livingston, Edinburgh, UK.

Lynch KS, Wilczynski W (2005) Gonadal steroids vary with reproductive stage in a tropically breeding female anuran. *General and Comparative Endocrinology* **143**, 51–56. doi:10.1016/j.ygcen.2005.02.023

Mannion P (2008) *Diagnostic Ultrasound in Small Animal Practice*. Blackwell Science, Oxford, UK.

Mansour N, Lahnsteiner F, Patzner RA (2011) Collection of gametes from live axolotl, *Ambystoma mexicanum*, and standardization of *in vitro* fertilization. *Theriogenology* **75**, 354–361. doi:10.1016/j.theriogenology.2010.09.006

Marcec R (2016) *Development of Assisted Reproductive Technologies for Endangered North American Salamanders*. Mississippi State University, Starkville MS, USA.

Mc Creery BR, Licht P (1983) Induced ovulation and changes in pituitary responsiveness to continuous infusion of Gonadotropin-releasing hormone during the ovarian cycle in the Bullfrog, *Rana catesbiana*. *Biology of Reproduction* **29**, 863–871. doi:10.1095/biolreprod29.4.863

Michael SF, Buckley C, Toro E, Estrada AR, Vincent S (2004) Induced ovulation and egg deposition in the direct developing anuran *Eleutherodactylus coqui*. *Reproductive Biology and Endocrinology* **2**, 6. doi:10.1186/1477-7827-2-6

Morrill GA, Schatz F, Kostellow A, Blocj E (2006) Gonadotropin stimulation of steroid synthesis and metabolism in the *Rana pipiens* ovarian follicle: sequential changes in endogenous steroids during ovulation, fertilization and cleavage stages. *The Journal of Steroid Biochemistry and Molecular Biology* **99**, 129–138. doi:10.1016/j.jsbmb.2006.01.008

Muths E, Scherer RD, Lambert BA (2010) Unbiased survival estimates and evidence for skipped breeding opportunities in females. *Methods in Ecology and Evolution* **1**, 123–130. doi:10.1111/j.2041-210X.2010.00019.x

Muths E, Scherer RD, Bosch J (2013) Evidence for plasticity in the frequency of skipped breeding opportunities in common toads. *Population Ecology* **55**, 535–544. doi:10.1007/s10144-013-0381-6

Narayan E, Hero JM (2011) Urinary corticosterone responses and haematological stress indicators in the endangered Fijian ground frog (*Platymantis vitiana*) during transportation and captivity. *Australian Journal of Zoology* **59**, 79–85. doi:10.1071/ZO11030

Narayan EJ, Molinia FC, Christi KS, Morley CG, Cockrem JF (2010) Annual cycles of urinary reproductive steroid concentrations in wild and captive endangered Fijian ground frogs (*Platymantis vitiana*). *General and Comparative Endocrinology* **166**, 172–179. doi:10.1016/j.ygcen.2009.10.003

Norris DO (2007) Reproduction in amphibians. In *Vertebrate Endocrinology*. 4th edn. pp. 371–428. Elsevier Academic Press, San Diego CA, USA.

Norris DO (2011) Hormones and reproductive patterns in urodele and gymnophionid amphibians. In *Hormones and Reproduction of Vertebrates*. Volume 2: Amphibians (Eds DO Norris and KH Lopez) pp. 187–202. Elsevier Inc. London, UK.

Norris DO, Lopez KH (2005) Anatomy of the Amphibian endocrine system. In *Amphibian Biology*. Volume 6: Endocrinology. (Ed. H Heatwole) pp. 2021–2044. Surrey Beatty and Sons, Chipping Norton.

Obringer AR, O'Brien JK, Saunders RL, Yamamoto K, Kikuyama S, Roth TL (2000) Characterization of the spermiation response, luteinizing hormone release and sperm quality in the American toad (*Bufo americanus*) and the endangered Wyoming toad (*Bufo baxteri*). *Reproduction, Fertility and Development* **12**, 51–58. doi:10.1071/RD00056

Ogawa A, Dake J, Iwashina YK, Tokumoto T (2011) Induction of ovulation in *Xenopus* without hCG injection: the effect of adding steroids into the aquatic environment. *Reproductive Biology and Endocrinology* **9**, 11. doi:10.1186/1477-7827-9-11

Ogielska M, Bartmanska J (2009) Oogenesis and female reproductive system in Amphibia-Anura. In *Reproduction of Amphibians*. Series: Biological Systems in Vertebrates. (Ed. M Ogielska) pp. 153–272. Science Publishers, Enfield NH, USA.

Oguchi A, Tanaka S, Aida T, Yamamoto K, Kikuyama S (1997) Enhancement by prolactin o the GnRH-induced release of LH from dispersed anterior pituitary cells of the bullfrog (*Rana catesbeina*). *General and Comparative Endocrinology* **107**, 128–135. doi:10.1006/gcen.1997.6904

Oterino J, Sanchez-Toranzo G, Zelarayan L, Ajmat MT, Bonilla F, Buhler MI (2006) Behaviour of the vitelline envelope in *Bufo arenarum* oocytes matured *in vitro* in blockade to polyspermy. *Zygote (Cambridge, England)* **14**, 97–106. doi:10.1017/S0967199406003662

Pagotto U, Marsicano G, Cota D, Lutz B, Pasquali R (2006) The emerging role of the endocannabinoid system in endocrine regulation and energy balance. *Endocrine Reviews* **27**, 73–100. doi:10.1210/er.2005-0009

Polzonetti-Magni AM, Mosconi G, Carnevali O, Yamamoto K, Hanaoka Y, Kikuyama S (1998) Gonadotropins and reproductive function in the anuran amphibian, *Rana esculenta*. *Biology of Reproduction* **58**, 88–93. doi:10.1095/biolreprod58.1.88

Polzonetti-Magni AM, Mosconi G, Soverchia L, Kikuyama S, Carnevali O (2004) Multihormonal control of vitellogenesis in lower vertebrates. *International Review of Cytology* **239**, 1–46. doi:10.1016/S0074-7696(04)39001-7

Popesku JT, Mennigen JA, Chang JP, Trudeau VL (2011) Dopamine D1 receptor blockage potentiates AMPA-stimulated luteinizing hormone release in the goldfish. *Journal of Neuroendocrinology* **23**, 302–309. doi:10.1111/j.1365-2826.2011.02114.x

Reyer HU, Bättig I (2004) Identification of reproductive status in female frogs – a quantitative comparison of nine methods. *Herpetologica* **60**, 349–357. doi:10.1655/03-77

Reyer HU, Frei G, Som C (1999) Cryptic female choice: frogs reduce clutch size when amplexed by undesired males. *Proceedings of the Royal Society of London* **266**, 2101–2107. doi:10.1098/rspb.1999.0894

Rorabaugh JC, Sredl MJ (2014) Herpetofauna of the 100-mile circle: Chiricahua leopard frog (*Lithobates chiricahuensis*). *Sonoran Herpetologist* **27**, 61–70.

Roth TL, Szymanski DC, Keyster ED (2010) Effects of age, weight, hormones, and hibernation on breeding success in boreal toads (*Bufo boreas boreas*). *Theriogenology* **73**, 501–511. doi:10.1016/j.theriogenology.2009.09.033

Rowlison T (2013) *The Comparative Effects of Arginine Vasotocin on Reproduction in the* Bufo boreas boreas *and* Bufo fowleri *toad*. Mississippi State University, Starkville MS, USA.

Ruiz-Fernández MJ, Jiménez S, Fernández-Valle E, García-Real MI, Castejón D, Moreno N, González-Soriano J (2020) Sex determination in two species of anuran amphibians by magnetic resonance imaging and ultrasound techniques. *Animals (Basel)* **10**, 2142. doi:10.3390/ani10112142

Santana F, Swaisgood RR, Lemm JM, Fisher R (2015) Chilled frogs are hot: hibernation and reproduction of the Endangered mountain yellow-legged frog *Rana mucosa*. *Endangered Species Research* **27**, 43–51. doi:10.3354/esr00648

Sato KI, Tokmakov AA (2020) Toward the understanding of biology of oocyte life cycle in *Xenopus laevis*: no oocytes left behind. *Reproductive Medicine and Biology* **19**, 114–119. doi:10.1002/rmb2.12314

Schildger B, Triet H (2001) Ultrasonography in amphibians. *Seminars in Avian Exotic Pet Medicine* **10**, 169–173. doi:10.1053/saep.2001.24673

Semlitsch RD, Gibbons JW (1990) Effects of egg size on success of larval salamanders in complex aquatic environments. *Ecology* **71**, 1789–1795. doi:10.2307/1937586

Seong JY, Wang L, Oh DY, Yun O, Maiti K, Li JH, Soh JM, Choi HS, Kim K, Vaudry H (2003) Ala/Thr201 in extracellular loop 2 and Leu/Phe290 in transmembrane domain 6 type 1 frog gonadotropin-releasing hormone receptor confer differential ligand sensitivity and signal transduction. *Endocrinology* **144**, 454–466. doi:10.1210/en.2002-220683

Sever DM, Rania LC, Krunz JD (1996) Reproduction of the salamander *Siren intermedia* LeConte with especial reference to oviductal anatomy and mode of fertilisation. *Journal of Morphology* **227**, 335–348. doi:10.1002/(SICI)1097-4687(199603)227:3<335::AID-JMOR5>3.0.CO;2-4

Sharon R, Degani G, Warburg MR (1997) Oogenesis and the ovarian cycle in *Salamandra salamandra infraimmaculata* Mertens (Amphibia; Urodela; Salamandridae) in fringe areas of the taxon's distribution. *Journal of Morphology* **231**, 149–160. doi:10.1002/(SICI)1097-4687(199702)231:2<149::AID-JMOR4>3.0.CO;2-9

Silla AJ (2011) Effect of multiple priming injections of luteinizing hormone-releasing hormone on spermiation and ovulation in Günther's Toadlet, *Pseudophryne guentheri*. *Reproductive Biology and Endocrinology* **9**, 68. doi:10.1186/1477-7827-9-68

Silla AJ, Byrne PG (2019) The role of reproductive technologies in amphibian conservation breeding programs. *Annual Review of Animal Biosciences* **7**, 499–519. doi:10.1146/annurev-animal-020518-115056

Silla AJ, Byrne PG (2021) Hormone-induced ovulation and artificial fertilisation in four terrestrial-breeding anurans. *Reproduction, Fertility and Development* **33**, 615–618. doi:10.1071/RD20243

Silla AJ, McFadden M, Byrne PG (2018) Hormone-induced spawning of the critically endangered northern corroboree frog *Pseudophryne pengilleyi*. *Reproduction, Fertility and Development* **30**, 1352–1358. doi:10.1071/RD18011

Silla AJ, McFadden M, Byrne PG (2019) Hormone-induced sperm-release in the critically endangered Booroolong frog (*Litoria booroolongensis*): effects of gonadotropin-releasing hormone and human chorionic gonadotropin. *Conservation Physiology* **7**, coy080. doi:10.1093/conphys/coy080

Silla AJ, Roberts JD (2012) Investigating patterns in the spermiation response of eight Australian frogs administered human chorionic gonadotropin (hCG) and luteinizing hormone-releasing hormone (LHRHa). *General and Comparative Endocrinology* **179**, 128–136. doi:10.1016/j.ygcen.2012.08.009

Snyder WE, Trudeau VL, Loskutoff NM (2012) A noninvasive, transdermal absorption approach for exogenous hormone induction of spawning in the northern cricket frog, *Acris crepitans*: a model for small, endangered amphibians. *Reproduction Fertility and Development* **25**, 232–233. doi:10.1071/RDv25n1Ab168

Soderstrom K, Leid M, Moore FL, Murray TF (2000) Behavioral, pharmacological, and molecular characterization of an amphibian cannabinoid receptor. *Journal of Neurochemistry* **75**, 413–423. doi:10.1046/j.1471-4159.2000.0750413.x

Soroori S, Molazem M, Rostami A, Nejad MRE, Nejad MK (2019) Radiographic and ultrasonographic evaluation of egg in healthy and egg-bound green iguana. *Journal of Veterinary Research* **73**, Pe491–Pe498.

Sotowska-Brochocka J, Martyńska L, Licht P (1994) Dopaminergic inhibition of gonadotropic release in hibernating frogs, *Rana temporaria*. *General and Comparative Endocrinology* **93**, 192–196. doi:10.1006/gcen.1994.1022

Sretarugsa P, Weerachatyanukul W, Chavedej J, Kruatrachue M, Sobhon P (2001) Classification of developing oocytes, ovarian developments and seasonal variation in *Rana tigrina*. *Science Asia* **27**, 1–14. doi:10.2306/scienceasia1513-1874.2001.27.001

Stamper DL, Licht P (1990) Effect of gonadotropin-releasing hormone on gonadotropin biosynthesis in pituitaries of the frog, *Rana pipiens*. *Biology of Reproduction* **43**, 420–426. doi:10.1095/biolreprod43.3.420

Stamper DL, Licht P (1994) Influence of androgen on the GnRH-stimulated secretion and biosynthesis of gonadotropins in the pituitary of juvenile female bullfrogs, *Rana catesbeiana*. *General and Comparative Endocrinology* **93**, 93–102. doi:10.1006/gcen.1994.1011

Stetter M (2001) Diagnostic imaging of amphibians. In *Amphibian Medicine and Captive Husbandry* (Eds KM Wright and BR Whitaker) pp. 253–272. Krieger Publishing Company, Malabar FL, USA.

Stoops MA, Campbell MK, Dechant CJ (2014) Successful captive breeding of *Necturus beyeri* through manipulation of environmental cues and exogenous hormone administration: a model for endangered *Necturus*. *Herpetological Review* **45**, 251–256.

Szymanski DC, Gist DH, Roth TL (2006) Anuran gender identification by fecal steroid analysis. *Zoo Biology* **25**, 35–46. doi:10.1002/zoo.20077

Takahashi S, Nakazawa H, Watanabe A, Onitake K (2006) The outermost layer of egg-jelly is crucial to successful fertilization in the newt, *Cynops pyrrhogster*. *The Journal of Experimental Zoology* **305A**, 1010–1017. doi:10.1002/jez.a.295

Titon SCM, Assis VR, Titon JB, Cassettari BO, Fernandes PACM, Gomes FR (2017) Captivity effects on immune response and steroid plasma levels of a Brazilian toad (*Rhinella schneideri*). *Journal of Experimental Zoology. Part A, Ecological and Integrative Physiology* **327**, 127–138. doi:10.1002/jez.2078

Titon SCM, Junior BT, Assis VR, Kinker GS, Fernandes PACM, Gomes FR (2018) Interplay among steroids, body condition and immunity in response to long-term captivity in toads. *Scientific Reports* **8**, 17168. doi:10.1038/s41598-018-35495-0

Toro E, Michael SF (2004) *In vitro* fertilization and artificial activation of eggs of the direct-developing anuran *Eleutherodactylus coqui*. *Reproductive Biology and Endocrinology* **2**, 60. doi:10.1186/1477-7827-2-60.

Toyoda F, Matasuda K, Yamamoto K, Kikuyama S (1996) Involvement of endogenous prolactin in the expression of courtship behavior in the newt, *Cynops pyrrhogaster*. *General and Comparative Endocrinology* **102**, 191–196. doi:10.1006/gcen.1996.0060

Trottier TM, Armstrong JB (1975) Hormonal stimulation as an aid to artificial insemination in *Ambystoma mexicanum*. *Canadian Journal of Zoology* **53**, 171–173. doi:10.1139/z75-021

Trudeau VL (2015) Kiss and tell: deletion of kisspeptins and receptors reveal surprising results. *Endocrinology* **156**, 769–771. doi:10.1210/en.2015-1019

Trudeau VL, Somoza GM, Natale GS, Pauli B, Wignall J, Jackman P, Doe K, Schueler FW (2010) Hormonal induction of spawning in 4 species of frogs by co-injection with a gonadotropin-releasing hormone agonist and a dopamine antagonist. *Reproductive Biology and Endocrinology* **8**, 36. doi:10.1186/1477-7827-8-36.

Trudeau VL, Schueler FW, Navarro-Martin L, Hamilton CK, Bulaeva E, Bennett A, Fletcher W, Taylor L (2013) Efficient induction of spawning of northern leopard frogs (*Lithobates pipiens*) during and outside the natural breeding season. *Reproductive Biology and Endocrinology* **11**, 14. doi:10.1186/1477-7827-11-14.

Tsai PS, Kessler AE, Jones JT, Wahr KB (2005) Alteration of the hypothalamic-pituitary-gonadal axis in estrogen-and androgen-treated adult male leopard frog, *Rana pipiens*. *Reproductive Biology and Endocrinology* **3**, 2. doi:10.1186/1477-7827-3-2.

Uribe MCA (2009) Oogenesis and female reproductive system in Amphibia-Urodela. In *Reproduction of Amphibians*. Series: Biological Systems in Vertebrates. (Ed M Ogielska) pp. 273–304. Science Publishers, Enfield NH, USA.

Uribe MCA (2011) Hormones and the Female Reproductive System of Amphibians. In *Hormones and Reproduction of Vertebrates*. Volume 2: Amphibians. (Eds DO Norris and KH Lopez) pp. 55–85. Elsevier Inc., London, UK.

Uteshev VK, Gakhova EN, Kramarova LI, Shishova NV, Kaurova SA, Browne RK (2018) Refrigerated storage of European common frog *Rana temporaria* oocytes. *Cryobiology* **83**, 56–59. doi:10.1016/j.cryobiol.2018.06.004

Vu M, Trudeau VL (2016) Neuroendocrine control of spawning in amphibians and its practical applications. *General and Comparative Endocrinology* **234**, 28–39. doi:10.1016/j.ygcen.2016.03.024

Vu M, Weiler B, Trudeau VL (2017) Time- and dose-related effects of a gonadotropin-releasing hormone agonist and dopamine antagonist on reproduction in the northern leopard frog (*Lithobates pipiens*). *General and Comparative Endocrinology* **254**, 86–96. doi:10.1016/j.ygcen.2017.09.023

Wake MH (2015) Fetal adaptations for viviparity in amphibians. *Journal of Morphology* **276**, 941–960. doi:10.1002/jmor.20271

Whittier JM, Crews D (1987) Seasonal reproduction: patterns and control. In *Hormones and Reproduction in Fishes, Amphibians and Reptiles*. (Eds DO Norris and RE Jones) pp. 385–409. Plenum Press, New York NY, USA.

Wlizla M, Falco R, Peshkin L, Parlow AF, Horb ME (2017) Luteinizing hormone is an effective replacement for hCG to induce ovulation in Xenopus. *Developmental Biology* **426**, 442–448. doi:10.1016/j.ydbio.2016.05.028

Wright KM, Whitaker BR (2001) *Amphibian Medicine and Captive Husbandry*. Krieger Publishing Company, Malabar FL, USA.

Zohar Y, Mylonas CC (2001) Endocrine manipulations of spawning in cultured fish: from hormones to genes. In *Reproductive Biotechnology in Finfish Aquaculture* (Eds CS Lee and EM Donaldson) pp. 99–136. Elsevier Inc., Amsterdam, The Netherlands.

7   Protocols for hormonally induced spermiation, and the cold storage, activation, and assessment of amphibian sperm

Aimee J. Silla and Cecilia J. Langhorne

INTRODUCTION

A variety of proactive conservation initiatives are being implemented in an attempt to arrest the alarming rate of global amphibian decline. *In situ* conservation strategies aim to conserve species in their native habitats while threats are being ameliorated. Amphibian declines often are driven by several deleterious factors operating simultaneously (Beebee and Griffiths 2005) and *in situ* efforts alone may not be sufficient to ensure the maintenance or recovery of genetically viable populations (Chapter 2). Captive survival-assurance colonies aim to maintain genetically representative populations *ex situ* and are most beneficial to the recovery of species when providing high numbers of genetically diverse individuals for release *in situ* as part of an integrated approach to managing conservation (Pritchard *et al.* 2012). Reproductive technologies can assist integrated conservation breeding programs to achieve propagation targets and to manage genetic diversity (Silla and Byrne 2019). In recent decades, amphibian reproductive technologies have been successfully utilised to overcome reproductive failure in captivity (Kouba *et al.* 2009; Silla and Byrne 2019); enhance the genetic diversity and adaptive potential of amphibians bred in captivity (Silla *et al.* 2018; Silla and Byrne 2019); investigate potential genetic incompatibilities between individuals from different source populations (Byrne and Silla 2020); and capture genetic diversity and prolong the reproductive lifespan of individual's through biobanking of biological material (Kouba *et al.* 2013; Clulow and Clulow 2016; Della Togna *et al.* 2020).

Fundamental to the application of many reproductive technologies is the efficient collection of viable suspensions of sperm. Once effective protocols for collection have been established for a species, suspensions of sperm can be collected from both captive and wild populations and, through effective storage techniques, transported between captive facilities to create a metapopulation aided by artificial fertilisation. The potential for reproductive technologies to assist amphibian conservation breeding programs is considerable and, with recent advances, the field is poised to make substantial contributions to species' recovery. This review discusses the state of current technologies to collect samples of sperm via hormonal induction, effectively hold these suspensions in cold storage, activate the motility of sperm, and assess the quality of sperm suspensions. The information presented in this chapter is seminal to the implementation of artificial fertilisation (also referred to as *in vitro* fertilisation) and sperm cryopreservation technologies reviewed elsewhere (Kouba *et al.* 2013; Clulow and Clulow 2016; Silla and Byrne 2019) (Chapters 8 and 9), and vital to the integration of *in situ* and *ex situ* conservation (Chapters 1 and 11).

HORMONAL INDUCTION OF SPERMIATION

The first routine use of the administration of exogenous hormones to induce the release of sperm (spermiation) in amphibians dates back several decades when Argentine toads (*Rhinella* [*Bufo*] *arenarum*) and northern leopard frogs (*Lithobates* [*Rana*] *pipiens*) were employed to diagnose pregnancy in humans (Bruehl 1952). Since then, the collection of amphibian sperm has facilitated several scientific discoveries in the fields of evolutionary biology, biotechnology, and physiology. However, despite a long history of the hormonal induction of spermiation in amphibians, the development of protocols beyond a small number of model species has only been achieved in recent decades (Silla and Byrne 2019). Research developing reproductive technologies for amphibians has increased substantially since the extent and severity of amphibian declines were formally recognised (Stuart *et al.* 2004) and the first Amphibian Conservation Action Plan (ACAP) released (Gascon *et al.* 2007). The application of hormonal therapy to induce spermiation is arguably the most successful and frequently employed reproductive technology in amphibian conservation breeding programs. Successful protocols have been refined for a diversity of species (Table 7.1) and provide a valuable foundation for the application of hormonal therapies to novel species. Importantly, protocols are species-specific and it is essential that trials are conducted to determine the optimal hormone and dose to administer, and the timing of the peak release of sperm for each target species.

Hormones and doses

Gonadotropin releasing-hormone analogue and human chorionic gonadotropin

The two most commonly employed hormones to stimulate the release of sperm in amphibians are the synthetic gonadotropin releasing-hormone analogue (GnRHa, also known as luteinizing-hormone-releasing-hormone [LHRHa]) and purified human chorionic gonadotropin (hCG). These hormones are widely available, typically as a lyophilised powder that is reconstituted in sterile saline, and can be administered either as a standardised dose or an adjusted one based on individual bodyweight (Table 7.1). The induction of spermiation has been achieved in a wide range of amphibian species through exogenous administration of hCG and GnRHa, either alone or in combination (Table 7.1). Both GnRHa and hCG act by manipulating the neuroendocrine pathways associated with the hypothalamic-pituitary-gonadal (HPG) axis that controls reproduction in amphibians (Chapter 4). The administration of GnRHa mimics the bioactivity of natural GnRH-1 molecules, operating at the level of the pituitary, to stimulate the endogenous synthesis and release of luteinizing hormone (LH) and follicle stimulating hormone (FSH) (reviewed by Clulow *et al.* 2014; Silla *et al.* 2021; Chapter 4). Purified hCG is an agonist of LH, targeting the gonads and attempting to mimic the natural endogenous surge in LH molecules responsible for the maturation and release of sperm (reviewed by Clulow *et al.* 2014; Silla *et al.* 2021; Chapter 4). Exogenous GnRHa is more commonly used to induce spermiation in a diversity of amphibian species. By comparison, hCG is required in higher concentrations and interspecific variability in the spermiation responses to hCG among amphibian species is high. While there are several species in which the spermiation response to hCG is suboptimal, there are several other species in which spermiation has been achieved successfully in response to the administration of this hormone (Table 7.1). It has been hypothesised that the high variability in spermiation responses to hCG is due to species-specific differences in gonadal LH-receptor affinities (Clulow *et al.* 2014; Silla *et al.* 2021), which may reflect the divergent evolution of LH receptors among amphibian families (Silla and Roberts 2012). Insufficient data currently exist directly comparing the spermiation response of amphibians from different lineages to the administration of GnRHa and hCG, at a range of doses, to determine whether responses are genus-specific or family-specific.

Central to the development of protocols to induce spermiation for novel species is the establishment of dose–response curves for GnRHa and hCG, and verification of peak collection times post-administration (discussed further below). If suboptimal or supraoptimal doses of hormone are administered, release of gametes will either not occur, or sperm will be released in lower quantities and with reduced viability and capacity for fertilisation (Silla and Byrne 2019; Silla *et al.* 2019). Importantly, optimal doses required to induce spermiation, and peak time for collecting sperm post-administration, have been shown to vary enormously among amphibian species (Table 7.1). Failure to quantify these appropriately may lead to incorrect conclusions when comparing the effectiveness of protocols to induce spermiation.

Table 7.1:   Examples of successful hormone-administration protocols used to induce spermiation in a diversity of amphibian species over the past two decades.

Species	Common name	Status	Hormone	Route	Dose	Response	Reference
Agalychnis callidryas	Red-eyed treefrog	LC	GnRHa	IP	4 µg/g GnRHa	95% (19/20)	Jacobs *et al.* (2016)
Ambystoma mexicanum	Mexican axolotl	CR	hCG	IM	200IU hCG	100% (10/10)	Mansour *et al.* (2011)
Ambystoma tigrinum	Tiger salamander	LC	GnRHa	IM	0.025 µg/g 'priming' + 0.1 µg/g GnRHa	73% (11/15)	Marcec (2016)
Anaxyrus americanus	American toad	LC	hCG	IP	300IU hCG	100% (16/16)	Kouba *et al.* (2012)
			GnRHa	T	100 µg/g GnRHa	75% (12/16)	Rowson *et al.* (2001)
Anaxyrus baxteri	Wyoming toad	EW	hCG	IP	300 IU hCG	80% (8/10)	Browne *et al.* (2006)
			GnRHa	IP	0.2 µg/g GnRHa	100% (30/30)	Poo *et al.* (2019)
Anaxyrus fowleri	Fowler's toad	LC	hCG	IP	300 IU hCG	95% (19/20)*	McDonough *et al.* (2016)
			GnRHa	N	10 µg GnRHa	93% (14/15)	Julien *et al.* (2019)
Atelopus zeteki	Panamanian golden toad	CR	GnRHa MET	IP	0.4 µg/g GnRHa + 10 µg/g MET	100% (24/24)	Della Togna *et al.* (2017)
Bufo lemur	Puerto Rican crested toad	EN	hCG	IP	4 IU/g hCG	100% (4/4)	Langhorne (2016)
Crinia glauerti	Glauert's froglet	LC	GnRHa	SC/DLS	2 µg/g GnRHa	100% (6/6)	Silla and Roberts (2012)
Crinia georgiana	Quacking frog	LC	GnRHa	SC/DLS	2 µg/g GnRHa	100% (8/8)	Silla and Roberts (2012)
Crinia pseudinsignifera	False western froglet	LC	GnRHa	SC/DLS	2 µg/g GnRHa	100% (8/8)	Silla and Roberts (2012)
Dendrobates auratus	Green poison frog	LC	hCG	DLS	100IU hCG 'priming' + 100IU hCG	100% (6/6)	Lipke (2008)
Epidalea calamita	Natterjack toad	LC	hCG	IP	10 IU/g hCG	100% (5/5)	Arregui *et al.* (2019)
Geocrinia rosea	Roseate frog	LC	GnRHa	SC/DLS	1 µg GnRHa	100% (10/10)	Silla *et al.* (2020)
			GnRHa	T	100 µg GnRHa	100% (9/9)	Silla *et al.* (2020)
Heleioporus albopunctatus	Western spotted frog	LC	hCG	SC/DLS	13 IU/g hCG	100% (8/8)	Silla and Roberts (2012)
Heleioporus eyrei	Moaning frog	LC	hCG	SC/DLS	13 IU/g hCG	100% (9/9)	Silla and Roberts (2012)
Lithobates catesbeianus	American bullfrog	LC	GnRHa	CC	0.4 µg GnRHa	100% (5/5)	Pereira *et al.* (2013)
Lithobates sevosus	Dusky gopher frog	CR	GnRHa hCG	IP	15 µg GnRHa + 500IU hCG	100% (17/17)	Langhorne (2016)
Litoria booroolongensis	Booroolong frog	CR	hCG	SC/DLS	40 IU/g hCG	100% (20/20)	Silla *et al.* (2019)
Litoria caerulea	Green tree frog	LC	hCG	SC/DLS	300IU hCG	100% (7/7)	Clulow *et al.* (2018)
Neobatrachus pelobatoides	Humming frog	LC	hCG	SC/DLS	13 IU/g hCG	100% (9/9)	Silla and Roberts (2012)

Species	Common name	Status	Hormone	Route	Dose	Response	Reference
Notophthalmus meridionalis	Black-spotted newt	EN	GnRHa	IM	0.025 µg/g 'priming' + 0.1 µg/g GnRHa	100% (10/10)	Guy *et al.* (2020)
Pelophylax lessonae	Pool frog	LC	GnRHa	IP	0.5 µg/g GnRHa	100% (9/9)	Uteshev *et al.* (2013)
Pleurodeles waltl	Sharp-ribbed newt	NT	GnRHa	IM	50 µg GnRHa	100% (4/4)	Uteshev *et al.* (2015)
Pseudophryne corroboree	Southern corroboree frog	CR	GnRHa	SC/DLS	5 µg/g GnRHa	82% (9/11)	Byrne and Silla (2010)
			GnRHa	SC/DLS	2 µg/g GnRHa	88% (7/8)ᵛ	A. J. Silla, unpublished data
Pseudophryne guentheri	Günther's toadlet	LC	GnRHa	SC/DLS	2 µg/g GnRHa	100% (18/18)	Silla (2010; 2011)
Pseudophryne pengilleyi	Northern corroboree frog	EN	GnRHa	SC/DLS	1 µg/g GnRHa	100% (5/5)	A. J. Silla, unpublished data
Rana arvalis	Moor frog	LC	GnRHa	SC/DLS	0.75 µg/g GnRHa	82% (79/96)**	Sherman *et al.* (2010)
Rana temporaria	European common frog	LC	GnRHa	IP	1.2 µg/g GnRHa	100% (4/4)	Uteshev *et al.* (2012)
Rhaebo guttatus	Smooth-sided toad	LC	hCG	IP	10 IU/g hCG	100% (5/5)	Hinkson *et al.* (2019)
Rhinella marina	Cane toad	LC	hCG	IP	2000 IU hCG	86% (12/14)^	Iimori *et al.* (2005)
Tylototriton kweichowensis	Kweichow newt	VU	GnRHa	IM	0.025 µg/g 'priming' + 0.1 µg/g GnRHa	67% (4/6)	Guy *et al.* (2020)

Abbreviations: Status (according to the IUCN Red List): DD – data deficient; LC – least concern; NT – near threatened; VU – vulnerable; EN – endangered; CR – critically endangered; EW – extinct in the wild. **Hormone:** E – oestradiol; GnRHaª – gonadotropin –releasing hormone agonist; hCG – human chorionic gonadotropin; MET – metoclopramide; PMSG – pregnant mare's serum; P4 – progesterone. **Route:** CC – coelomic cavity; DLS – dorsal lymph sac; IM – intramuscular; IP – intraperitoneal; N – nasal; SC – subcutaneous; T – topical/epicutaneous.
For species in which a taxonomic revision has occurred, contemporary nomenclature is displayed.
Note: Several forms of synthetic GnRH agonists exist with differing biological potencies, original research articles should be consulted for specific details. * Calculated as the total number of toads spermiating in response to all treatments after the initial administration of hormone (week 0). ᵛ A. J. Silla unpublished data showed 2 µg/g GnRHa induced the release of higher sperm concentrations compared to those in response to 5 µg/g GnRHa reported by Byrne and Silla (2010). ** Response rates from C. D. Sherman, unpublished data. ^ Total number of toads responding (*n* = 6/treatment) not reported; data provided are number of spermic urine samples recovered/total urine samples collected.

Dopamine antagonists

In addition to the administration of GnRHa and hCG, there is growing interest in the combined administration of GnRHa with a dopamine antagonist (e.g. domperidone, pimozide, or metoclopramide). Dopaminergic inhibition of LH in fish has been described (Zohar and Mylonas 2001; Dufour *et al.* 2005) and recent research indicates that a similar mechanism might exist in amphibians (Vu *et al.* 2017). Essentially, dopamine is thought to inhibit the effects of LH when stimulating the maturation and release of gametes. Co-injection with a dopamine-antagonist has therefore been suggested to enhance the stimulatory effects of GnRHa administration (Zohar and Mylonas 2001; Vu and Trudeau 2016). While the combined administration of GnRHa with metaclopromide (a protocol also referred to as AMPHIPLEX) has been used successfully to induce spawning in several amphibian species (Trudeau *et al.* 2010; Trudeau *et al.* 2013), the effectiveness of this protocol remains equivocal. The rates of spawning success of neither the northern leopard frog (*Lithobates pipiens*) nor the American bullfrog (*Lithobates catesbeianus*) are enhanced by the combined administration of GnRHa and dopamine antagonists, with the administration of GnRHa alone at optimal doses reported to be equally effective (do Nascimento *et al.* 2015; Vu *et al.* 2017). To date, only a single study of an amphibian species (Panamanian golden toad, *Atelopus zeteki*), has reported a benefit of the administration of GnRHa in combination with a dopamine antagonist for inducing spermiation (Della Togna *et al.* 2017), thus highlighting a clear need for further research.

Priming hormones

The administration of a low dose of priming hormone before a higher resolving dose is a common protocol used to prime the ovary and promote ovulation in female amphibians (Browne *et al.* 2006; Silla 2011; Silla and Byrne 2021). In male anurans, spermiation is frequently achieved in response to a single application of hormone (Table 7.1), which can stimulate sperm release for up to 24 h post-administration (Roth and Obringer 2003; Kouba and Vance 2009). Research on priming hormones in males is limited and equivocal. It has been reported in male Günther's toadlets (*Pseudophryne guentheri*) that priming doses administered at short intervals (1–2 h) have a negative effect on spermiation (Silla 2011). In contrast, research on the green poison frog (*Dendrobates auratus*) reported an improvement in the spermiation response of males administered two doses of hCG 1 h apart, compared to males receiving a single hCG dose (Lipke 2008). In the common toad (*Bufo bufo*) the administration of a priming dose of GnRHa 24 h before a resolving dose of hCG led to an increase in the concentration of sperm expelled but the difference was not significant (Uteshev *et al.* 2012). In urodeles, several species, including the sharp-ribbed newt (*Pleurodeles waltl*) and the Mexican axolotl (*Ambystoma mexicanum*) respond favourably to a single injection of hormone (Mansour *et al.* 2011; Uteshev *et al.* 2015) (Table 7.1). However, other species such as the black-spotted newt (*Notophthalmus meridionalis*), Kweichow newt (*Tylototriton kweichowensis*) and tiger salamander (*Ambystoma tigrinum*) benefit from the administration of a priming dose administered 24 h before a resolving dose of GnRHa (Marcec 2016; Guy *et al.* 2020; Gillis *et al.* 2021) (Table 7.1). Although currently untested, it has recently been suggested that priming male amphibians with a low dose of hormones before, or early in, the breeding season may promote the maturation of sperm and enhance its quality (Silla *et al.* 2021), and this will be an important avenue for future research.

Frequency of induced spermiation and reproductive trade-offs

In order to collect replicate samples of sperm for storage or artificial fertilisation, captive managers may hormonally induce spermiation repeatedly within short intervals of days to weeks (Roth and Obringer 2003). Empirical research quantifying the effects of repeatedly inducing spermiation on male fitness-determining traits, such as the quality of sperm, its capacity for fertilisation, and male health or longevity, is scarce. However, it has been suggested that the frequency of hormonally induced gamete release both for male and female amphibians should be limited, with appropriate recovery periods imposed, in order to maintain the quality of gametes and to promote animal welfare (Silla *et al.* 2021). Recent research on the natterjack toad (*Epidalea calamita*) reported a decline in percentage sperm motility in males induced to spermiate weekly compared with those induced biweekly or monthly (Arregui *et al.* 2019). Similarly, a study in the Fowler's toad (*Anaxyrus* [*Bufo*] *fowleri*) recommended a minimum recovery period of 2–3 weeks, having shown that shorter intervals between hormonal inductions led to a significant reduction in the concentration of sperm released and in the body condition of males (McDonough *et al.* 2016). Of note, both aforementioned species are classified as prolonged seasonal breeders. The recovery period required for a species to avoid the depletion of sperm quantity and maintenance of sperm quality is likely to be related to the species' reproductive mode and spermatogenetic cycle.

Beyond the potential immediate impacts of repeated spermiation, it is important to consider the effect that the frequency of induced spermiation events within an annual cycle may have on the long-term reproductive output and lifespan of amphibians. A trade-off between current reproductive effort and future reproductive output is one of the fundamental principles of life-history theory (Stearns 1992). Therefore inducing multiple spermiation events on an annual cycle may result in amphibians paying a reproductive cost in the form of reduced fertility (lower quality or concentration of sperm), reproductive longevity, or lifespan. Such reproductive trade-offs have been traditionally explained within the context of resource limitation (Stearns 1992). In captivity, where resource limitation is relaxed compared with that occurring in natural environments, one might predict that the trade-off between current and future reproduction may be less pronounced or nonexistent. However, recent research investigating the proximate mechanisms underpinning reproductive trade-offs suggests that there may be other core mediators such as reactive oxygen species (ROS) production (and the costs of oxidative stress) that operate independently of resource-limitation (Dowling and Simmons 2009). Given that data are limited on the long-term impacts of repeated hormonal induction of spermiation in amphibians, it is recommended that the natural reproductive ecologies of a given species (e.g. the average number of spawning events observed per male, per breeding season, in natural populations of the species) are considered to avoid over-stimulating reproductive effort beyond natural levels. While avoiding over-stimulation is considered 'best practice', for certain species or individuals, there may be a need for urgent action and repeated induction of spermiation over short periods. In these cases, conservation practitioners should weigh the potential conservation benefits against the risks posed to individual animals.

Methods for the administration of hormones

Traditionally, the administration of reproductive hormones to amphibians has been achieved safely and effectively via injection using ultra-fine (30–31 gauge) single-use sterile syringes (Silla *et al.* 2021). Injections given to anurans are typically administered intraperitoneally (IP, injection into the peritoneal cavity of the abdomen) or subcutaneously (SC, usually in the vicinity of dorsal lymph sac), whereas urodeles are typically injected intramuscularly (IM) (Table 7.1).

Intraperitoneal, subcutaneous, and intramuscular injections

Intraperitoneal injections are a common technique but are considered to be of moderate risk (particularly for amphibians of small body size, or when injections are to be administered frequently), and care must be taken to avoid perforation of the urinary bladder or intestine (Roth and Obringer 2003; Turner *et al.* 2011). Injections should be administered at a 45° angle to the skin. Substances administered intraperitoneally are absorbed into the mesenteric vessels and may undergo hepatic metabolism before reaching the systemic circulation (Turner *et al.* 2011). When intraperitoneal delivery is intended, caution should be taken not to puncture the skin superficially or at too acute an angle, as doing so will result in subcutaneous rather than intraperitoneal administration, a common mistake when performing this procedure (Turner *et al.* 2011). With appropriate training and competency in monitoring, accurate delivery of intraperitoneal injections can be readily achieved (Turner *et al.* 2011) and this method of administration has been successfully employed to induce spermiation in a variety of anurans (Table 7.1).

An alternative route of administering hormones that is commonly used to induce spermiation in anurans is subcutaneous injection (Table 7.1); this method of delivery is considered safe and reliable in anurans of any body size. It is important to note, however, that the location of subcutaneous delivery is paramount to the speed and efficiency of the hormone's absorption. A previous study comparing the efficacy of different routes of GnRHa administration in the American toad, *Anaxyrus [Bufo] americanus*, reported reduced spermiation in toads that were administered the hormone subcutaneously compared to intraperitoneal injection (Obringer *et al.* 2000). It is important to note that subcutaneous injections were delivered into the hindlimb in this study, which may have reduced efficacy. Subcutaneous injections should be delivered by locating one of several dorsal lymph sacs located just beneath the skin, allowing for rapid absorption via the lymphatic system and subcutis capillaries (Whitaker and Wright 2001). More comprehensive empirical research is required to directly compare the speed and efficacy of intraperitoneal and subcutaneous injections for inducing spermiation in anurans. Nevertheless, subcutaneous injections offer several technical advantages: they are considered relatively easy to master with minimal training and have the advantage of allowing for the administration of larger volumes of fluid (Turner *et al.* 2011). Importantly, while subcutaneous injections are one of the preferred methods of administering hormones for anurans, this route is not recommended for urodeles due to added adherence of the skin to underlying muscle (Whitaker and Wright 2001).

Intramuscular injections can be reliably administered to the hindlimb (or in some cases the forelimb) of amphibians large enough to provide a sufficient mass of muscle (Whitaker and Wright 2001). Intramuscular injections can also be administered between the shoulder blades (Gillis *et al.* 2021) and this route has been employed for the administration of hormones to large-bodied urodeles (Table 7.1). Due to the rich vascular supply in these regions, intramuscular injections are expected to result in uniform and rapid absorption (Turner *et al.* 2011). Of note, when administering hormones intramuscularly, consideration must be given to the volume of fluid, as clinical guidelines recommend that substantially smaller volumes be administered intramuscularly compared with either intraperitoneal or subcutaneous injection (Turner *et al.* 2011).

Alternative routes of administration

Injections can be safely and reliably administered, even to very small species, provided that there has been adequate training (Browne *et al.* 2019; Della Togna *et al.* 2020; Silla *et al.* 2021). However, there has been a recent impetus to develop alternative protocols of administration due to the perceived animal-welfare benefits of such techniques, including reduced stress, reduced risk of the transmission of disease, and/or reduced risk of injury (to the animal or clinician) (Silla *et al.* 2020). The most commonly studied alternative is the topical (also referred to as transdermal or epicutaneous) administration of reproductive hormones directly to the skin (Table 7.1). The permeability and hypervascularisation of amphibian skin allow the absorption of many chemicals rapidly into circulation, particularly when applied to the ventral abdominal surface (Llewelyn *et al.* 2016). Percutaneous absorption of GnRHa has been shown to effectively induce spermiation in American and Gulf Coast toads (*Anaxyrus americanus* and *Incilius* [*Bufo*] *valliceps*; Rowson *et al.* 2001) and roseate frogs (*Geocrinia rosea*; Silla *et al.* 2020), as well as spawning in northern corroboree frogs (*Pseudophryne pengilleyi*; Silla *et al.* 2018). A recent study in the Australian roseate frog reported that 100% of males spermiated in response to the topical application of GnRHa (applied dropwise to the ventral abdomen), with comparable sperm viability and over 1.5 times the mean sperm concentration of sperm samples collected in response to the optimal injected dose (*Pseudophryne pengilleyi*; Silla *et al.* 2020). Importantly, GnRHa doses successfully applied to anurans topically are substantially higher than those administered via injection, suggesting that transdermal absorption of GnRHa, while effective, may be less efficient than hormone injection (Rowson *et al.* 2001; Silla *et al.* 2018; Silla *et al.* 2020) (Table 7.1).

Another alternative procedure to induce spermiation, recently investigated in the Fowler's toad (*Anaxyrus fowleri*), is the intranasal delivery of GnRHa (Julien *et al.* 2019). Suspensions of hormone were pipetted directly into the nares of toads and resulted in 93% of males spermiating in response to the optimal dose (Julien *et al.* 2019) (Table 7.1). It is important to note that sperm concentrations were lower than those previously reported for Fowler's toads administered GnRHa via intraperitoneal injections (McDonough *et al.* 2016; Julien *et al.* 2019) although high concentrations of sperm can still be obtained via intranasal delivery if multiple collections are pooled. Finally, in a recent landmark success, the dropwise topical application of GnRHa to the nasolabial groove was applied to the endangered Texas blind salamander (*Eurycea rathbuni*), resulting in 92% of male-female pairs spawning (Glass Campbell *et al.* 2021). This alternate

approach has also been applied to the vulnerable San Marcos salamander (*Eurycea nana*) to successfully stimulate spawning (R. Marcec-Greaves, *pers. comm.*). Salamanders from the family Plethodontidae exhibit distinctive nasolabial grooves (cutaneous depressions, or folds, that extend from the upper lip to the nares) that are functionally important for chemical communication during courtship and territoriality (Dawley 1992). Topical application of GnRHa directly to the nasolabial groove may offer a non-invasive alternative for inducing spermiation in other plethodontid species (the majority of living salamanders; Dawley 1992) and warrants further investigation.

Further research will be required before these alternative methods of administering hormones can be used routinely in a wide number of species. Importantly, these methods may offer several advantages, including reduced handling of animals, decreased risk of transmitting disease, elimination of discomfort caused by puncture during injection, and reduced consumable waste and disposal of sharps (Byrne and Silla 2020; Della Togna *et al.* 2020). However, it must be emphasised that responses to hormones may be more variable, and the hormone concentrations needed may be higher (which in some cases may be financially restrictive) compared to traditional methods of injection; therefore, cost–benefit analyses of different methods of administration need to be considered on a case-by-case basis (Byrne and Silla 2020; Della Togna *et al.* 2020).

COLLECTION OF SPERM SAMPLES
Methods for collecting sperm

Once anurans have been administered an effective dose of reproductive hormones to stimulate spermiation, the collection of spermic urine is relatively straightforward. Urination is a common defensive behaviour for many anurans (Toledo *et al.* 2011) so males of most species will urinate quickly when handled, allowing sperm to be collected as they are flushed through the cloaca in a mixture of urine and cloacal secretions (Kouba *et al.* 2013; Silla *et al.* 2015; Silla *et al.* 2020). Collection of spermic urine from anurans can be facilitated by suspending the male above a dry Petri dish and waiting for passive urination (Kouba *et al.* 2009; 2012; Della Togna *et al.* 2017) (Figure 7.1A). Alternatively, spermic urine can be collected by gently inserting the end of a smoothed (fire-polished and cooled) microcapillary tube or small catheter into the cloaca (Byrne and Silla 2010; Silla 2010;

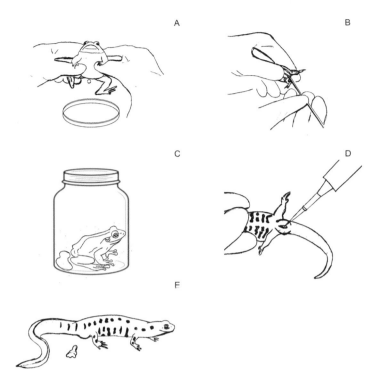

Figure 7.1: Collection of sperm from amphibians, showing (A) collection of spermic urine from an anuran over a dry Petri dish, (B) collection of spermic urine from the cloaca of an anuran using a microcapillary tube or catheter, (C) collection of spermic urine excreted into a dry holding receptacle, (D) collection of semen/milt from the cloaca of a urodele using a micropipette, and (E) passive collection of a spermatophore from a urodele. Note that males must be isolated from females during the collection of sperm.

Della Togna *et al.* 2017; Silla *et al.* 2019) (Figure 7.1B). The tip of the tube can be gently inserted a few millimetres into the cloaca, relaxing pressure from the urogenital sphincter, thereby promoting urination; urine enters the tube by capillary action (Silla 2010). To access the cloaca during the collection of spermic urine (Figure 7.1B), small to medium-sized anurans can be restrained using a gentle, but firm, hold around the trunk (mid-section below the forearms), or by cupping them in a fist. Securing an animal's anterior body (head and trunk) in a gentle, but firm, fist has the advantage of shielding its eyes and restricting vision, which may alleviate stress and reduce movement, thereby facilitating access to the cloaca (Silla *et al.* 2021). Spermic urine may also be collected by placing the animal in a dry container and collecting the urine that is excreted into the holding receptacle (Sherman *et al.* 2008) (Figure 7.1C). If using this method, caution must be taken to check for urine regularly, as the quality of sperm will decline rapidly at room temperature. This approach is not recommended for experimental studies as variation in the time since urination between individuals may influence data on the quality of sperm. Also, this method is unsuitable when repeat sampling is required over a prolonged period, as animals may become dehydrated if remaining in the holding container without the provision of a moist substrate. Sperm from urodeles can be collected as either semen (also referred to as milt) or spermatophores. Semen from urodeles can be collected by suspending the male above a dry Petri dish (Uteshev *et al.* 2015), or a more common method is to collect semen directly from the cloacal opening using a micropipette (Figure 7.1D). As with anurans, small to medium-sized urodeles can be restrained using a secure hold around the trunk or in a gentle fist to shield the eyes, but care should be taken to avoid handling the tail because autonomy can occur (Wright 2001). Alternatively, sperm can be obtained via the deposition of spermatophores (Figure 7.1E). Most urodeles exhibit internal fertilisation, with males transferring sperm to females via the deposition of external spermatophores, collected by the female via the cloaca, and subsequently transferred to the sperm storage organ (spermatheca) (Houck and Arnold 2003). Collection of sperm passively via the deposition of spermatophores reduces handling of the animal; however, this method is generally avoided when sperm are to be cryopreserved because the gelatinous matrix encapsulating the sperm is difficult to macerate and may interfere with effective cryopreservation (Guy *et al.* 2020).

The importance of peak periods for collecting sperm

Collecting sperm samples of the highest quality is an important target for maximising the success of subsequent reproductive technologies. In particular, the percentage fertilisation achieved during artificial fertilisation has been shown to be dependent on the concentration of sperm applied (Cabada 1975; Edwards *et al.* 2004; Silla 2013) and on sperm motility (Dziminski *et al.* 2009). Optimising the quality of sperm is therefore critical, particularly when samples are to be cryopreserved, as current freeze–thaw protocols typically result in a significant loss of viability and capacity for fertilisation (Pearl *et al.* 2017; Della Togna *et al.* 2020). An important step in ensuring that suspensions of sperm are of the highest quality is the quantification of time-dependent spermiation in order to identify optimal collection periods following the administration of hormones (Silla and Byrne 2019; Silla *et al.* 2019; Della Togna *et al.* 2020). Importantly, species-specificity in peak collection periods has been previously reported (Silla and Roberts 2012; Silla and Byrne 2019; Della Togna *et al.* 2020). These differences are likely to reflect variation in reproductive physiology among amphibian species, in particular the structure of mating systems and associated differences in species' basal androgen levels and capacity for the production of sperm (Silla and Byrne 2019). As a result, peak collection periods need to be established on a species-specific basis. In order to effectively identify targeted collection periods, it is recommended that the quality of sperm be intensively quantified over time, commencing within an hour of administering the hormone, and continuing every 1–2 h for 24 h (Kouba *et al.* 2012; Della Togna *et al.* 2017; Silla *et al.* 2019). Maintaining adequate hydration to facilitate repeated urination is essential during the collection of sperm; consequently males should be held in individual holding tanks containing a clean, saturated substrate (wetted sponge or paper towel), or in a small reservoir of filtered water during collection (Uteshev *et al.* 2012; Della Togna *et al.* 2017; Silla *et al.* 2019; Guy *et al.* 2020).

COLD STORAGE OF SPERM
Benefits of the short-term storage of sperm

The temporary storage of gametes is an important tool routinely employed before cryopreservation and/or artificial fertilisation. Subjugating issues associated with the asynchronous release of gametes from male and female amphibians is imperative to the success of artificial fertilisation. The sexes differ considerably with regard to how quickly they respond to hormone treatment, with females exhibiting a protracted response compared to conspecific males. Cold-storage protocols that prolong the viability of gametes are therefore extremely useful in coordinating the simultaneous availability of male and female gametes. While the fertilisation capacity of oocytes declines within just a few hours, the longevity of sperm can be extended over several days to weeks at low temperatures (Table 7.2), thereby ensuring that viable sperm are available during the narrow fertilisation window following oviposition (Kouba *et al.* 2009; Silla and Byrne 2019). The storage of sperm not only can maximise the success of artificial fertilisation, but may also facilitate genetic exchange between breeding colonies, thereby providing a cost-effective alternative to the transportation of live animals without compromising biosecurity or animal welfare. Additional benefits include the ability to increase effective population sizes, create meta-populations by linking captive colonies, as well as the potential to incorporate genetic diversity from *in situ* populations without depleting wild breeding stocks. A review article by Kouba *et al.* (2013) described the production of offspring of the Mississippi gopher frog (*Lithobates sevosus*) via artificial fertilisation using cold-stored sperm transported between breeding facilities, thereby validating the potential of cold storage as a viable genetic management tool.

Storage temperature, medium osmolality, and the addition of antibiotics

Temperature and medium osmolality are important factors affecting the longevity of stored sperm, and suspensions are best maintained in a quiescent state in isotonic diluents (220–280 mOsmol kg^{-1}) at temperatures of 0–5°C (Kouba *et al.* 2009; Browne *et al.* 2019). The sperm of amphibians and fish (unlike those of mammals) are tolerant to cold shock, and it is therefore recommended that suspensions of sperm be immediately cooled in a refrigerator/cooling chamber or plunged into an ice bath (Kouba *et al.* 2009; Contreras *et al.* 2020). The cold-storage longevity of amphibian sperm obtained as spermic urine,

Table 7.2: Longevity of amphibian sperm during cold storage (0–5°C).

Species	Common name	Suspension type	Viability assay	Initial mean viability	Days until ≤ 50% viability	Days until ≤ 25% viability	Reference
Ambystoma mexicanum	Mexican axolotl	Spermatophores	Live/dead	45%	< 1 day	> 28	Figiel (2020)
Ambystoma tigrinum	Tiger salamander	Milt	Motility	55%	< 1	< 1	Gillis *et al.* (2021)
		Spermic urine	Motility	85%	< 1	< 1	Gillis *et al.* (2021)
Anaxyrus boreas boreas	Boreal toad	Spermic urine	Motility	88%	5	> 14	Langhorne *et al.* (2021)
Anaxyrus fowleri	Fowler's toad	Spermic urine	Live/dead	89%	> 4	> 4	Germano *et al.* (2013)*
			Motility	83%	2	> 4	
Epidalea calamita	Natterjack toad	Spermic urine	Motility	78%	2	–	Arregui *et al.* (2019)
Limnodynastes peronii	Striped marsh frog	Testes macerate	Motility	50%	3	< 6	Browne *et al.* (2002)
Litoria booroolongensis	Booroolong frog	Spermic urine	Motility	58%	2	4	Keogh *et al.* (2017b)*
		Testes macerate	Motility	78%	21	> 21	Silla *et al.* (2015)*
Litoria ewingii	Southern brown tree frog	Testes macerate	Live/dead	97%	14	> 14	Love (2011)
Litoria verreauxii	Whistling tree frog	Testes macerate	Motility	90%	6	> 9	Browne *et al.* (2002)
Pelophylax shqipericus	Albanian water frog	Testes macerate	Live/dead	98%	3	4	Turani *et al.* (2019)
			Motility	99%	3	4	
Pleurodeles waltl	Sharp-ribbed newt	Spermic urine	Motility	72%	2	3	Uteshev *et al.* (2015)
Pseudophryne guentheri	Günther's toadlet	Testes macerate	Live/dead	94%	11	13	Silla (2013)
			Motility	83%	7	13	
Rhinella marina	Cane toad	Whole testes	Motility	81%	10	12	Browne *et al.* (2001)

Species for which a taxonomic revision has occurred, contemporary nomenclature is displayed.
* Indicates studies that applied extender treatments; data reported are for control treatments only to allow interspecific comparisons. Discussion of the effectiveness of extender treatments can be found in text.

testes macerates, spermatophores, or milt has been investigated in a small number of species (Table 7.2), highlighting a need for further research. Overall the effective duration of storage is highly variable between species and largely dependent on the type of suspension (Table 7.2). For most species, the viability of suspensions of spermic urine drops to below 50% within 2 days and below 25% within 4–5 days of storage (with the exception of Boreal Toads, Table 7.2). Testes macerates and spermatophores display improved longevity during cold storage, with the sperm of certain species retaining viability for up to 3–4 weeks (Table 7.2). A comparison of two studies of the booroolong frog (*Litoria booroolongensis*) highlights the marked difference in the longevity of sperm between types of suspension within a species, with testes macerates retaining viability at least three-and-a-half times longer than spermic urine (Silla *et al.* 2015; Keogh *et al.* 2017b) (Table 7.2). Several factors may contribute to the reduced longevity of stored spermic urine compared with testes macerates, including comparatively lower osmolality, lower sperm concentration, and higher bacterial abundance within spermic-urine suspensions. A study in the booroolong frog (*Litoria booroolongensis*) reported a 100-fold increase in bacterial abundance in spermic urine samples compared with those of testes macerates (Keogh *et al.* 2017b). In fish, the presence of bacteria in semen has been associated with reduced longevity during storage, sperm motility, and fertility, and the treatment of sperm suspensions with antibiotics is therefore used routinely during cold storage (Contreras *et al.* 2020). Antibiotic treatment is also used as an important tool for preventing the transfer of disease agents between aquaculture facilities when sperm are transported for the purpose of biobanking and artificial fertilisation. In amphibians, the addition of antibiotics has been shown to be detrimental to sperm longevity in two studies – this may have been the result of antibiotic toxicity at supraoptimal doses (Germano *et al.* 2013; Silla *et al.* 2015). Research on two Australian hylids (*Litoria booroolongensis* and *Litoria ewingii*) showed that the addition of gentamicin to sperm suspensions can significantly reduce bacterial abundance without impacting the longevity of stored sperm, highlighting the potential of this method to be used for amphibian biosecurity (Love 2011; Keogh *et al.* 2017b). Further research will be required to determine whether alternative antibiotics and/or lower doses benefit the longevity of amphibian sperm during storage.

Aeration and gaseous environment

The gaseous environment in which both fish and amphibian sperm are stored has been shown to influence the longevity of sperm in cold storage (Bencic *et al.* 2001; Silla *et al.* 2015; Contreras *et al.* 2020). Aerating sperm suspensions with carbon dioxide has been shown to be detrimental to the longevity of amphibian sperm (Silla *et al.* 2015), whereas aeration with atmospheric air or oxygen has been shown to have beneficial effects on the storage of suspensions both of spermic urine and testes macerates (Germano *et al.* 2013; Silla *et al.* 2015). Importantly, opening storage containers to replace atmospheric air, without physical agitation, appears insufficient to improve sperm storage (Browne *et al.* 2001). Indeed, surface renewal theory for gas absorption highlights the importance of agitating samples during aeration, as mechanisms of gas transfer require the movement of fluid from the sample core to the interface (Sinha and De 2012). Physical agitation of suspensions of sperm at regular intervals during storage in the presence of either atmospheric air or oxygen is therefore recommended to alleviate the harmful effects of anoxia (Germano *et al.* 2013; Silla *et al.* 2015).

ACTIVATION OF SPERM MOTILITY

Identifying and refining properties of the activation medium required for optimising the initiation of sperm motility following storage (either cold storage or cryopreservation) is fundamental to achieving successful artificial fertilisations. Sperm motility in externally fertilising amphibians is activated by a drop in osmolality between that of the testis and the external fertilisation medium. This decrease in osmotic strength, or hypotonic shock, stimulates the activation of flagellar movement by promoting an increase in intracellular cyclic adenosine monophosphate (cAMP), which, in turn, promotes the phosphorylation of protein kinase A (PKA) substrates (O'Brien *et al.* 2011). The most significant characteristic of the activation medium to refine is therefore its osmolality. Importantly, the optimal osmolality for the activation of sperm motility is species-specific and may reflect differences in natural fertilisation ecologies among species. While activation of sperm motility and artificial fertilisation typically are achieved in media of very low osmolality in aquatic-breeding anurans (< 25 mOsm/kg) (Wolf and Hedrick 1971; Cabada 1975; Edwards *et al.* 2004), recent research suggests that optimal osmolality of the activation media may be broader than expected in aquatic species adapted to reproducing in stochastic environments (Byrne *et al.* 2015). Optimal osmolalities have been shown to range from 10 to 50 mOsm/kg, and 10 to 75 mOsm/kg in the aquatic-breeding species *Crinia signifera* and *Litoria booroolongensis*, respectively (Byrne *et al.* 2015; Silla *et al.* 2017). Additionally, terrestrial-breeding anurans are adapted to achieve the activation of sperm motility and subsequent fertilisations in the absence of fresh, free-standing water (Wells 2007). For terrestrial-breeding anurans, activation of sperm motility probably occurs in a fluid medium comprised of a mixture of urine, cloacal, and reproductive-tract secretions typically higher in osmolality than that of bodies of freshwater (Silla 2013). Optimal medium osmolalities ranged from 25 to 100 mOsm/kg for the terrestrially breeding anuran *Pseudophryne guentheri* (Silla 2013). Given this interspecific diversity, sperm-motility curves should be established to determine the optimal osmolality of media used to activate the motility of sperm for each individual species.

Phosphodiesterase inhibitors and antioxidants

The addition of phosphodiesterase (PDE) inhibitors (such as caffeine, theophylline, and pentoxifylline; also referred to as stimulants) to sperm-activation media has been shown to inhibit PDE activity, increase levels of cAMP, and result in the activation of sperm motility in a variety of taxa ranging from marine invertebrates to mammals

(Yoshida *et al.* 1994; Yunes *et al.* 2005; Stephens *et al.* 2013). Phosphodiesterase inhibitors have seldom been applied to the sperm suspensions of amphibians, but there is evidence for both the immediate and delayed beneficial effects of caffeine and theophylline on percentage sperm motility and sperm velocity (Fitzsimmons *et al.* 2007; Silla *et al.* 2017). It has been suggested that, in addition to their PDE-inhibition properties, caffeine and theophylline may improve the duration of sperm motility by acting as anti-oxidants (Silla *et al.* 2017). However, a study in the booroolong frog (*Litoria booroolongensis*) found that the addition of other known antioxidants, vitamins C (ascorbic acid) and E (α-tocopherol), were either detrimental or had no effect on the activation or duration of sperm motility (Keogh *et al.* 2017a).

Duration of motility

The activation of amphibian sperm motility is initiated within seconds in media of appropriate osmolality and is then followed by a rapid to moderate decline in motility and velocity when sustained at ambient temperatures (Browne *et al.* 2015). Browne *et al.* (2015) reviewed data on the duration of sperm motility following activation at room temperature in 21 amphibian species and reported a minimum of 2.8 min in the Chinese giant salamander (*Andrias davidianus*) to a maximum duration of 4.5 h in the broad-palmed frog (*Litoria latopalmata*) (mean = 68.3 min, *n* = 21 species). A recent study in the booroolong frog (*Litoria booroolongensis*) observed sperm motility 10 h post-activation at 22°C (Silla *et al.* 2017). The study by Silla *et al.* (2017) reported unprecedented sperm endurance at an ecologically relevant temperature that more than doubled the maximum value previously recorded in an anuran (Browne *et al.* 2015). It is unclear why such extreme variation in the duration of motility exists in amphibian sperm but theory suggests that it may be associated with the evolution of mating systems (levels of polyandry), egg size, or the turbulence of water where fertilisation occurs (Stockley *et al.* 1996; Snook 2005).

ASSESSMENT OF SPERM QUALITY

There are several characteristics, or biomarkers, that can be used to assess the quality of sperm suspensions and indicate their potential for fertilisation. Sperm traits that are typically reported include the concentration of sperm, motility parameters (percentage motility, forward progression, and velocity), viability (proportion live/dead and DNA fragmentation), and morphology (morphological deformities, head length, tail length, and total length).

Concentration of sperm

The concentration of sperm is an important initial measure of the spermiation response and is generally quantified using specialised manual counting chambers, such as a haemocytometer. The same basic technique applies, regardless of the specific counting chamber used, whereby a predetermined volume of sample (the exact volume will differ depending on the type of slide) is loaded into a cell chamber of a known consistent depth and the number of sperm cells within a designated grid system are counted and used to calculate the number of sperm cells per millilitre (sperm/mL) (Kouba *et al.* 2012; Della Togna *et al.* 2017; Silla *et al.* 2019). Sperm counts should be repeated two to four times per sample and averaged to improve accuracy. It is also important to note that precision decreases with decreasing concentration of sperm ($< 1 \times 10^4$ sperm/mL), and the counting area may need to be increased to improve accuracy. Alternatively, if working with species with very low concentrations of sperm, it may be necessary to count the total number of sperm in a set volume, or the entire sample, to obtain accurate data (e.g. Silla *et al.* 2020). Importantly, in order to obtain an accurate count manually, it is essential that sperm be immotile. Sperm obtained from testes macerates or from spermatophores are inactive, but motility may be activated upon release as spermic urine, and dilution in a solution of high osmolality (i.e. simplified amphibian ringer, SAR) may be required to inhibit motility. If dilution is required, all samples must be treated in a standardised manner and it is essential that dilution factors are incorporated into the final calculation. Alternatively, the concentration of motile sperm can be determined with high accuracy and reproducibility using a computer-assisted sperm-analysis system (described below) (Centola 1996).

Motility of sperm

Traditionally, sperm motility has been evaluated manually via light microscopy using wet slide preparations. Typically, 100 sperm in randomly selected fields are assessed for percentage motility (the proportion of sperm exhibiting flagellar movement) and percentage forward progressive motility (the proportion of motile sperm exhibiting forward motion). The quality of motion may also be ranked on a qualitative scale (Obringer *et al.* 2000; Kouba *et al.* 2012; McDonough *et al.* 2016; Arregui *et al.* 2019). Although the manual assessment of motility is standardised practice in many laboratories, it is important to note that this procedure is prone to reproducibility error and observer bias (Centola 1996) and is also time-consuming when processing a large number of samples. An alternative approach is to quantify the motility of sperm using computer-assisted sperm-analysis (CASA) (Byrne *et al.* 2015; Silla *et al.* 2015; Keogh *et al.* 2017b; Silla *et al.* 2019). Computer-assisted sperm-analysis systems objectively and rapidly (within seconds) measure a wide range of motility parameters, including percentage motility, velocity, and straightness of the swimming path. When economic constraints inhibit the use of a complete CASA system, CASA analysis is possible by recording video footage (using a camera and microscope with c-section mount) and analysing motility using free software developed to assess the motility of fish sperm (see O'Brien *et al.* 2011). Of note, CASA systems need to be calibrated on a species-specific basis (e.g. to detect an appropriate cell size, which is dependent on the length of a species' sperm), and playback functions should be used to validate technical settings. Finally, osmolality and pH of sperm suspensions should also be quantified if assessing the motility of spermic urine, as these parameters are known to influence the activation of sperm motility.

Viability of sperm

In addition to sperm motility, assays of the viability of sperm provide a valuable measure of the quality of sperm. Commercial assay kits have been employed to quantify the integrity of the plasma membrane of sperm cells (Fitzsimmons *et al.* 2007; Sherman *et al.* 2008; Byrne and Silla 2010; Silla 2010; Silla and Roberts 2012; Jacobs *et al.* 2016; Figiel 2020) and fragmentation of DNA (Della Togna *et al.* 2017; Della Togna *et al.* 2018; Arregui *et al.* 2020). Commonly used live/dead viability assays employ SYBR-14, a membrane-permeant dye that stains the cell nuclei of living (membrane-intact) sperm bright green, followed by propidium iodide (PI) to penetrate membrane-compromised (dead) cells that fluoresce red (Sherman *et al.* 2008; Silla *et al.* 2020). Viability is reported as the proportion of living sperm to total sperm, counted under fluorescent microscopy. Viability should be assessed by counting 100 sperm per

sample (Rowson *et al.* 2001; Della Togna *et al.* 2017; Figiel 2020). Of note, the viability and percentage motility of sperm often are correlated in fresh samples of sperm (see Silla 2013); however, viability is generally higher than motility post-storage (either cold storage or cryopreservation: e.g. Mansour *et al.* 2010; Silla 2013). This higher level of viability is particularly evident post-thaw as the flagellum and mitochondrial vesicle often are damaged, possibly inhibiting motility in living sperm (Kouba *et al.* 2003).

Sperm morphology and fertilisation capacity of sperm

When the volume of sample permits, additional aliquots from the sperm suspension can be fixed for later assessment of sperm morphology. Computer-assisted sperm morphometry analysis systems (CASMA or CASA-morph) objectively measure a range of morphometric parameters including geometrics of the head, nucleus, acrosome, midpiece, and flagellum (Yániz *et al.* 2015). Such systems provide a powerful tool for morphometric analysis, initially developed for the assessment of mammalian sperm in the 1990s (Yániz *et al.* 2015) but are yet to be employed for amphibian sperm. Although less sophisticated, manual assessment of the length of the head and tail of sperm is also highly informative and may be used to investigate correlations between the size, speed, and fertilisation capacity of sperm (Dziminski *et al.* 2010; Simpson *et al.* 2014). Identification of morphological deformities and quantification of the proportion of normal/abnormal sperm within suspensions is another valuable indicator of the quality of sperm. Morphological deformities reported for amphibian sperm include abnormalities of the acrosome and head, swelling or rupture of the plasma membrane, absent or deformed flagellum, and loss of accessory structures (Kouba *et al.* 2003; Della Togna *et al.* 2017; Browne *et al.* 2019; Arregui *et al.* 2020; Guy *et al.* 2020). Finally, the quality of sperm suspensions may be assessed by quantifying their capacity for fertilisation, although this has rarely been conducted due to the need for fresh oocytes (Browne *et al.* 2019). Caution should also be taken in using fertilisation success or offspring fitness as proxies for sperm quality given the potential for maternal effects and genetic incompatibilities (Byrne *et al.* 2019) (see Chapter 8). For this reason, fertilisations must be replicated, with each suspension of sperm utilised to fertilise the eggs of several replicate females.

CONCLUSIONS

Hormonal therapies to induce spermiation, as well as protocols for the collection, cold storage, and activation of motility of sperm, have been successfully developed for a diversity of model species. In recent years, these protocols have been applied to a small but growing number of threatened amphibians to assist integrated conservation breeding programs. Despite such achievements, it has become increasingly apparent that optimal protocols vary considerably across species, which in part may be driven by the extreme diversity of reproductive modes exhibited by amphibians. Research reviewed in this chapter provides a solid framework from which species-specific protocols can be refined for novel species. Additionally, the standardisation of objective assessment of the quality of sperm will strengthen comparisons between research projects, thereby assisting in advancing the field.

ACKNOWLEDGEMENTS

The authors acknowledge the input of Gina Della Togna and Phillip Byrne during the preparation of this chapter. The support of the following funding grants is also acknowledged, of which Aimee Silla was in receipt during the preparation of this manuscript: Australian Research Council; Linkage Projects (LP140100808 & LP170100351), Discovery Early Career Researcher Award (DE210100812), and Zoo and Aquarium Association's Wildlife Conservation Fund.

REFERENCES

Arregui L, Diaz-Diaz S, Alonso-López E, Kouba AJ (2019) Hormonal induction of spermiation in a Eurasian bufonid (*Epidalea calamita*). *Reproductive Biology and Endocrinology* 17, 92. doi:10.1186/s12958-019-0537-0

Arregui L, Bóveda P, Gosálvez J, Kouba AJ (2020) Effect of seasonality on hormonally induced sperm in *Epidalea calamita* (Amphibia, Anura, Bufonidae) and its refrigerated and cryopreservated storage. *Aquaculture* 529, 735677. doi:10.1016/j.aquaculture.2020.735677

Beebee TJ, Griffiths RA (2005) The amphibian decline crisis: a watershed for conservation biology? *Biological Conservation* 125, 271–285. doi:10.1016/j.biocon.2005.04.009

Bencic DC, Ingermann RL, Cloud JG (2001) Does CO_2 enhance short-term storage success of chinook salmon (*Oncorhynchus tshawytscha*) milt? *Theriogenology* 56, 157–166. doi:10.1016/S0093-691X(01)00551-9

Browne R, Clulow J, Mahony M (2001) Short-term storage of cane toad (*Bufo marinus*) gametes. *Reproduction (Cambridge, England)* 121, 167–173. doi:10.1530/rep.0.1210167

Browne RK, Clulow J, Mahony M (2002) The short-term storage and cryopreservation of spermatozoa from hylid and myobatrachid frogs. *Cryo Letters* 23, 129–136.

Browne R, Seratt J, Vance C, Kouba A (2006) Hormonal priming, induction of ovulation and in-vitro fertilization of the endangered Wyoming toad (*Bufo baxteri*). *Reproductive Biology and Endocrinology* 4, 34. doi:10.1186/1477-7827-4-34

Browne R, Kaurova S, Uteshev V, Shishova N, Mcginnity D, Figiel C, Mansour N, Agnew D, Wu M, Gakhova E (2015) Sperm motility of externally fertilizing fish and amphibians. *Theriogenology* 83, 1–13. doi:10.1016/j.theriogenology.2014.09.018

Browne RK, Silla AJ, Upton R, Della-Togna G, Marcec-Greaves R, Shishova NV, Uteshev VK, Proaño B, Pérez OD, Mansour N (2019) Sperm collection and storage for the sustainable management of amphibian biodiversity. *Theriogenology* 133, 187–200. doi:10.1016/j.theriogenology.2019.03.035

Bruehl FS (1952) The development of pregnancy tests. *The American Journal of Nursing* 52, 591–593.

Byrne P, Silla A (2020) An experimental test of the genetic consequences of population augmentation in an amphibian. *Conservation Science and Practise* 2, e194. doi:10.1111/csp2.194

Byrne PG, Silla AJ (2010) Hormonal induction of gamete release, and *in-vitro* fertilisation, in the critically endangered Southern Corroboree Frog, *Pseudophryne corroboree*. *Reproductive Biology and Endocrinology* 8, 144. doi:10.1186/1477-7827-8-144

Byrne PG, Dunne C, Munn AJ, Silla AJ (2015) Environmental osmolality influences sperm motility activation in an anuran amphibian. *Journal of Evolutionary Biology* 28, 521–534. doi:10.1111/jeb.12584

Byrne P, Gaitan Espitia J, Silla A (2019) Genetic benefits of extreme sequential polyandry in a terrestrial-breeding frog. *Evolution* 73, 1972–1985. doi:10.1111/evo.13823

Cabada MO (1975) Sperm concentration and fertilization rate in *Bufo arenarum* (Amphibia: Anura). *The Journal of Experimental Biology* 62, 481–486. doi:10.1242/jeb.62.2.481

Centola G (1996) Comparison of manual microscopic and computer-assisted methods for analysis of sperm count and motility. *Archives of Andrology* 36, 1–7. doi:10.3109/01485019608987878

Clulow J, Clulow S (2016) Cryopreservation and other assisted reproductive technologies for the conservation of threatened amphibians and reptiles: bringing the ARTs up to speed. *Reproduction, Fertility and Development* 28, 1116–1132. doi:10.1071/RD15466

Clulow J, Trudeau VL, Kouba AJ (2014) Amphibian declines in the twenty-first century: why we need assisted reproductive technologies. IN *Reproductive Sciences in Animal Conservation* (Eds WV Holt, JL Brown and P Comizzoli) pp. 275–316. Springer, New York NY, USA.

Clulow J, Pomering M, Herbert D, Upton R, Calatayud N, Clulow S, Mahony MJ, Trudeau VL (2018) Differential success in obtaining gametes between male and female Australian temperate frogs by hormonal induction: a review.

General and Comparative Endocrinology **265**, 141–148. doi:10.1016/j.ygcen.2018.05.032

Contreras P, Dumorné K, Ulloa-Rodríguez P, Merino O, Figueroa E, Farías JG, Valdebenito I, Risopatrón J (2020) Effects of short-term storage on sperm function in fish semen: a review. *Reviews in Aquaculture* **12**, 1373–1389.

Dawley EM (1992) *Correlation of salamander Vomeronasal and Main Olfactory System Anatomy with Habitat and Sex: Behavioral Interpretations.* Chemical Signals in Vertebrates 6. Springer, Boston MA, USA.

Della Togna G, Trudeau VL, Gratwicke B, Evans M, Augustine L, Chia H, Bronikowski EJ, Murphy JB, Comizzoli P (2017) Effects of hormonal stimulation on the concentration and quality of excreted spermatozoa in the critically endangered Panamanian golden frog (*Atelopus zeteki*). *Theriogenology* **91**, 27–35. doi:10.1016/j.theriogenology.2016.12.033

Della Togna G, Gratwicke B, Evans M, Augustine L, Chia H, Bronikowski E, Murphy JB, Comizzoli P (2018) Influence of extracellular environment on the motility and structural properties of spermatozoa collected from hormonally stimulated Panamanian Golden Frog (*Atelopus zeteki*). *Theriogenology* **108**, 153–160. doi:10.1016/j.theriogenology.2017.11.032

Della Togna G, Howell LG, Clulow JC, Langhorne CJ, Marcec-Greaves R, Calatayud NE (2020) Evaluating amphibian biobanking and reproduction for captive breeding programs according to the Amphibian Conservation Action Plan objectives. *Theriogenology* **150**, 412–431. doi:10.1016/j.theriogenology.2020.02.024

Dowling DK, Simmons LW (2009) Reactive oxygen species as universal constraints in life-history evolution. *Proceedings. Biological Sciences* **276**, 1737–1745. doi:10.1098/rspb.2008.1791

Dufour S, Weltzien FA, Sebert ME, Le Belle N, Vidal B, Vernier P, Pasqualini C (2005) Dopaminergic inhibition of reproduction in teleost fishes: ecophysiological and evolutionary implications. *Annals of the New York Academy of Sciences* **1040**, 9–21. doi:10.1196/annals.1327.002

Dziminski MA, Roberts JD, Beveridge M, Simmons LW (2009) Sperm competitiveness in frogs: slow and steady wins the race. *Proceedings. Biological Sciences* **276**, 3955–3961. doi:10.1098/rspb.2009.1334

Dziminski MA, Roberts J, Simmons LW (2010) Sperm morphology, motility and fertilisation capacity in the myobatrachid frog *Crinia georgiana*. *Reproduction, Fertility and Development* **22**, 516–522. doi:10.1071/RD09124

Edwards DL, Mahony MJ, Clulow J (2004) Effect of sperm concentration, medium osmolality and oocyte storage on artificial fertilisation success in a myobatrachid frog (*Limnodynastes tasmaniensis*). *Reproduction, Fertility and Development* **16**, 347–354. doi:10.1071/RD02079

Figiel CR (2020) Cold storage of sperm from the axolotl, *Ambystoma Mexicanum*. *Herpetological Conservation and Biology* **15**, 367–371.

Fitzsimmons C, Mclaughlin E, Mahony M, Clulow J (2007) Optimisation of handling, activation and assessment procedures for *Bufo marinus* spermatozoa. *Reproduction, Fertility and Development* **19**, 594–601. doi:10.1071/RD06124

Gascon C, Collins JP, Moore RD, Church DR, Mckay JE, Mendelson III JR (2007) *Amphibian Conservation Action Plan.* IUCN SSC Amphibian Specialist Group. Gland, Switzerland and Cambridge, UK.

Germano JM, Arregui L, Kouba AJ (2013) Effects of aeration and antibiotics on short-term storage of Fowler's toad (*Bufo fowleri*) sperm. *Aquaculture* **396**, 20–24. doi:10.1016/j.aquaculture.2013.02.018

Gillis AB, Guy EL, Kouba AJ, Allen PJ, Marcec-Greaves RM, Kouba CK (2021) Short-term storage of tiger salamander (*Ambystoma tigrinum*) spermatozoa: The effect of collection type, temperature and time. *PLoS One* **16**, e0245047. doi:10.1371/journal.pone.0245047

Glass Campbell L, Anderson KA, Marcec-Greaves R (2021) Topical application of hormone gonadotropin-releasing hormone (GnRH-A) simulates reproduction in the endangered Texas blind salamander (*Eurycea rathbuni*). *Conservation Science and Practice* **2021**, e609. doi:10.1111/csp2.609

Guy EL, Gillis AB, Kouba AJ, Barber D, Poole V, Marcec-Greaves RM, Kouba CK (2020) Sperm collection and cryopreservation for threatened newt species. *Cryobiology* **94**, 80–88. doi:10.1016/j.cryobiol.2020.04.005

Hinkson KM, Baecher JA, Poo S (2019) Cryopreservation and hormonal induction of spermic urine in a novel species: the smooth-sided toad (*Rhaebo guttatus*). *Cryobiology* **89**, 109–111. doi:10.1016/j.cryobiol.2019.05.007

Houck LD, Arnold SJ (2003) Courtship and mating behavior. *Reproductive Biology and Phylogeny of Urodela* **1**, 383–424.

Iimori E, D'occhio M, Lisle A, Johnston S (2005) Testosterone secretion and pharmacological spermatozoal recovery in the cane toad (*Bufo marinus*). *Animal Reproduction Science* **90**, 163–173. doi:10.1016/j.anireprosci.2005.01.010

Jacobs LE, Robertson JM, Kaiser K (2016) Variation in male spermiation response to exogenous hormones among divergent populations of Red-eyed Treefrogs. *Reproductive Biology and Endocrinology* **14**, 83. doi:10.1186/s12958-016-0216-3

Julien AR, Kouba AJ, Kabelik D, Feugang JM, Willard ST, Kouba CK (2019) Nasal administration of gonadotropin releasing hormone (GnRH) elicits sperm production in Fowler's toads (*Anaxyrus fowleri*). *BMC Zoology* **4**, 3. doi:10.1186/s40850-019-0040-2

Keogh L, Byrne PG, Silla AJ (2017a) The effect of antioxidants on sperm motility activation in the Booroolong frog. *Animal Reproduction Science* **183**, 126–131. doi:10.1016/j.anireprosci.2017.05.008

Keogh LM, Byrne PG, Silla AJ (2017b) The effect of gentamicin on sperm motility and bacterial abundance during chilled sperm storage in the Booroolong frog. *General and Comparative Endocrinology* **243**, 51–59. doi:10.1016/j.ygcen.2016.11.005

Kouba AJ, Vance CK (2009) Applied reproductive technologies and genetic resource banking for amphibian conservation. *Reproduction, Fertility and Development* **21**, 719–737. doi:10.1071/RD09038

Kouba AJ, Vance CK, Frommeyer MA, Roth TL (2003) Structural and functional aspects of *Bufo americanus* spermatozoa:

effects of inactivation and reactivation. *Journal of Experimental Zoology. Part B, Molecular and Developmental Evolution* **295A**, 172–182. doi:10.1002/jez.a.10192

Kouba A, Vance C, Willis E (2009) Artificial fertilization for amphibian conservation: current knowledge and future considerations. *Theriogenology* **71**, 214–227. doi:10.1016/j.theriogenology.2008.09.055

Kouba AJ, Delbarco-Trillo J, Vance CK, Milam C, Carr M (2012) A comparison of human chorionic gonadotropin and luteinizing hormone releasing hormone on the induction of spermiation and amplexus in the American toad (*Anaxyrus americanus*). *Reproductive Biology and Endocrinology* **10**, 59. doi:10.1186/1477-7827-10-59

Kouba AJ, Lloyd RE, Houck ML, Silla AJ, Calatayud N, Trudeau VL, Clulow J, Molinia F, Langhorne C, Vance C (2013) Emerging trends for biobanking amphibian genetic resources: the hope, reality and challenges for the next decade. *Biological Conservation* **164**, 10–21. doi:10.1016/j.biocon.2013.03.010

Langhorne C (2016) *Developing Assisted Reproductive Technologies for Endangered North American Amphibians*. PhD Thesis, PhD Thesis, Mississippi State University, USA.

Langhorne CJ, Calatayud NE, Kouba CK, Willard ST, Smith T, Ryan PL, Kouba AJ (2021) Efficacy of hormone stimulation on sperm production in an alpine amphibian (*Anaxyrus boreas boreas*) and the impact of short-term storage on sperm quality. *Zoology* **146**, 125912. doi:10.1016/j.zool.2021.125912

Lipke C (2008) Induced spermiation and sperm morphology in the green poison frog, *Dendrobates auratus*. PhD thesis. University of Veterinary Medicine, Germany.

Llewelyn V, Berger L, Glass B (2016) Percutaneous absorption of chemicals: developing an understanding for the treatment of disease in frogs. *Journal of Veterinary Pharmacology and Therapeutics* **39**, 109–121. doi:10.1111/jvp.12264

Love E (2011) Hormonal induction of spermiation and short-term sperm storage in two Australian tree frogs. BSC (Honours) thesis. Monash University, Australia.

Mansour N, Lahnsteiner F, Patzner RA (2010) Motility and cryopreservation of spermatozoa of European common frog, *Rana temporaria*. *Theriogenology* **74**, 724–732. doi:10.1016/j.theriogenology.2010.03.025

Mansour N, Lahnsteiner F, Patzner RA (2011) Collection of gametes from live axolotl, *Ambystoma mexicanum*, and standardization of in vitro fertilization. *Theriogenology* **75**, 354–361. doi:10.1016/j.theriogenology.2010.09.006

Marcec R (2016) Development of assisted reproductive technologies for endangered North American salamanders. PhD thesis. Mississippi State University, USA.

McDonough CE, Martin MW, Vance CK, Cole JA, Kouba AJ (2016) Frequency of exogenous hormone therapy impacts spermiation in male Fowler's toad (*Bufo fowleri*). *Reproduction, Fertility and Development* **28**, 995–1003. doi:10.1071/RD14214

do Nascimento NF, Silva RC, Valentin FN, Paes MDCF, De Stéfani MV, Nakaghi LSO (2015) Efficacy of buserelin acetate combined with a dopamine antagonist for spawning induction in the bullfrog (*Lithobates catesbeianus*). *Aquaculture Research* **46**, 3093–3096. doi:10.1111/are.12461

O'Brien ED, Krapf D, Cabada MO, Visconti PE, Arranz SE (2011) Transmembrane adenylyl cyclase regulates amphibian sperm motility through protein kinase A activation. *Developmental Biology* **350**, 80–88. doi:10.1016/j.ydbio.2010.11.019

Obringer AR, O'brien JK, Saunders RL, Yamamoto K, Kikuyama S, Roth TL (2000) Characterization of the spermiation response, luteinizing hormone release and sperm quality in the American Toad (*Bufo americanus*) and the endangered Wyoming Toad (*Bufo baxteri*). *Reproduction, Fertility and Development* **12**, 51–58. doi:10.1071/RD00056

Pearl E, Morrow S, Noble A, Lerebours A, Horb M, Guille M (2017) An optimized method for cryogenic storage of *Xenopus* sperm to maximise the effectiveness of research using genetically altered frogs. *Theriogenology* **92**, 149–155. doi:10.1016/j.theriogenology.2017.01.007

Pereira MM, Ribeiro Filho OP, Zanuncio JC, Navarro RD, Seixas Filho JT, De Lima Ribeiro CD (2013) Evaluation of the semen characteistics after induced spermiation in the bullfrog *Lithobates catesbeianus*. *Acta Scientiarum. Biological Sciences* **35**, 305–310.

Poo S, Hinkson KM, Stege E (2019) Sperm output and body condition are maintained independently of hibernation in an endangered temperate amphibian. *Reproduction, Fertility and Development* **31**, 796–804. doi:10.1071/RD18073

Pritchard DJ, Fa JE, Oldfield S, Harrop SR (2012) Bring the captive closer to the wild: redefining the role of ex situ conservation. *Oryx* **46**, 18–23. doi:10.1017/S0030605310001766

Roth TL, Obringer AR (2003). Reproductive research and the worldwide amphibian extinction crisis. In *Reproductive Science and Integrated Conservation*. (Eds WV Holt, AR Pickard, JC Rodger and DE Wildt) pp. 359–374. Cambridge University Press, Cambridge, UK.

Rowson AD, Obringer AR, Roth TL (2001) Non-invasive treatments of luteinizing hormone-releasing hormone for inducing spermiation in American (*Bufo americanus*) and Gulf Coast (*Bufo valliceps*) toads. *Zoo Biology* **20**, 63–74. doi:10.1002/zoo.1007

Sherman CD, Uller T, Wapstra E, Olsson M (2008) Within population variation in ejaculate characteristics in a prolonged breeder, Peron's tree frog, *Litoria peronii*. *Naturwissenshaften*. **95**, 1055–1061. doi:10.1007/s00114-008-0423-7

Sherman CD, Sagvik J, Olsson M (2010) Female choice for males with greater fertilization success in the swedish Moor frog, *Rana arvalis*. *PLoS One* **5**, e13634. doi:10.1371/journal.pone.0013634

Silla AJ (2010) Effects of luteinizing hormone-releasing hormone and arginine-vasotocin on the sperm-release response of Günther's Toadlet, *Pseudophryne guentheri*. *Reproductive Biology and Endocrinology* **8**, 139. doi:10.1186/1477-7827-8-139

Silla AJ (2011) Effect of priming injections of luteinizing hormone-releasing hormone on spermiation and ovulation in Günther's Toadlet, *Pseudophryne guentheri*. *Reproductive Biology and Endocrinology* **9**, 68. doi:10.1186/1477-7827-9-68

Silla AJ (2013) Artificial fertilisation in a terrestrial toadlet *Pseudophryne guentheri*: effect of medium osmolality, sperm concentration and gamete storage. *Reproduction, Fertility and Development* **25**, 1134–1141. doi:10.1071/RD12223

Silla AJ, Byrne PG (2019) The role of reproductive technologies in amphibian conservation breeding programs. *Annual Review of Animal Biosciences* **7**, 499–519. doi:10.1146/annurev-animal-020518-115056

Silla AJ, Byrne PG (2021) Hormone-induced ovulation and artificial fertilisation in four terrestrial-breeding anurans. *Reproduction, Fertility and Development* **33**, 615–618. doi:10.1071/RD20243

Silla AJ, Roberts JD (2012) Investigating patterns in the spermiation response of eight Australian frogs administered human chorionic gonadotropin (hCG) and luteinizing hormone-releasing hormone (LHRHa). *General and Comparative Endocrinology* **179**, 128–136. doi:10.1016/j.ygcen.2012.08.009

Silla AJ, Keogh LM, Byrne PG (2015) Antibiotics and oxygen availability affect the short-term storage of spermatozoa from the critically endangered booroolong frog, *Litoria booroolongensis*. *Reproduction, Fertility and Development* **27**, 1147–1153. doi:10.1071/RD14062

Silla AJ, Keogh LM, Byrne PG (2017) Sperm motility activation in the critically endangered booroolong frog: the effect of medium osmolality and phosphodiesterase inhibitors. *Reproduction, Fertility and Development* **29**, 2277–2283. doi:10.1071/RD17012

Silla AJ, Mcfadden M, Byrne PG (2018) Hormone-induced spawning of the critically endangered northern corroboree frog *Pseudophryne pengilleyi*. *Reproduction, Fertility and Development* **30**, 1352–1358. doi:10.1071/RD18011

Silla AJ, Mcfadden MS, Byrne PG (2019) Hormone-induced sperm-release in the critically endangered Booroolong frog (*Litoria booroolongensis*): effects of gonadotropin-releasing hormone and human chorionic gonadotropin. *Conservation Physiology* **7**, coy080. doi:10.1093/conphys/coy080

Silla AJ, Roberts JD, Byrne PG (2020) The effect of injection and topical application of hCG and GnRH agonist to induce sperm-release in the roseate frog, *Geocrinia rosea*. *Conservation Physiology* **8**, coaa104. doi:10.1093/conphys/coaa104

Silla AJ, Calatayud NE, Trudeau VL (2021) Amphibian reproductive technologies: approaches and welfare considerations. *Conservation Physiology* **9**, coab011. doi:10.1093/conphys/coab011

Simpson JL, Humphries S, Evans JP, Simmons LW, Fitzpatrick JL (2014) Relationships between sperm length and speed differ among three internally and three externally fertilizing species. *Evolution* **68**, 92–104. doi:10.1111/evo.12199

Sinha AP, De P (2012) Interphase mass transfer. In *Mass Transfer: Principles and Operations*. (Eds AP Sinha and DE Parameswar) pp. 104–122. PHI Learning, New Delhi, India.

Snook RR (2005) Sperm in competition: not playing by the numbers. *Trends in Ecology & Evolution* **20**, 46–53. doi:10.1016/j.tree.2004.10.011

Stearns SC (1992) *The Evolution of Life Histories*. Oxford University Press, Oxford, UK.

Stephens TD, Brooks RM, Carrington JL, Cheng L, Carrington AC, Porr CA, Splan RK (2013) Effects of pentoxifylline, caffeine, and taurine on post-thaw motility and longevity of equine frozen semen. *Journal of Equine Veterinary Science* **33**, 615–621. doi:10.1016/j.jevs.2012.10.004

Stockley P, Gage MJG, Parker GA, Moller AP (1996) Female reproductive biology and the coevolution of ejaculate characteristics in fish. *Proceedings of the Royal Society of London. Series B, Biological Sciences* **263**, 451–458. doi:10.1098/rspb.1996.0068

Stuart SN, Chanson JS, Cox NA, Young BE, Rodrigues AS, Fischman DL, Waller RW (2004) Status and trends of amphibian declines and extinctions worldwide. *Science* **306**, 1783–1786. doi:10.1126/science.1103538

Toledo LF, Sazima I, Haddad CF (2011) Behavioural defences of anurans: an overview. *Ethology Ecology and Evolution* **23**, 1–25. doi:10.1080/03949370.2010.534321

Trudeau VL, Somoza GM, Natale GS, Pauli B, Wignall J, Jackman P, Doe K, Schueler FW (2010) Hormonal induction of spawning in 4 species of frogs by coinjection with a gonadotropin-releasing hormone agonist and a dopamine antagonist. *Reproductive Biology and Endocrinology* **8**, 36. doi:10.1186/1477-7827-8-36

Trudeau VL, Schueler FW, Navarro-Martin L, Hamilton CK, Bulaeva E, Bennett A, Fletcher W, Taylor L (2013) Efficient induction of spawning of Northern leopard frogs (*Lithobates pipiens*) during and outside the natural breeding season. *Reproductive Biology and Endocrinology* **11**, 14. doi:10.1186/1477-7827-11-14

Turani B, Aliko V, Shkembi E (2019) Characterization of Albanian water frog, *Pelophylax shqipericus*, sperm traits and morphology, by using phase contrast microscopy. *Microscopy Research and Technique* **82**, 1802–1809. doi:10.1002/jemt.23346

Turner PV, Brabb T, Pekow C, Vasbinder MA (2011) Administration of substances to laboratory animals: routes of administration and factors to consider. *Journal of the American Association for Laboratory Animal Science; JAALAS* **50**, 600–613.

Uteshev V, Shishova N, Kaurova S, Browne R, Gakhova E (2012) Hormonal induction of spermatozoa from amphibians with *Rana temporaria* and *Bufo bufo* as anuran models. *Reproduction, Fertility and Development* **24**, 599–607. doi:10.1071/RD10324

Uteshev V, Shishova N, Kaurova S, Manokhin A, Gakhova E (2013) Collection and cryopreservation of hormonally induced sperm of pool frog (*Pelophylax lessonae*). *Russian Journal of Herpetology* **20**, 105–109.

Uteshev V, Kaurova S, Shishova N, Stolyarov S, Browne R, Gakhova E (2015) In vitro fertilization with hormonally induced sperm and eggs from sharp-ribbed newts *Pleurodeles waltl*. *Russian Journal of Herpetology* **22**, 35–40.

Vu M, Trudeau VL (2016) Neuroendocrine control of spawning in amphibians and its practical applications. *General and Comparative Endocrinology* **234**, 28–39. doi:10.1016/j.ygcen.2016.03.024

Vu M, Weiler B, Trudeau VL (2017) Time- and dose-related effects of a gonadotropin-releasing hormone agonist and dopamine antagonist on reproduction in the northern leopard frog (*Lithobates pipiens*). *General and Comparative Endocrinology* **254**, 86–96. doi:10.1016/j.ygcen.2017.09.023

Wells KD (2007) The natural history of amphibian reproduction. In *The Ecology and Behavior of Amphibians*. (Ed. KD Wells) pp. 451–515. University of Chicago Press, Chicago IL, USA.

Whitaker BR, Wright KM (2001) Clinical techniques. In *Amphibian Medicine and Captive Husbandry*. (Eds KM Wright and BR Whitaker) pp. 89–110. Krieger Publishing Company, Malabar FL, USA.

Wolf DP, Hedrick JL (1971) A molecular approach to fertilization 11. Viability and artificial fertilization of *Xenopus laevis* gametes. *Developmental Biology* **25**, 360–376. doi:10.1016/0012-1606(71)90037-6

Wright KM (2001) Restraint techniques and euthanasia. In *Amphibian Medicine and Captive Husbandry*. (Eds KM Wright and BR Whitaker) pp. 111–122. Krieger Publishing Company, Malabar FL, USA.

Yániz J, Soler C, Santolaria P (2015) Computer assisted sperm morphometry in mammals: a review. *Animal Reproduction Science* **156**, 1–12. doi:10.1016/j.anireprosci.2015.03.002

Yoshida M, Inaba K, Ishida K, Morisawa M (1994) Calcium and cyclic AMP mediate sperm activation, but Ca^{2+} alone contributes sperm chemotaxis in the ascidian, *Ciona savignyi*: (ascidian/sperm motility/chemotaxis/calcium/cAMP). *Development, Growth & Differentiation* **36**, 589–595. doi:10.1111/j.1440-169X.1994.00589.x

Yunes R, Fernández P, Doncel GF, Acosta AA (2005) Cyclic nucleotide phosphodiesterase inhibition increases tyrosine phosphorylation and hyper motility in normal and pathological human spermatozoa. *Biocell* **29**, 287–293. doi:10.32604/biocell.2005.29.287

Zohar Y, Mylonas CC (2001) Endocrine manipulations of spawning in cultured fish: from hormones to genes. *Aquaculture* **197**, 99–136.

8   Genetic management of threatened amphibians: using artificial fertilisation technologies to facilitate genetic rescue and assisted gene flow

Phillip G. Byrne and Aimee J. Silla

INTRODUCTION

Genetic considerations should be at the forefront of plans to manage threatened species. Without careful consideration of the potential for genetic issues (such as loss of genetic diversity and inbreeding), even the most robust conservation plans are likely to fail. Consequently, there is an urgent need to integrate genetics and evolutionary theory into planning (Ralls *et al.* 2018). Critical to the success of this integration will be the acquisition of knowledge concerning the genetic underpinnings of phenotypic variation and adaptive variation. Such quantitative genetic information is directly relevant to most of the genetic issues in conservation biology as it provides a basis for genetic management (Frankham 1999). For animals characterised by external fertilisation, artificial fertilisation (also referred to as *in vitro* fertilisation) offers a powerful tool enabling controlled-breeding experiments and quantitative genetic analyses aimed at elucidating the genetic basis of phenotypic variation within and between populations. Amphibians are prime candidates for this research because almost all species (> 90%) are characterised by external fertilisation, and, remarkably, artificial fertilisation was first achieved in anuran models over 240 years ago (artificial fertilisation was first achieved by Spallanzani in 1776) (Capanna 1999). As an outcome, artificial fertilisation protocols have been developed for a diversity of species globally, with an excellent fundamental knowledge of general procedures required to optimise the release of gametes and promote successful fertilisation (Silla and Byrne 2019). In recent years there has been increasing recognition of the power of artificial fertilisation to assist with amphibian conservation (Silla and Byrne 2019), yet application of artificial fertilisation to threatened amphibian species remains limited (Browne *et al.* 2006; Byrne and Silla 2010; Turani *et al.* 2015, 2020; Upton *et al.* 2021). In this chapter we highlight the enormous, although largely untapped, potential for artificial fertilisation to assist with the genetic management of threatened anurans. We begin by identifying loss of genetic diversity and evolutionary potential as a pervasive threat to amphibian biodiversity. We then discuss how artificial fertilisation can address this concern by facilitating controlled breeding experiments aimed at testing the feasibility of various contemporary conservation actions. We focus our attention on discussing how artificial fertilisation can be used to test (1) the genetic compatibility and combining ability of populations targeted for population augmentation; (2) the genetic compatibility of animals integral to conservation breeding programs; and (3) the potential for assisted gene flow to improve the evolutionary potential of wild populations. We draw on quantitative-genetic models of modern evolutionary theory, and a small, although highly informative, empirical literature to assess the feasibility of each approach. Our goal is to present a set of case studies whose approaches and findings can be synthesised to provide a conceptual basis for future work. Extending from this foundation, our overarching goal is to provide a valuable resource for conservation practitioners. To this end, we propose a series of methodological and investigative frameworks to aid current and future amphibian conservation programs. We centre our attention on (1) outlining powerful, yet user friendly, experimental designs (and associated statistical and quantitative genetic analyses); and (2) identifying essential anuran artificial fertilisation techniques and protocols required to ensure the collection of robust data.

A growing need to maintain and preserve genetic diversity

Fundamental to recent advances in conservation genetics has been a general recognition of the pressing need to maintain and preserve genetic diversity, a goal endorsed by the IUCN (Coates *et al.* 2018). Loss of genetic variation represents a major threat to the persistence of populations, with irrefutable links between the loss of genetic variation and the extirpation of populations (Frankham 2005). When populations become small and genetically isolated, the inevitable outcome is that they will lose genetic diversity (through genetic drift and directional selection). In turn, with each passing generation they will become increasingly inbred, and more susceptible to a loss of fitness (inbreeding depression) (Frankham 2015; Weeks *et al.* 2017). This problem will be compounded by reduced adaptive potential, which is expected to elevate the risk of extinction, particularly in environments exposed to rapid environmental change (Frankham 2015; Weeks *et al.* 2017). In these situations, it is essential for conservation managers to develop effective strategies to preserve genetic variation. Implementing these strategies may only be possible after other priorities for management have been met, but it is important to factor genetic management into any long-term plans for conservation.

Preservation of genetic variation through population augmentations and genetic rescue

Arguably, the simplest way to elevate genetic variation in small, isolated populations is through population augmentations. Also referred to as supplementation or restoration, this approach involves supplementing a declining population with conspecific individuals from one or more genetically distinct source populations (Frankham *et al.* 2011; Weeks *et al.* 2011). In principle, providing a declining population with new genes (or more specifically alleles) sourced from populations with higher genetic diversity stands to overcome the problem of inbreeding depression (Weeks *et al.* 2011; 2017). Essentially, the introgression of novel genes is expected to counter the expression and accumulation of deleterious alleles (Keller and Waller 2002; Hedrick and Fredrickson 2010; Kronenberger *et al.* 2018). In turn, this should improve the recipient population's resilience and persistence by reducing vulnerability to demographic or environmental stochasticity (Ficetola and De Bernardi 2005; Weeks *et al.* 2011; Kronenberger *et al.* 2018). These presumed evolutionary advantages, termed 'genetic rescue', have led to population augmentation being widely recommended as a conservation action. Strong support for implementing this approach has come from the demonstration of genetic rescue in several iconic endangered species, such as the Florida panther (Pimm *et al.* 2006) and the Australian mountain pygmy possum (Hedrick and Fredrickson 2010; Frankham 2015; Weeks *et al.* 2017). Importantly, the broad-scale relevance of population augmentation continues to face scrutiny, fuelled mostly by concerns that there could be long-term negative genetic consequences resulting from matings between individuals from evolutionarily distinct populations (outbreeding depression) (Tallmon *et al.* 2004; Ficetola and De Bernardi 2005; Edmands 2007; Frankham *et al.* 2011; Huff *et al.* 2011; Weeks *et al.* 2011). Indeed, studies in a diversity of taxa have reported evidence for outbreeding depression (reviewed by Whitlock *et al.* 2013).

Why might population augmentation have negative genetic outcomes?

Negative genetic outcomes can result from outbreeding when parental populations are adapted to different environmental conditions (and are genetically distinct), and hybrids possess intermediate phenotypes that make them less fit in one or both environments (termed 'extrinsic outbreeding depression') (Whitlock *et al.* 2013). Furthermore, hybrids can suffer reduced fitness if there is a high level of genetic divergence between donor and recipient populations. Specifically, matings between locally adapted animals (from different populations and environments) can result in the disruption of co-adapted gene complexes, leading to parental genetic incompatibility, reduced fitness of the offspring, and compromised viability of the population (termed 'intrinsic outbreeding depression') (Goldberg *et al.* 2005; Frankham *et al.* 2011). For amphibians, the risk of outbreeding depression is a legitimate concern and should not be overlooked in the push to implement the augmentation of amphibian populations (Sagvik *et al.* 2005). Amphibians typically show extreme levels of genetic differentiation between populations due to strong philopatry, low dispersal abilities, and uneven distribution of breeding habitats (Beebee 2005; Sagvik *et al.* 2005; Watts *et al.* 2015). Consequently, for any given species, there is an inherent risk of outbreeding that necessitates empirical assessment of the genetic consequences of mixing populations before a program of population augmentation is initiated.

ARTIFICIAL FERTILISATION TO THE RESCUE: A TOOL FOR TESTING THE GENETIC CONSEQUENCES OF AUGMENTING POPULATIONS

For species characterised by external fertilisation, a direct way to test the suitability of population augmentation as a conservation action is to use artificial fertilisation to perform controlled-breeding experiments. For these experiments, gametes can be obtained from both sexes (see 'General methodological considerations' below), and artificial fertilisation can be used to make crosses between individuals from putative donor and recipient populations. Subsequent assays for the fitness of progeny from each family that is generated allows direct evaluation of the genetic consequences of outbreeding. Anurans provide excellent models for this type of investigation because, for most species, large numbers of gametes can be obtained from both sexes, allowing for comparison between tens of individuals across multiple populations (see discussion of experimental design below). Surprisingly, only four amphibian studies (spanning three species) have used artificial fertilisation to run breeding experiments that allow the genetic outcomes of augmenting populations to be evaluated. Below, we discuss each study in detail, with the goal of comparing and contrasting experimental approaches, and highlighting experimental deficiencies that need to be overcome in future studies.

Case studies of amphibian population augmentation

European common frogs (*Rana temporaria*)

This species is Europe's most abundant and widely distributed anuran, and is currently listed by the IUCN Red List as a species of Least Concern. Sagvik *et al.* (2005) used this species as a model to explore the influence of outbreeding on fitness and employed artificial fertilisation to cross 88 individuals (22 males and 22 females per population) from two disconnected populations separated by 130 km. One population was large (> 4000 frogs) whereas the other was relatively small (~300 frogs), isolated, and presumably inbred. The eggs of females belonging to each experimental group were fertilised by two experimental males, one from the same population and one from the different population. This experimental design provided some important insights into the potential genetic consequences of outbreeding. For females from the large population, crosses with males from the small population (simulating outbreeding) resulted in a higher incidence of malformed hatchlings compared to crosses with a male from the same population. For the small population, the patterns were similar, although not significant. Considered together, these findings were suggestive of population-specific outbreeding depression. In contrast, in a follow-up experiment, Uller *et al.* (2006) found that outbreeding resulted in an increase in the size of metamorphs when females from the small population were crossed with males from the large population. However, in the reciprocal cross the pattern was in the opposite direction, suggesting strong genetic or non-genetic maternal affects. Nevertheless, the fitness benefit reported was in line with the notion that hybrid offspring generated via outbreeding will have elevated fitness (heterosis) (Frankham 2015). When considered within the context of the earlier study, this finding suggests that the incidence of outbreeding depression may depend on the life stage at which the offspring's performance is measured. This knowledge has important implications for the design of fitness assays (see 'General methodological considerations' below).

Rice frogs (*Fejervarya limnocharis*)

Rice frogs have a broad distribution extending from India to Japan and the species is currently listed by the IUCN Red List as of Least Concern. With the goal of exploring genetic divergence in rice frogs, Sumida *et al.* (2007) made 39 crosses using a mix of individuals taken from populations in Japan (seven females, three males), Northern Thailand (three females, two males), Central Thailand (two females, three males) and Sri Lanka (one male). Because sample sizes were so low, the study suffered several major deficiencies. Most notably, the mating design did not use a balanced and systematic crossing program. Clutches were not evenly split between sires, frogs from each population were not crossed in every possible combination, and an equal number of males and females were not used from each population. This created the potential for individual pairs, frogs, or even eggs to have a disproportionate influence on average fitness values. In turn, this could have inflated or deflated estimates of variance, and dramatically altered the perceived risk of outbreeding depression. Moreover, some population crosses were not well replicated, only involving one or two mating pairs, providing an extremely limited representation of 'within population' genetic variation, and thereby limiting the potential for detecting heterotic effects. Nevertheless, by providing robust estimates of genetic distance, the study was able to infer that the degree of genetic similarity between populations predicted inter-population compatibility. Specifically, crosses between more genetically divergent populations resulted in higher levels of abnormal development and mortality both in embryonic and larval stages. Interestingly, while some crosses failed to produce any viable offspring, most crosses produced at least some individuals that metamorphosed, indicating that population augmentation potentially could be carried out over large geographical distances in this species, as well as in other broadly distributed species free from barriers to gene flow. Nevertheless, given the deficiencies of the experimental design, the data should be interpreted with caution, and additional research would be needed to assess the feasibility of population augmentation in this species.

Brown toadlets (*Pseudophryne bibronii*)

Brown toadlets were once widely distributed along the eastern seaboard of Australia, but in recent years have suffered rapid declines and are currently listed by the IUCN Red List as Near Threatened. With the goal of exploring the potential for population augmentation, Byrne and Silla (2020) used artificial fertilisation and a cross-classified breeding design (North Carolina type II) to conduct a preliminary investigation into the risk of outbreeding depression. In three discrete experimental blocks (involving either five females and five males, or six females and five males), a group of males ($n = 15$) and females ($n = 14$) were crossed in every possible pairwise combination. The study involved frogs from three discrete populations, with each experimental block involving frogs from the same population, as well as a single foreign female. The final combined design involved 85 families of paternal and maternal half-siblings, with 70 within-population crosses and 15 between-population crosses. The main finding from the study was that outbreeding consistently had a negative effect on offsprings' fitness, with outcrossed pairings having significantly lower fertilisation success, hatching success, and lower survival of larvae to metamorphosis (Byrne and Silla 2020). Overall, more than half of the between-population crosses (i.e. outbreed pairings) failed to produce viable offspring (compared to the within-population crosses), indicating that brown toadlets may be highly susceptible to genetic incompatibility and outbreeding depression. Moreover, because populations were only ~20 km apart, the study provided novel evidence that outbreeding depression in anurans can occur over very narrow geographical scales. A limitation of this study, however, was that all of the 'between-population' crosses involved females from the same population; consequently, genetic factors specific to this population may have been responsible for the 'between-population' incompatibilities that were reported. Future studies could avoid such issues by employing full factorial breeding designs that involve an equal number of parents from multiple populations (for further discussion see 'General methodological considerations' below).

Future directions for amphibian population-augmentation programs: general advice for conservation practitioners

Taken together, findings produced by the aforementioned studies provide two important take-home messages relevant to the development of genetic rescue programs for amphibians. First, because the studies spanned a taxonomically diverse set of species, the possibility exists that outbreeding depression is a real issue for various amphibian groups. Accordingly, when working with a new target species, and beginning to discuss population augmentation as a recovery action, it would be prudent for conservation managers to prioritise investigation into the risk of outbreeding depression. Second, the potential for outbreeding depression (complete failure of clutches in some cases) makes it very clear that managers should implement a program of population augmentation with extreme caution, particularly if poor fertilisation rates are observed in controlled-breeding studies. If there are high levels of genetic incompatibility detected, releasing large numbers of frogs into a threatened population could compromise the viability of the population and accelerate its decline. In these cases, alternative actions, such as *in situ* management, would need to be considered. By contrast, if low to moderate levels of genetic incompatibility are detected, managers should weigh the risk of outbreeding depression against the immediate risk of extirpating the population using a structured decision-making framework (Weeks *et al.* 2011; Ralls *et al.* 2018). If intervention is essential, and *in situ* management is unsuitable, the best approach may be to introduce a relatively small number of individuals each breeding season. In theory, so long as crosses are not 100% incompatible, and the level of gene flow from the donor population does not exceed 20%, incremental introgression of new genes may lead to the recovery of population fitness without compromising locally adapted gene complexes (Weeks *et al.* 2011). Under this scenario, artificial fertilisation could be used to screen for incompatibilities, with the subsequent release of animals restricted to only the most viable individuals (Uller *et al.* 2006). Importantly, because the F1 hybrids will be more genetically (and presumably phenotypically) similar to frogs in the recipient population, this approach should also reduce the risk of behavioural incompatibility, and expedite the introgression of new alleles (and manifestation of beneficial heterotic effects). Without question, artificial fertilisation is a highly effective way to conduct controlled-breeding experiments that permit a direct assessment of the risk of outbreeding and the feasibility of augmenting populations. Moreover, these assessments can be performed relatively quickly and inexpensively, particularly when compared to the practice of using genetic markers to infer suitability for augmentation based on levels of genetic divergence. The feasibility of doing this continues to be debated, given a lack of consistent empirical evidence for an association between the genetic distance of parental populations and outbreeding depression (Whitlock *et al.* 2013). Given the aforementioned benefits of using artificial fertilisation to directly assess the risk of outbreeding depression, we encourage the use of artificial fertilisation for this purpose in amphibian conservation programs worldwide. When designing breeding experiments to test for outbreeding depression, there are several experimental details that require careful consideration. Here we provide some recommendations for running these types of experiments in order to avoid generating spurious findings that misdirect management practices. We outline an experimental design, and associated set of methodologies, that will allow an effective test of the feasibility of augmenting populations for the purpose of genetic rescue.

Experimental design and analysis to test for heterosis and outbreeding depression

What is the best mating design to use?

How well parental genomes combine during hybridisation to transmit beneficial (or desirable) traits to offspring is termed 'combining ability' (Fasahat *et al.* 2016). The most balanced and systematic biometrical approach to deduce the combining ability of populations earmarked for an augmentation program is to employ a full-diallel cross-mating design. In this context, the design involves taking a set of males and females (in equal numbers) from two genetically distinct populations and generating hybrid offspring from all possible parental genotypes (i.e. crossing every parent/genotype in every possible pairwise combination) (Figure 8.1B). Critically, the full design includes reciprocal crosses, whereby crosses are made between females from population 1 and males from population 2, as well as between females from population 2 and males from population 1 (Figure 8.1A and B). These reciprocal crosses make it possible to test for population-specific paternal and maternal effects. Understanding the relative importance of these effects may be critical in anurans given the potential for maternal effects, and the findings of Uller *et al.* (2006) that estimates of outbreeding depression in European common frogs were strongly influenced by which sex (and population) was tested. In the final mating design we describe, 1/4 of the crosses represent 'within-population crosses' from population A, 1/4 of the crosses represent 'within-population crosses' from population B, 1/4 of the crosses are 'between-population' crosses involving females from population A and males from population B, and 1/4 of the crosses are 'between-population' crosses involving females from population B and males from population A (Figure 8.1B). This fully balanced design enables the fitness of offspring from non-outbred crosses within both parental populations to be quantified as comparators for hybrid performance.

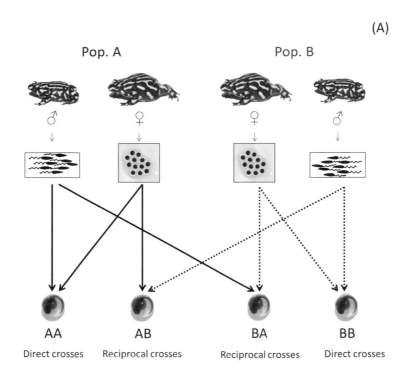

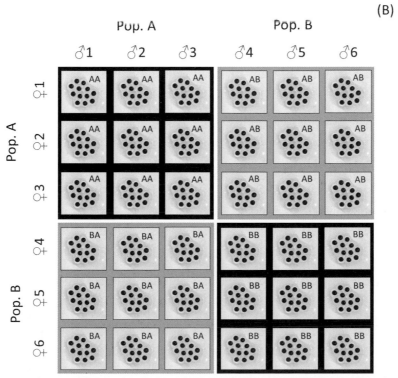

Figure 8.1: Schematic diagram showing (A) a mating design incorporating direct and reciprocal crosses between a male and female from one population (Pop. A), and a male and a female from a second population (Pop. B), and (B) a full diallel cross mating design for three males and three females (represented by numbers 1, 2, and 3) from one population (Pop. A) crossed in every pairwise combination with three males and three females (represented by numbers 4, 5, and 6) from a second population (Pop. B). Cells framed in black represent purebred crosses while cells framed in grey represent hybrid crosses.

Experimental blocks

When running a diallell cross experiment it is essential to keep in mind that there is an inherent risk of introducing temporal confounds. Specifically, crosses need to be performed very quickly after collecting the gametes to avoid any decline in the gametes' quality during short-term storage, which could influence fertilisation capacity and offsprings' performance (Gasparini *et al.* 2017). This time restriction limits the number of crosses that can be made over a set period. To maximise sample size, while ensuring effective management of time, we recommend that the full diallel cross design be used in combination with a randomised block design. Specifically, we recommend that each diallel cross experiment be replicated over several experimental blocks, using a new set of randomly sampled individuals for each block.

Estimating combining ability

Two important concepts associated with the diallel cross design, and commonly used when evaluating the level of compatibility between individuals from different populations are generalised combining ability (GCA) and specific combining ability (SCA). GCA refers to the generalised combining ability of any given parental genotype (average performance of a parental genotype in a series of hybrid crosses), while SCA refers to the specific combining ability of two parents (performance of a parental genotype in one specific hybrid cross) (Fasahat *et al.* 2016). Critically, GCA and SCA provide information on gene action involved in the expression of quantitative characters (the fitness-determining traits under investigation) expressed by offspring in the F1 generation. Generalised combining ability effectively reflects total additive genetic effects, whereas SCA reflects total non-additive genetic effects (dominance and epistasis) that result from interactions between parental genomes. Analysis of combining ability therefore allows practitioners to identify suitable parents for hybridisation, and superior cross combinations that may be used to exploit heterosis (hybrid vigour) to improve the population's viability through augmentation (Fasahat *et al.* 2016).

Statistical analyses

Data obtained from a diallel cross design is typically analysed using general linear models (GLMs) with fixed effects to estimate (1) GSA and SCA (to allow the viability of parental combinations to be ranked); (2) reciprocal effects; (3) environmental effects (if offspring are reared in multiple environments); and (4) any interactive effects of interest (e.g. GSA × environment). If a randomised block design is used, block effects can also be included in the model as a random effect. GLMs also can be used to specifically test for paternal and maternal effects, and estimate genetic components of variation (additive genetic effects, dominant genetic effects, epistatic genetic effects). In recent years, Bayesian approaches also have been developed to deal with the problem of estimating parameters from data that are incomplete, imbalanced, or contain numerous outliers (Lenarcic *et al.* 2012).

Sample sizes: number of parents

A fundamental question to ask before running a diallel cross experiment is: how many parents should be used? There is no simple answer to this question because it will depend on the type of information being sought. If the goal is to estimate combining ability, then a minimum of four parental crosses is needed to generate enough data to make the required calculations. Whether these data are meaningful or not will depend on the variance surrounding the estimates of GCA and SCA. If the variance is high, the sample size will need to be increased to gain more precise estimates. If the goal is to gain estimates of genetic parameters, the minimum number of parents to include is much less clear, largely because this topic has received very little empirical attention. Some general insights have come from early theoretical work and a small number of computer simulations, which together have indicated that eight to 10 parents would be needed to obtain reliable estimates (Hayman 1960; Pederson 1971; Hayward 1979). An informal review of the literature on plant and animal breeding indicates that the number of parents involved in diallel cross-mating designs is highly variable, although typically it ranges between four and 20 parents per population/strain. In reality, sample sizes largely will be decided, based on biological constraints imposed by the species (such as clutch size), as well as availability of resources such as space to rear offspring. With these limitations in mind, a logical approach when investigating the feasibility of population augmentation would be to run as many crosses as manageable. Even if the variance surrounding the genetic estimates is high, experimenters should at least gain some insight into whether crosses between populations generate viable hybrids (e.g. Persson *et al.* 2007). These data can then be used to run power analyses to determine the scale of work required to gain deeper insights into the genetic basis of variation in specific traits, or to allow initial decisions to be made about the value of trialling the release of a small number of individuals annually (Weeks *et al.* 2011).

Sample sizes: number of offspring per family

Another important question is how many offspring should be tested per family? Again, there is no simple answer. Saying this, using very low numbers of individuals (e.g. one to four) per family greatly increases the risk of generating imprecise and spurious estimates of mean fitness. Stated simply, as sample size increases the variance of the sample mean will decrease, and precision will increase. Therefore, once again, a logical approach will be to maximise sample size based on biological and practical constraints. For most anuran species, clutch sizes are large (> 100 eggs), so, in principle, it should be possible to have at least 10 offspring per family in a diallel involving eight to 10 females.

What generation should be tested?

Theoretically, reductions of fitness associated with outbreeding depression are expected to be manifest in the F1 generation, and then become more evident in the F2 generation and beyond. This is because between-population heterozygosity that increased in the F1 (and may have caused heterosis) will effectively be halved by segregation. Moreover, recombination will disrupt locally adapted combinations of parental genes present in each source population (Frankham 2016). Indeed, there is empirical evidence from various taxa to support such generational shifts (Whitlock *et al.* 2013). While tests in the F1 generation are informative, the full effects of outbreeding might only be elucidated by extending fitness assays into the F2 generation, and beyond (Frankham 2016). If the fitness of the F2 hybrids differs from the mid-parent value, then outbreeding can be considered to have had an absolute impact (for better or worse). It should be mentioned that in nature F1 hybrids are likely to backcross to parental lines. Offspring from these mating's will have levels of heterozygosity equivalent to F2 and later-generation hybrids, but with a reduced epistatic cost (Whitlock *et al.* 2013). Given this expectation, it may be informative for practitioners to also explore changes in fitness that result from backcrossing. If benefits were identified, this would support the managed immigration of hybrids.

USING ARTIFICIAL FERTILISATION AND BREEDING EXPERIMENTS TO ASSESS GENETIC COMPATIBILITY WITHIN POPULATIONS

Within populations, artificial fertilisation can be used to facilitate quantitative genetic investigation into the influence of parental genotypes on the fitness of offspring. Knowledge gained from this approach has considerable potential to assist with the recovery of threatened species by improving the genetic management of captive populations deemed integral to the success of conservation breeding programs.

Genetic quality and the viability of captive populations

In an effort to maximise genetic diversity and minimise deleterious effects resulting from inbreeding and genetic drift, vertebrate conservation breeding programs traditionally have adopted the approach of random-pairing. However, there is a growing body of evidence to indicate that this approach fails to capture the complexity of genetic quality within populations (Neff and Pitcher 2005; Neff *et al.* 2011). Theoretically, genetic quality consists of two components: (1) the independent effects of parental genotypes (good genes); and (2) the interaction between parental genotypes (compatible genes) (Byrne *et al.* 2019). If 'good genes' determine offspring fitness, certain individuals will consistently produce offspring of superior viability, irrespective of the quality of their mating partner. By contrast, if 'compatible genes' are important, resulting from allelic interactions within loci (dominance) or between loci (epistasis), pairings between certain individuals (genotypes) are required to produce viable offspring. From an evolutionary perspective, 'good genes' will come under directional selection, and, because they are heritable, the effects of 'good genes' will be revealed as additive-genetic variance in fitness. By contrast, 'compatible genes' will not be targets of directional selection, and the effects of 'compatible genes' will maintain non-additive genetic variation (Byrne *et al.* 2019).

When working with species characterised by external fertilisation, it is possible to use artificial fertilisation to construct fully replicated cross-classified breeding designs that control (or remove) non-genetic sources of variation and simultaneously estimate the relative contributions of additive and non-additive sources of genetic variation to the fitness of offspring. Understanding the importance of genetic quality (arising from 'good genes' and 'compatible genes') for the fitness of offspring has substantially improved the success of captive breeding programs for a diversity of endangered fish by improving the number and survival of offspring generated (Neff *et al.* 2011). Therefore, we recommend that amphibian conservation breeding programs move in this direction to better understand how variance in genetic quality might influence the long-term viability of captive populations. Quantitative genetic studies in a small number of non-threatened species of frogs (conducted in the context of sexual selection so as to better understand the genetic benefits of mate choice and polyandry) provide a basis for applying this approach to the management of threatened species. We discuss several case studies below, highlighting subtle differences in experimental design that can be integrated into future conservation-focused programs.

Case studies of amphibian genetic quality

Quacking frogs (*Crinia georgiana*)

This small to medium-sized ground-dwelling frog is characterised by high levels of simultaneous polyandry, but with no obvious material benefit derived from this behaviour. Therefore, Dziminski *et al.* (2008) ran a cross-classified experiment to test whether multiple paternity might function as a genetic bet-hedging strategy to improve the genetic quality of offspring. Using artificial fertilisation and a North Carolina type II mating design, a set of four males and four females were crossed in every pairwise combination and this crossing design was repeated in four experimental blocks, producing 16 families of maternal and paternal half-siblings in each block, and a total of 64 families. Unexpectedly, there was no evidence that fertilisation success or offspring fitness was influenced by additive genetic effects (i.e. there was no evidence that sires can supply their offspring with good genes). In contrast, the presence of strong and consistent non-additive genetic effects on fertilisation success and on various fitness-related traits indicated that parental genetic compatibility is critical for the quality of the offspring. A novel aspect to the study was that it controlled for variation in maternal provisioning by measuring eggs, and including egg size (yolk volume) as a covariate in the statistical analyses. Because variation in the provisioning of yolk in amphibians is known to influence the size, performance, and viability of offspring, future studies should routinely photograph eggs (for later measurement) immediately before (or immediately after) artificial fertilisation.

Brown toadlets (*Pseudophryne bibronii*)

The brown toadlet, a terrestrial-breeding Australian frog, has the highest known incidence of sequential polyandry of any vertebrate, with females routinely dividing their eggs between the nests of up to eight males (Byrne and Keogh 2009). To test for the genetic benefits of polyandry, Byrne *et al.* (2019) used artificial fertilisation to produce 70 families of paternal and maternal half-siblings. This was achieved by conducting three experimental blocks. In the first block, four females were crossed with five males, and, in the second and third blocks, five females were crossed with five males. Quantitative genetic analysis revealed no significant additive genetic effects on the viability of embryos and larvae, although highly significant non-additive genetic effects (sire-by-dam interactive effects) were observed. Negative effects were not apparent at fertilisation but occurred in the embryo and larval stages, indicating that negative impacts of parental genetic incompatibility are increasingly manifest during development. This change highlights the importance of testing for genetic effects on the viability of offspring across multiple life stages.

Red-backed toadlets (*Pseudophryne coriacea*)

The red-backed toadlet is an Australian terrestrial-breeding anuran in which females are highly discriminant in mate choice and prefer to mate with more closely related males (O'Brien *et al.* 2019). To test whether females produced more viable offspring by mating with close relatives, Byrne *et al.* (2021), used artificial fertilisation to produce 64 families of paternal and maternal half-siblings. This was achieved by making three experimental blocks. In the first block, four females were crossed with four males, and, in the second and third, six females were crossed with four males. The study also took the novel approach of genotyping all the parents using single nucleotide polymorphisms (SNPs), and including an estimate of parental genetic similarity as a fixed effect in the analyses. The study found no significant additive genetic effects, but highly significant non-additive genetic effects (sire-by-dam interactive effects), on fertilisation success and on the viability of offspring. The implication of these results was that genetic compatibility can strongly influence female mate choice in frogs. Parental genetic similarity (estimated using SNPs) had no effect on the success of fertilisation or the viability of offspring, yet this probably reflected the fact that the parents used in the study were not overly genetically variable. Nevertheless, inclusion of genetic similarity in the analyses represents a methodological advance that should be considered for future studies.

Future directions for conservation breeding programs: general advice for conservation practitioners

The studies described above provide consistent evidence for compatible genetic effects across species, highlighting the profound impact that parental genetic compatibility can have on the quality of offspring, and the value of testing for such effects in captive populations. Interestingly, although the case studies described above provide no evidence for 'good genes' effects, studies investigating the influence of parental genetic quality on the offsprings' quality in the context of environmental change have provided evidence for 'good genes'. Therefore, the relative importance of 'good genes' versus 'compatible genes' may vary considerably across species. Irrespective of which genetic mechanism is most important, the critical point is that artificial fertilisation can facilitate quick experiments that provide key insights into the influence of parental genetic quality and interactions on the quality of offspring. This has important implications for the management of captive populations of threatened anurans. For example, a better understanding of these effects could allow more rapid generation of viable offspring, which could help maintain the size and viability of captive colonies. In turn, this could increase the number and quality of individuals available for reintroduction, which could improve recovery success (see Armstrong and Seddon 2008). The approach we describe will also allow the fine-scale genetic management of captive populations to avoid genetic issues, such as inbreeding depression and adaptation to captivity. To the best of our knowledge no amphibian conservation breeding program is routinely using artificial fertilisation to manage the genetic quality of a threatened amphibian; thus we strongly encourage conservation practitioners to consider doing so. Below we provide some guidelines to assist with this integration.

Experimental design and analysis relevant to amphibian captive breeding programs

What is the best mating design to use?

The most effective quantitative genetic approach to test for 'good-genes' or 'compatible-genes' effects on the viability of offspring is the North Carolina type II mating design. This cross-classified design involves taking a set of sires and dams from the same population and crossing them in every possible pairwise combination (Lynch and Walsh 1998) (Figure 8.2). By splitting the clutch of each female between multiple males it is possible to hold maternal effects constant so that 'half-sibs' differ only in paternally inherited genes. Across multiple crosses (families), variance in the fitness of offspring can then be precisely partitioned among additive genetic effects (good genes), non-additive effects (compatible genes), and

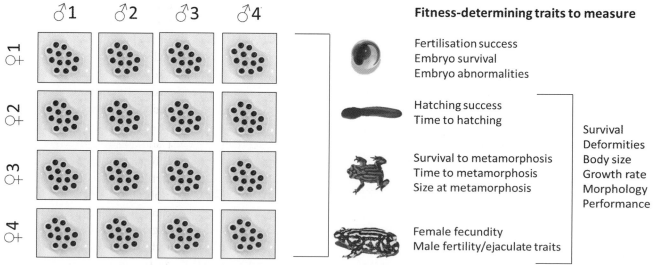

Figure 8.2: Schematic diagram showing a North Carolina type II mating design whereby a set of sires (here four males) and dams (here four females) from the same population are crossed in every possible pairwise combination, and a suite of key offspring traits to measure that represent different components of fitness in anuran amphibians.

maternal effects (encompassing maternal genetic effects and environmental effects) (Lynch and Walsh 1998; Byrne *et al.* 2019; Byrne *et al.* 2021). Importantly, in this design the influence of environmental effects can also be minimised (or manipulated) by raising half-sibs in a controlled environment (Rudin-Bitterli *et al.* 2018). In most cases, the best approach is to simulate conditions experienced by the species in the wild to ensure that the expression of traits is ecologically relevant. In some cases there may be value in challenging offspring (e.g. exposing them to extreme developmental conditions), as this is when the impacts of 'good genes' and 'compatible genes' on fitness will be most apparent (Dziminski *et al.* 2008; Rudin-Bitterli *et al.* 2018). If offspring are reared in benign non-challenging environments such effects might be masked. This advice regarding choice of environments for rearing amphibians also applies to investigations of the feasibility of augmenting populations for genetic rescue.

Sample sizes: number of parental crosses and number of offspring per family

As discussed earlier in regard to inter-population crosses, the number of crosses made and the number of offspring per family will be dictated by biological and practical constraints, but care should still be taken to avoid very low numbers that would generate spurious findings. Data can be analysed with as few as one individual per family, but increasing replicates will reduce variance and increase the capacity to draw firm conclusions.

Statistical analysis and quantitative genetic analysis

To estimate effects of sires, dams, and their interaction on the fitness of offspring, generalized linear mixed effects models (GLMM) typically are used to partition sources of phenotypic variation in fitness-determining traits; this analysis normally implements restricted maximum-likelihood methods (REML) (Byrne *et al.* 2019; Byrne and Silla 2020). In the model, sire ID, dam ID, and the interaction between sire and dam are treated as random effects. If the experiment involves blocks, the block ID is also included as a random effect. If the experiment involves an environmental treatment (e.g. offspring reared in two contrasting environments) this is included in the model as a fixed effect. In this instance, sire and dam can also be independently crossed with the environment to test for genotype-by-environment interactions underlying the expression of traits. In addition, the model allows for inclusions of covariates (included as random effects) such

as eggs' size (to control for maternal provisioning) and genetic similarity. For quantitative genetic analyses, components of variance are extracted from the GLMMs or as VS (sire), VD (dam), and VI (sire × dam interaction). Based on the assumption that epistasis is negligible, additive (VA), non-additive (VNA, including dominance) and maternal variance (VM) are typically calculated as follows: VA = 4VS; VNA = 4VI; VM = VD – VS (Lynch and Walsh 1998). The sire variance components (covariance among paternal half-siblings) provide an estimate of additive genetic effects, and dam variance components (the covariance between maternal half-siblings) provide an estimate of genetic and environmental maternal effects. The sire × dam interaction variance provides an estimate of the genetic variance due to non-additive nuclear gene action (dominance and extranuclear interactions).

USING ARTIFICIAL FERTILISATION AND BREEDING EXPERIMENTS TO EXPLORE EVOLUTIONARY POTENTIAL

Rapid environmental change is creating a myriad of stressors threatening amphibians globally. In the short term, many species can respond to environmental change by shifting distributions or expressing phenotypic plasticity (Hoffmann and Sgro 2011). In the long term, if directional change persists and populations become isolated, a genetic response will be necessary to avoid extinction (Sgrò *et al.* 2011). From a conservation perspective, our capacity to predict the potential for adaptive genetic responses will be greatly improved through an increased understanding of the genetic architecture of threatened populations (Merilä and Hendry 2014). In turn, acquiring this knowledge will provide the foundations for evaluating the feasibility of manipulation of genomes (assisted evolution) and population genetic structure to improve adaptive capacity and resilience (Aitken and Whitlock 2013). Within the framework of assisted evolution, one strategy receiving growing attention is assisted gene flow, the movement of individuals (or genotypes) into targeted populations from source populations, which are better adapted to a specific environmental stressor (e.g. high temperature) (Aitken and Whitlock 2013). The expectation is that beneficial pre-adapted genes will spread and improve a population's resilience and adaptive potential (Aitken and Whitlock 2013).

As assisted gene flow becomes increasingly considered for more threatened species, quantitative genetic studies will play an important role in assessing the feasibility of this action. Specifically, by using artificial fertilisation to run diallel crosses within populations it is possible to gain direct insight into the extent that variance in fitness-related traits are explained by additive genetic variance, which is the portion of genetic variance through which selection operates (Blows and Hoffmann 2005). Importantly, the extent to which genetically determined traits are expressed, and respond to selection, can depend on genotype-by-environment interactions (Charmantier and Garant 2005). Moreover, strong genotype-by-environment interactions will indicate genetic variation in phenotypic plasticity, which itself can be heritable and a target of selection (Tedeschi *et al.* 2016). Therefore, quantitative genetic studies will be most informative when they investigate sources of additive genetic variation in multiple populations (originating from contrasting environments) under a range of environmental conditions (discussed further below).

Rapid environmental change has played a major role in amphibian decline, with the main drivers being climatic change, disease, pollution, and destruction and fragmentation of habitats (see Chapter 2). Reflecting the widely held view that these threats will worsen at an exponential rate, a small but rapidly growing number of studies have employed artificial fertilisation and quantitative genetic approaches to investigate patterns of local adaptation in anurans and the genetic architecture of various traits related to viability. These studies, which to date have focused on understanding genetic variance in tolerance to anthropogenic acidification and drying of habitats, have delivered some important insights into anuran evolutionary potential and the feasibility of using assisted gene flow as a management strategy. Below we review some of the most influential studies conducted to date, with a focus on comparing the types of experimental approaches used, and assessing the potential for genetic responses to environmental change. Because assisted gene flow is yet to be implemented as a conservation tool for amphibians, we then propose an investigative framework to help direct future conservation efforts.

Case studies of environmental acidification and tolerance of acid-stress

Wood frogs (*Lithobates [Rana] sylvaticus*)

Wood frogs are the most widespread North American anuran species, and have an IUCN Red List status of Least Concern. Wood frogs have been exposed to the acidification of their habitat in various parts of their range, with largely unknown impacts on their populations. With the aim of investigating adaptations to environmental acidity, Pierce and Harvey (1987) studied four to eight populations, representing a range of environmental pH conditions. Artificial fertilisation was used to make within-population crosses, and embryos and larvae were reared across a range of pH. Comparison among populations showed that differences in tolerance of embryos to acid were unrelated to the extent of acidification in the source populations, but for larvae there were positive relationships. This result provided evidence for life stage–specific local adaptation in tolerance to acidity. In a set of follow-up experiments, the genetic basis of tolerance to acid was explored. Frogs from two populations were crossed (using a diallel cross-mating design) and the tolerance of embryos and larvae to acid was quantified. In the first experiment, six frogs from one population (three males and three females) were crossed with six frogs (three males and three females) from a different population in every possible pairwise combination, and 20–30 embryos per cross (family) were reared under neutral and acidic conditions. The second experiment used a similar design, although embryos were reared until hatching under neutral conditions, at which point six individuals per cross were subjected to a 24-h test of tolerance to acid. Geographical origin of the sires had no significant effect on the tolerance of embryos, a result that provided no evidence that the observed differences in tolerance among populations had any genetic basis. However, there was an association between maternal origin and embryos' tolerance, thereby providing evidence for maternal effects. Interestingly, the patterns were reversed for larvae because there was evidence for paternal-origin effects, with larvae of sires from more acidic sites having higher tolerances to acid. These results provided good evidence that both maternal affects and genetic effects contribute to wood frogs' tolerance to acid; this finding corroborated the results of an earlier study (Pierce and Sikand 1985). More broadly, the findings demonstrate that the expression of genetic effects can differ between life stages, reiterating the importance of testing for tolerance to stress across multiple life stages.

European common frogs (*Rana temporaria*)

In order to investigate the effects of acid on this common European ranid, Pakkasmaa *et al.* (2003) estimated genetic and maternal effects on the viability of tadpoles exposed to neutral and acidic conditions. Using artificial fertilisation, and following a factorial breeding design (North Carolina type 1), 60 families of paternal half-sibs were generated by crossing 30 males with two different females, with eggs from each family split between neutral and acidic pH. For several of the viability-related traits examined, maternal effects were strongly linked to variation in egg size, and these effects were mediated by pH, thereby providing evidence for environmentally mediated maternal effects (Pakkasmaa *et al.* 2003). Nevertheless, across both pH environments, viability and viability-related traits were significantly heritable (although heritability was low), and both additive and non-additive effects significantly contributed to variation in viability (at similar magnitudes). Although viability was significantly reduced under acidic conditions, there was no evidence that acid-stress altered estimates of heritability, or changed the relative contributions of additive and non-additive effects to variance. These findings indicate that tolerance to acid stress by European common frogs is underpinned to some extent by genetic effects, and that acid stress does not heavily impact the genetic architecture of fitness-related traits.

Moor frogs (*Rana arvalis*)

This species has a wide distribution throughout Europe (and is particularly abundant in eastern Europe and West Siberia) and has a status of Least Concern. Merilä *et al.* (2004) used artificial fertilisation to conduct a factorial mating experiment (North Carolina type II) in moor frogs to investigate the genetics of tolerance to acid-stress. Earlier studies of this species indicated that local adaptation in tolerance was highly likely, and also suggested a strong role for maternal effects (SäNEN *et al.* 2003a,b). For each of two populations (an acid-origin population and a neutral-origin population), three males and three females were crossed in every possible pairwise combination, with this design repeated three times (creating three blocks per population). The design did not involve crosses between populations. Offspring from each family were raised under five different pH treatments, with each treatment replicated three times per population (one per block). The most salient findings from the study were that both viability and viability-related traits had relatively low heritability, compared to the importance of dominance effects (reflecting parental genetic incompatibility) and maternal effects. Furthermore, the expression of additive genetic variance was independent of pH,

indicating limited additive genetic variation in tolerance to acid, and that variation in tolerance was underpinned by non-genetic maternal effects. Nevertheless, across all treatments, heritability was consistently significant (albeit low) for all traits, providing evidence that environmental acidification is unlikely to reduce the species' capacity to respond to selection. In a follow-up study, Persson *et al.* (2007) tested whether the findings in the laboratory held true when embryos were reared under more stressful field conditions. Artificial fertilisation was used to make reciprocal crosses between the 'acid-origin' population and the 'neutral-origin' population (involving six 2 × 2 crosses between females and males originating from acidic and neutral sites), with embryos then transplanted to an acidic field site. This design suffered from low sample size, which greatly reduced the capacity for meaningful quantitative genetic analysis. Nevertheless, it did allow for comparisons between populations, which revealed that acid-origin females had much higher survival than did neutral-origin females. Overall, the findings provided evidence for adaptive divergence in acid tolerance, and provided further support that maternal effects can be an important determinant of local adaptation to acidification. The most likely explanation for the strong maternal effects is that maternally derived differences in the quality of egg capsules underpins variation in tolerance of acid (SäNEN *et al.* 2003a).

Case studies on environmental drying and of tolerance of desiccation

European common frogs (*Rana temporaria*)

Laurila *et al.* (2002) used artificial fertilisation to conduct a factorial mating experiment (North Carolina type II) designed to investigate the genetic basis of plastic responses to the drying of ponds by two populations (a northern population with low risk of desiccation and a southern population with high risk of desiccation). In each population, there were two experimental blocks, the first involving all possible pairwise crosses between five females and five males, and the second involving all possible pairwise crosses between four females and five males. This resulted in 45 crosses per population. No crosses were made between populations. From each family, a subsample of 18 tadpoles were taken and evenly split between three treatments: no change in water level (control), slow decrease in water level, and fast decrease in water level. Comparisons between populations revealed that there were differences in plastic responses (indicating inter-population divergence in life-history traits' importance for tolerance to desiccation), but within each population there was no evidence for the existence of genetic variation in plastic responses. Importantly, most of the measured traits exhibited significant additive genetic variation, as well as weak maternal effects, indicating that life-history traits important in desiccation tolerance have considerable potential for adaptive micro-evolutionary response. More broadly, the evidence gained for additive genetic effects for developmental traits is congruent with the findings of pioneering studies investigating the genetic basis of anuran development (Travis *et al.* 1987; Newman 1988; Blouin 1992; Semlitsch 1993). These studies used artificial fertilisation and basic breeding designs to estimate additive genetic effects. However, the power of these studies to estimate non-additive genetic effects and maternal effects was very limited due to the use of incomplete factorial designs and/or very low numbers of females.

Günther's toadlet (*Pseudophryne guentheri*)

Günther's toadlet, a Western Australian terrestrial-breeding frog, has a status of Least Concern, although it has shown local population declines in recent years and is facing a rapidly drying climate throughout its range. In response to this presumed threat, it has become a model species for investigating the potential for assisted gene flow. Eads *et al.* (2012) used artificial fertilisation to conduct a factorial mating experiment (North Carolina type II) to explore patterns of genetic variation in embryos' tolerance of desiccation in a single population. The design involved 54 families produced by crossing three males and three females in every pairwise combination, in a series of six blocks. Eggs from each family were raised under three environmental conditions (wet, intermediate, or dry), with two or three replicates per treatment (depending on clutch size and availability of eggs). Although a range of viability and viability-related traits was tested, significant additive genetic effects were found only for larval shape at hatching. Moreover, there were no sire-by-environment interactions detected, indicating no significant genetic variation in the plasticity of traits. Together, these findings suggest that the capacity for adaptive genetic responses to the drying of habitats might be restricted to certain traits. Significant maternal effects were detected for all traits, and significant non-additive effects were detected for embryo's survival (indicative of

genetic incompatibility). Moreover, for several viability and viability-related traits, there were significant sire-by-dam-by-environment interactions, indicating that negative effects of genetic incompatibility only became apparent in challenging environments. Extending from this work, Rudin-Bitterli *et al.* (2020) investigated intraspecific variation in the tolerance of embryos and larvae to desiccation (derived from within-population parental crosses) from four populations across a rainfall gradient (expected to cause differential selection on tolerance to desiccation). Formal quantitative genetic analyses were not performed, presumably because some sires were used multiple times, and the experimental design was unbalanced. Nevertheless, as expected, tolerance of desiccation was higher in animals from drier sites, indicating that the species has the potential to show adaptive genetic responses to aridity (although phenotypic plasticity could not be ruled out). Based on this finding, it was concluded that frogs from xeric populations (adapted to drier conditions) could be used as a source for assisted gene flow into threatened mesic populations. In support of this conclusion, a genome-wide scan (using single nucleotide polymorphism data) across 12 central and range-edge populations provided evidence for local adaptation, and identified hundreds of loci that were putatively under directional selection (Cummins *et al.* 2019). Many of these loci were correlated with climatic variables, with one locus homologous to a gene known to play a role in the maturation of oocytes (which might influence fitness under dry conditions) (Cummins *et al.* 2019). More recently, Rudin-Bitterli *et al.* (2021) used artificial fertilisation to create reciprocal within and between population crosses using frogs sourced from two populations at the northern edge of the species' range (a region characterised by low rainfall), and two populations closer to the centre of the species' range (a region characterised by higher rainfall). Overall, embryonic survival was lower (and malformation rates were higher) for crosses between populations separated by the greatest geographical distance, indicative of outbreeding depression. However, for populations separated by a relatively short geographical distance there was limited evidence for outbreeding depression (outcomes were mostly neutral) and in one instance offspring from mixed crosses had improved hatching rates under dry-rearing conditions compared to pure crosses, indicative of either hybrid vigour or the introduction of desiccation-tolerant genes. These findings underscore the potential for targeted gene flow to be used as a strategy to improve anuran adaptive capacity, though such benefits should be weighed against fitness costs tied to outbreeding depression.

Quacking frogs (*Crinia georgiana*)

The quacking frog has a distribution restricted to the south-western corner of Western Australia, and while it is listed as being of Least Concern, its range overlaps with Gunther's toadlet, and it faces similar threats from drying habitats. Rudin-Bitterli *et al.* (2018) used artificial fertilisation to conduct a factorial mating experiment (North Carolina type II) to explore patterns of genetic variation in the desiccation tolerance of larvae from a single population. The experiment involved 90 families (produced by crossing three males with three females in a series of 10 blocks), with tadpoles from each family reared in two environments (low water or normal water levels). It was found that all viability and viability-related traits examined exhibited limited additive genetic variation, although there was evidence for significant non-additive genetic variation, and these effects were even stronger when offspring were reared under stressful conditions simulating the rapid drying of shallow breeding pools (Rudin-Bitterli *et al.* 2018). Although the study showed limited potential for the species to respond genetically to increased risk of desiccation, the findings showed that patterns of non-additive genetic variance were dependent on environmental context. More specifically, the fitness costs of genetic incompatibility may be masked in benign environments. Again, this highlights the critical need to test for genetic effects in contrasting (although ecologically relevant) environments.

Future directions for work on adaptive potential: general advice for practitioners

Taken together, the case studies discussed above provide several important insights. First, across species there was pervasive evidence for strong maternal effects on viability and viability-related traits under challenging developmental conditions. The prevalence of these effects is noteworthy because there is growing theoretical and empirical evidence that maternal effects can provide a medium for non-genetic inheritance (see Bonduriansky and Day 2009; O'Dea *et al.* 2016). This raises the exciting possibility that maternal effects could elevate adaptive potential. At present, the mechanistic basis of non-genetic inheritance is not well understood, so we do not discuss this topic further. However, this field is advancing rapidly, so there are likely to be exciting new options for management in the near future (see Bonduriansky and Day 2009; O'Dea *et al.* 2016). Second, across species there was evidence for environmentally mediated maternal effects, as well as sire-by-dam-by-environment interactions. These interactions highlight the potential for the expression of genetic variation to be environmentally dependent, and once again reiterate the importance of testing for determinants of fitness under various environmental conditions. Third, and most relevant to future work on assisted evolution, significant amounts of additive genetic variation in tolerance to different stressors was evident for various species, and for various viability and viability-related traits. Genetic effects were generally small, but this is to be expected for traits that have a strong impact on fitness, and that have been subjected to the erosive force of directional selection (Falconer and Mackay 1996). Therefore, we conclude that various anuran species are likely to be capable of adaptive micro-evolutionary responses to environmental change, and encourage increased consideration of the feasibility of assisted gene flow as a management strategy for threatened species. Bolstering this view, a recent modelling study reported that targeted gene flow can effectively reduce the risk of extinction, while maintaining very high levels of genetic diversity (and the survival of genomes) in recipient populations (Kelly and Phillips 2019). This finding was particularly encouraging because the study addressed two key aspects of making decisions about conservation (i.e. the timing and size of introductions) in response to a novel threat under various evolutionary scenarios (achieved by altering population dynamics, recombination rates, and levels of outbreeding depression). With a view towards developing a logical and cohesive framework to guide future conservation efforts concerning assisted gene flow, we suggest the following investigative process.

Step 1. Assess the level of local adaptation

To investigate whether there is 'within-species' geographical variation in tolerance to an environmental stressor, manipulative laboratory experiments should be performed. This should follow a 'common garden experiment' approach. Specifically, under controlled laboratory conditions artificial fertilisation should be used to make 'within-population' crosses (rather than collecting gametes from the wild, which can inflate environmental effects) and offspring from each 'within-population' cross should be raised under treatments that reflect ecologically relevant levels (or forecast levels) of the stressor.

Ideally, populations should be selected along an environmental gradient, or from sites representing contrasting environments. Including range-edge populations can be highly informative as these are expected to harbour the bulk of a species' adaptive variation (see Macdonald *et al.* 2017). If individuals from within a population, but from different localities, exposed to higher levels of the stressor perform or survive better at higher levels, then there is evidence for local adaptation, and the investigation can move to step 2.

Step 2. Investigate the genetic architecture of fitness-determining traits in putative donor and recipient populations

After gaining evidence for local adaptation, the next step is to determine whether target traits have a genetic basis. For putative donor and recipient populations, artificial fertilisation should be used in combination with a North Carolina type II mating design to conduct 'within-population' crosses. Studies by Laurila *et al.* (2002) and Merilä *et al.* (2004) provide examples of model experimental designs. After rearing offspring under a range of environmental treatments to allow genotype-by-environment effects to be examined, quantitative genetic analyses should be undertaken to assess the degree of additive genetic variance for all traits of interest. If high levels of additive genetic variance are detected, the target populations should be considered as candidates for assisted evolution. Experiments to assess combining ability should then be undertaken (step 3). At this stage it may be useful to conduct a genome-wide scan of individuals from the target populations. This could help identify genes or combinations of genes that might underpin tolerance (see Cummins *et al.* 2019) and allow more targeted selection of genomes for assisted gene flow. To this end, quantitative trait locus mapping (QTL) and/or genomic selection are emerging as particularly useful approaches.

Step 3. Test the combining ability of putative donor and recipient populations and the potential for effective introgression

For populations displaying moderate to high levels of additive genetic variance in traits expected to have a positive influence on tolerance, artificial fertilisation should be used to conduct inter-population crosses. The design (and associated methodology) we described earlier for studies testing the feasibility of augmenting populations (Figure 8.1) is suitable for this purpose. For all crosses, the offspring that are generated should be reared under relatively benign environments as well as under stressful ones (simulating prevailing environmental conditions in the donor and recipient populations) to test for gene × environment effects. This approach is necessary to confirm that genetic factors contribute to differences in tolerance among the populations, and evaluate whether hybrids in the F1 generation have improved tolerance limits. Ideally, evaluation should also extend into the F2 generation because heterozygosity (and the potential for heterosis) is expected to peak in the F1 generation (Frankham 2016). Without testing for effects across multiple generations it is not possible to know whether improvements in fitness will persist at beneficial levels.

Step 4. Commence a program of assisted gene flow

If there is no evidence for outbreeding depression, and there is evidence for heterosis, there is justification for beginning a program of targeted assisted gene flow. Following the introduction of foreign animals, *in situ* monitoring of populations should take place to ensure that there are no detrimental impacts on the population's viability that requires intervention. Ongoing testing of tolerances (as per step 1) should also be conducted to determine whether the introgression of new genes has improved the target population's evolutionary potential. Critically, the empirical framework we describe above can be applied to a broad spectrum of environmental stressors linked to amphibian decline, including pollution, disease, and climatic change. Increased work in this area, with a specific focus on threatened species, stands to make significant contributions to amphibian conservation.

GENERAL METHODOLOGICAL CONSIDERATIONS

For the remainder of the chapter, we draw attention to practical issues essential to consider when collecting gametes, conducting artificial fertilisations, and rearing offspring. Additionally, we identify the critical fitness-determining traits to measure throughout different amphibian life stages (Figure 8.2).

Collection of gametes and conducting artificial fertilisations

The process of conducting artificial fertilisations in amphibians has been reviewed previously (Clulow *et al.*

2014; Silla and Byrne 2019). Here, we highlight the importance of optimising and standardising methods for the collection and storage of gametes, as well as aspects of fertilisation (including the concentration of sperm and osmolality of the fertilisation medium), and the spatial arrangement of developing embryos, to guarantee consistency across 'crosses' in order to ensure that differences in fitness-determining traits are a result of genetic differences between crosses and not influenced by variability in experimental conditions.

Collection of gametes

It is essential that gametes of maximum viability are collected and that the quality of gametes does not vary among crosses. Such variance could skew measures of fitness such as the success of fertilisation. For several anuran species, females ovulate before entering the breeding chorus; thus, females may be intercepted as they enter the breeding site and the collection of oocytes is achieved by gently applying pressure to the abdomen (a technique referred to as 'stripping') (Dziminski *et al.* 2008; Sherman *et al.* 2008). In these situations, females should be stripped of oocytes within 24 h of collection to avoid spontaneous oviposition and/or the over-ripening of oocytes (over-ripening results from the aging and deterioration of oocytes retained for an extended period within the coelomic cavity after ovulation; Silla 2011; Silla *et al.* 2018). Ultrasound may also be used as a tool to evaluate whether females have ovulated or not (see Chapter 6). For anuran species that do not ovulate before entering the breeding site, collection of oocytes is facilitated through the hormonal induction of ovulation (see Byrne *et al.* 2019; Byrne *et al.* 2021; Byrne and Silla 2020). Protocols for hormonally inducing ovulation must be optimised on a species-specific basis to ensure a reliable and predictable response (Silla and Byrne 2019; Silla and Byrne 2021). Importantly, females that are hormonally induced to ovulate should be stripped once eggs are present in the oviduct or visible at the cloaca, and fertilisations conducted immediately. Females should not be allowed to spontaneously ovulate and oocytes should not be stored before fertilisation as their capacity for fertilisation declines within minutes of oviposition (Silla 2013; Silla and Byrne 2019). Such declines will result in variability in the fertilisation rate of oocytes between females, which will falsely inflate maternal effects. For this reason, it is also recommended that more females than required to complete the full suite of crosses are induced to ovulate simultaneously to ensure

that clutches of oocytes can be collected and subsequently fertilised in quick succession (excluding females that exhibit a delayed ovulatory response).

To ensure that fertilisations are conducted quickly once oocytes are expelled, suspensions of sperm should be pre-prepared and placed in cold storage (0–4°C; see Chapter 7) until use. Typically, studies conducting artificial fertilisation for the purpose of investigating genetic architecture have utilised suspensions generated from testes that have been removed from the body and macerated (see Dziminski *et al.* 2008; Byrne *et al.* 2019; Byrne *et al.* 2021). This procedure offers several advantages, including the capacity to generate suspensions of high concentration and ability to control the osmolality and pH of the suspension fluid. Also, given the high storage capacity of macerated sperm, suspensions can be prepared and refrigerated for several hours without a loss of the capacity for fertilisation ($\leq$ 12 h storage is recommended) (Silla and Byrne 2019; Chapter 7). With practice, euthanasia, dissection, and removal and maceration of the testes from a single male can all be achieved in less than 3 min, resulting in the generation of suspensions of sperm with minimal temporal variability that would otherwise inflate non-genetic paternal effects. Alternatively, suspensions of sperm can be generated non-invasively via hormonal induction (Silla and Byrne 2019; Chapter 7). Optimal protocols for hormonally inducing spermiation (collected as spermic urine), and determination of the peak period for collecting sperm after administration of hormones (Silla *et al.* 2019; Chapter 7), must first be quantified to determine the feasibility of conducting artificial fertilisations using hormonally induced suspensions of sperm. Additionally, if utilising spermic urine, the osmolality of sperm suspensions must be standardised, as the osmolality of urine can vary considerably depending on a male's state of hydration. Fertilisation should also be conducted as soon as possible following the collection of fresh spermic urine as the viability of these suspensions declines at a greater rate during cold storage compared with sperm suspensions obtained by macerating the testes (Chapter 7). Finally, while multiple samples may be collected from living individual males to conduct artificial fertilisation for propagative purposes, using live males may be inappropriate for investigations of genetic architecture because a single suspension of sperm must be generated and split between several sub-clutches produced by individual females.

Concentration of sperm

Ensuring that the concentration of sperm applied to each sub-clutch is standardised is of paramount importance because of the relationship between the concentration of sperm and the success of fertilisation. Typically, fertilisations are achieved by sperm concentrations ranging from 10^4 to 10^6 sperm/mL, with terrestrial and foam-nesting species generally exhibiting lower optimal concentrations compared with aquatically breeding species (Silla and Byrne 2019). A fertilisation curve should be established to determine the concentration of sperm required to reach an optimal or asymptotic rate of fertilisation (Silla 2013). The concentration of each sperm suspension should be quantified in advance, such that a pre-calculated volume of suspension containing a set concentration of sperm can be allocated to each fertilisation tray. Irrespective of natural fertilisation ecologies, artificial fertilisations should be conducted in a small volume of medium without water (known as dry fertilisation) to increase the probability of sperm–egg collisions (Silla and Byrne 2019). Additional fluid can be added after a few minutes to ensure adequate hydration or to flood embryos according to a species' natural fertilisation ecology.

Osmolality of the fertilisation medium

Conducting artificial fertilisations in a medium of optimal osmolality is imperative for maximising the activation of the motility of sperm, one important functional characteristic that enables sperm to reach and penetrate the egg (Silla and Byrne 2019) (Chapter 7). Anuran artificial fertilisation typically is conducted in fertilisation media of very low osmolality (0–25 mOsm/kg), as sperm motility is activated in response to hypoosmotic shock. Current research suggests that fertilisation success in terrestrially breeding species is optimised in media of higher osmolality (Silla 2013), likely reflecting the higher osmolality of the fertilisation environment. Thus, the natural fertilisation ecologies of targeted species should be considered when designing a fertilisation strategy, and when possible, fertilisation curves should be established to determine the optimal osmolality of medium at which to conduct artificial fertilisations (Silla and Byrne 2019).

The importance of the allocation of oocytes, order of fertilisations, and spatial arrangement of developing embryos

To avoid potential differences in the capacity for fertilisation between the first and the last oocytes expelled from each female, it is important to ensure that each sub-clutch has a mixture of early and late oocytes. In principle, this can be achieved by systematically stripping a small number of eggs into each fertilisation tray and repeating the process until the female is spent. Alternatively, an entire clutch of eggs can be stripped, gently mixed and randomly allocated to fertilisation trays. If employing the latter method (which may not be possible in species whose eggs are tightly bound by egg jelly), utmost care must be taken not to damage or over-manipulate the oocytes before fertilisation, as the protective jelly layers of the eggs will not yet have reached full thickness. Although rare, over-manipulation of eggs also can potentially cause auto-activation (parthenogenetic cleavage) (Rugh 1951), which can obscure estimates of fertilisation success. In species in which auto-activation appears to be an issue, the extent of auto-activation can be quantified by including a procedural control, whereby a sub-clutch of eggs is treated without the addition of sperm to the fertilisation tray. Once a female's clutch has been evenly divided among fertilisation trays, the order that sub-clutches are fertilised with each males' sperm must be randomised, and accurate records kept to track and monitor the crosses. Following artificial fertilisations, embryos should be separated and reared individually in ventilated containers (e.g. ventilated Petri dishes). Developing embryos should be randomly allocated to a position on the shelf, and throughout development should be regularly rotated through positions on the shelf to avoid micro-environmental differences (e.g. differences in temperature and humidity). Following these guidelines minimises micro-environmental variations among sub-clutches that might otherwise influence estimates of genetic variance.

What environment should be used for rearing offspring?

Studies investigating the genetic basis of variation in traits are typically conducted in controlled laboratory environments to minimise the effects of environmental variation on estimates of phenotypic variation. In general, rearing offspring under standardised laboratory conditions will be the preferred approach so that genotype-by-environment interactions can be accurately assessed. However, if the goal is to gain a more accurate picture of the full complexity of a selection regime experienced in nature, it may be necessary to raise offspring under natural or semi-natural conditions (Reznick and Travis 1996). Ultimately, complementary laboratory and field approaches may be necessary to fully understand the extent and ecological relevance of the genetic underpinnings of variations in traits. Irrespective of whether rearing occurs in the laboratory or in the field, immediately after artificial fertilisation has occurred offspring should be reared individually. Individual rearing is essential to avoid introducing microenvironmental confounds (including density-dependent effects) that can arise when families are raised together (Byrne *et al.* 2019). It should be noted that for several aquatically breeding anurans the adherent nature of the egg jelly (forming strings of eggs) may require the jelly to be cut to allow separation and individual rearing.

What fitness-determining traits should be measured?

Multiple traits contribute to total fitness (i.e. the probability that a phenotype or genotype is represented in future generations). Therefore, practitioners should aim to measure a suite of traits that represent different components of fitness, such as survival or fecundity, as well as traits that are less certainly related to fitness (e.g. body size or growth rate) (Figure 8.2). The first trait to quantify is the proportion of eggs fertilised, known as fertilisation success. Not all studies score fertilisation success, but this is a mistake because it can lead to inaccurate estimates of early mortality of the embryos (Sagvik *et al.* 2005). Genetic incompatibilities that reduce the offsprings' viability are expected to manifest, and compound, during early development; therefore it is important to quantify the survival and abnormalities of embryos, hatching success, and time to hatching (Byrne and Silla 2020) (Figure 8.2). Following hatching of larvae (tadpoles) from the egg capsule, important measures of fitness include larval survival, deformity, growth rate, and morphology (body size and shape) (Figure 8.2). Next, metamorphosis is a physiologically demanding transformation and is typically accompanied by a spike in mortality, so survival to metamorphosis should be measured, along with time to metamorphosis and size at metamorphosis (Byrne *et al.* 2019; Byrne and Silla 2020) (Figure 8.2). In the adult life stage, adult survival and body size, female fecundity, and male fertility are important indicators of fitness (Figure 8.2). Finally, assays of performance (e.g. escape, exercise, foraging), general-movement behaviour, and sexual display can all provide estimates of fitness and can be quantified throughout larval and adult life stages (Rudin-Bitterli *et al.* 2020) (Figure 8.2).

CONCLUSIONS

While artificial fertilisation has been widely used in the propagation of anurans, it is yet to be well integrated into amphibian genetic management on a broad scale. We highlighted that artificial fertilisation is a powerful tool enabling controlled-breeding experiments that provide fundamental insights into the genetic underpinnings of viability and fitness-related traits. We also discussed how application of this knowledge promises to greatly improve the implementation and operation of three conservation actions: population augmentation and genetic rescue, conservation breeding, and assisted gene flow. While augmentation of populations has enormous potential to aid in the genetic rescue of small, inbred populations, widespread implementation has been held back by concerns that mixing populations could cause outbreeding depression. Data obtained from breeding experiments in a small number of model frog species indicate that these concerns are founded, based on strong evidence for genetic incompatibility between populations of certain species. In contrast, for other species mixing populations does not appear to have major detrimental effects, and for some fitness-related traits there is even evidence for heterotic effects (hybrid vigour). Reasons for interspecific variation probably relates to differences in capacity for long-distance dispersal and the extent of uniformity and connectivity of habitats. Expecting high species-specificity, we encourage conservation practitioners to use artificial fertilisation–mediated breeding experiments as a tool to inform decision-making surrounding the use of population augmentation. Compared to augmentation of populations, the value of conservation breeding programs remains largely uncontested, although there is certainly room for improvement. To date, most conservation breeding programs have genetically managed populations by working to maximise genetic variation. While effective at reducing the risk of inbreeding depression, and curbing adaptation to captivity, this approach neglects the potential influence of parental genetic quality on the viability of offspring. Artificial fertilisation–facilitated breeding experiments have shown high levels of genetic incompatibility between randomly paired parents in various species of frogs. We argue that a deeper understanding of the genetic architecture of fitness-related traits in captive species, and the specific influence of 'compatible genes' and 'good genes' on the viability of offspring, could significantly elevate the viability and sustainability of 'assurance' populations. Finally, of the three management strategies we discussed, by far the most cutting edge and topical is assisted evolution through assisted gene flow. While currently being implemented for various plant and animal groups, assisted evolution is yet to be applied to any amphibian. To this end, artificial fertilisation-facilitated breeding experiments have a major role to play by allowing tests of local adaptation, and assessment of the genetic consequences of moving pre-adapted genotypes into threatened populations. Studies with various species of frogs facing rapid environmental changes (e.g. acidification and drying of habitats) have revealed heritability in various traits shown to influence tolerance, and have also provided evidence for geographical variation in tolerance. This demonstrates real potential for assisted gene flow to improve the adaptive potential and resilience of threatened amphibian populations, thereby justifying a push for increased investment in this field of research. Here, we provided a 'road map' for the collection, analysis, and interpretation of data to address our three main points, and help answer some of the poignant questions we raised above. We conclude that increased use of artificial fertilisation to control and understand the outcomes of breeding within and between anuran populations will improve our capacity to genetically manage threatened species globally.

ACKNOWLEDGEMENTS

We thank Natalie Rosser for providing feedback on the chapter, and David Hunter and Michael McFadden for discussions regarding the importance of genetic management in amphibian conservation. This work was supported by grants from the Australian Research Council (LP140100808, LP170100351 & DE210100812).

REFERENCES

Aitken SN, Whitlock MC (2013) Assisted gene flow to facilitate local adaptation to climate change. *Annual Review of Ecology, Evolution, and Systematics* **44**, 367–388. doi:10.1146/annurev-ecolsys-110512-135747

Armstrong DP, Seddon PJ (2008) Directions in reintroduction biology. *Trends in Ecology & Evolution* **23**, 20–25. doi:10.1016/j.tree.2007.10.003

Beebee T (2005) Conservation genetics of amphibians. *Heredity* **95**, 423–427. doi:10.1038/sj.hdy.6800736

Blouin MS (1992) Genetic correlations among morphometric traits and rates of growth and differentiation in the green tree frog, *Hyla cinerea*. *Evolution* **46**, 735–744. doi:10.1111/j.1558-5646.1992.tb02079.x

Blows MW, Hoffmann AA (2005) A reassessment of genetic limits to evolutionary change. *Ecology* **86**, 1371–1384. doi:10.1890/04-1209

Bonduriansky R, Day T (2009) Nongenetic inheritance and its evolutionary implications. *Annual Review of Ecology, Evolution, and Systematics* **40**, 103–125.

Browne RK, Seratt J, Vance C, Kouba A (2006) Hormonal priming, induction of ovulation and *in-vitro* fertilization of the endangered Wyoming toad (*Bufo baxteri*). *Reproductive Biology and Endocrinology* **4**, 34. doi:10.1186/1477-7827-4-34

Byrne PG, Keogh JS (2009) Extreme sequential polyandry insures against nest failure in a frog. *Proceedings. Biological Sciences* **276**, 115–120. doi:10.1098/rspb.2008.0794

Byrne PG, Silla AJ (2010) Hormonal induction of gamete release, and *in-vitro* fertilisation, in the critically endangered Southern Corroboree Frog, *Pseudophryne corroboree*. *Reproductive Biology and Endocrinology* **8**, 144. doi:10.1186/1477-7827-8-144

Byrne PG, Silla AJ (2020) An experimental test of the genetic consequences of population augmentation in an amphibian. *Conservation Science and Practise* **2**, e194. doi:10.1111/csp2.194

Byrne PG, Gaitan Espitia JD, Silla AJ (2019) Genetic benefits of extreme sequential polyandry in a terrestrial-breeding frog. *Evolution* **73**, 1972–1985. doi:10.1111/evo.13823

Byrne PG, Keogh SJ, O'brien DM, Gaitan Espitia JD, Silla AJ (2021) Evidence that genetic compatability underpins female mate choice in a monandrous amphibian. *Evolution* **75**, 529–541.

Capanna E (1999) Lazzaro Spallanzani: at the roots of modern biology. *The Journal of Experimental Zoology* **285**, 178–196. doi:10.1002/(SICI)1097-010X(19991015)285:3<178::AID-JEZ2>3.0.CO;2-#

Charmantier A, Garant D (2005) Environmental quality and evolutionary potential: lessons from wild populations. *Proceedings. Biological Sciences* **272**, 1415–1425. doi:10.1098/rspb.2005.3117

Clulow J, Trudeau VL, Kouba AJ (2014) Amphibian declines in the twenty-first century: why we need assisted reproductive technologies. In *Reproductive Sciences in Animal Conservation*. (Eds WV Holt, JL Brown and P Comizzoli) pp. 275–316. Springer, New York NY, USA.

Coates DJ, Byrne M, Moritz C (2018) Genetic diversity and conservation units: dealing with the species-population continuum in the age of genomics. *Frontiers in Ecology and Evolution* **6**, 165. doi:10.3389/fevo.2018.00165

Cummins D, Kennington WJ, Rudin-Bitterli T, Mitchell NJ (2019) A genome-wide search for local adaptation in a terrestrial-breeding frog reveals vulnerability to climate change. *Global Change Biology* **25**, 3151–3162. doi:10.1111/gcb.14703

Dziminski MA, Roberts JD, Simmons LW (2008) Fitness consequences of parental compatibility in the frog *Crinia georgiana*. *Evolution* **62**, 879–886. doi:10.1111/j.1558-5646.2008.00328.x

Eads AR, Mitchell NJ, Evans JP (2012) Patterns of genetic variation in desiccation tolerance in embryos of the terrestrial-breeding frog, *Pseudophryne guentheri*. *Evolution* **66**, 2865–2877. doi:10.1111/j.1558-5646.2012.01616.x

Edmands S (2007) Between a rock and a hard place: evaluating the relative risks of inbreeding and outbreeding for conservation and management. *Molecular Ecology* **16**, 463–475. doi:10.1111/j.1365-294X.2006.03148.x

Falconer D, Mackay T (1996) *Introduction to Quantitative Genetics.* 4th edn. Longman, Harlow, UK.

Fasahat P, Rajabi A, Rad J, Derera J (2016) Principles and utilization of combining ability in plant breeding. *Biometrics and Biostatistics International Journal* **4**, 1–24. doi:10.15406/bbij.2016.04.00085

Ficetola GF, De Bernardi F (2005) Supplementation or in situ conservation? Evidence of local adaptation in the Italian agile frog *Rana latastei* and consequences for the management of populations. *Animal Conservation* **8**, 33–40.

Frankham R (1999) Quantitative genetics in conservation biology. *Genetical Research* **74**, 237–244. doi:10.1017/S001667239900405X

Frankham R (2005) Genetics and extinction. *Biological Conservation* **126**, 131–140. doi:10.1016/j.biocon.2005.05.002

Frankham R (2015) Genetic rescue of small inbred populations: meta-analysis reveals large and consistent benefits of gene flow. *Molecular Ecology* **24**, 2610–2618. doi:10.1111/mec.13139

Frankham R (2016) Genetic rescue benefits persist to at least the F3 generation, based on a meta-analysis. *Biological Conservation* **195**, 33–36. doi:10.1016/j.biocon.2015.12.038

Frankham R, Ballou JD, Eldridge MD, Lacy RC, Ralls K, Dudash MR, Fenster CB (2011) Predicting the probability of outbreeding depression. *Conservation Biology* **25**, 465–475. doi:10.1111/j.1523-1739.2011.01662.x

Gasparini C, Dosselli R, Evans JP (2017) Sperm storage by males causes changes in sperm phenotype and influences the reproductive fitness of males and their sons. *Evolution Letters* **1**, 16–25. doi:10.1002/evl3.2

Goldberg TL, Grant EC, Inendino KR, Kassler TW, Claussen JE, Philipp DP (2005) Increased infectious disease susceptibility resulting from outbreeding depression. *Conservation Biology* **19**, 455–462. doi:10.1111/j.1523-1739.2005.00091.x

Hayman B (1960) The theory and analysis of diallel crosses. III. *Genetics* **45**, 155–172. doi:10.1093/genetics/45.2.155

Hayward M (1979) The application of the diallel cross to outbreeding crop species. *Euphytica* **28**, 729–737. doi:10.1007/BF00038941

Hedrick PW, Fredrickson R (2010) Genetic rescue guidelines with examples from Mexican wolves and Florida panthers. *Conservation Genetics* **11**, 615–626. doi:10.1007/s10592-009-9999-5

Hoffmann AA, Sgro CM (2011) Climate change and evolutionary adaptation. *Nature* **470**, 479–485. doi:10.1038/nature09670

Huff DD, Miller LM, Chizinski CJ, Vondracek B (2011) Mixed-source reintroductions lead to outbreeding depression in second-generation descendents of a native North American fish. *Molecular Ecology* **20**, 4246–4258. doi:10.1111/j.1365-294X.2011.05271.x

Keller LF, Waller DM (2002) Inbreeding effects in wild populations. *Trends in Ecology & Evolution* **17**, 230–241. doi:10.1016/S0169-5347(02)02489-8

Kelly E, Phillips B (2019) How many and when? Optimising targeted gene flow for a step change in the environment. *Ecology Letters* **22**, 447–457. doi:10.1111/ele.13201

Kronenberger JA, Gerberich JC, Fitzpatrick SW, Broder ED, Angeloni LM, Funk WC (2018) An experimental test of alternative population augmentation scenarios. *Conservation Biology* **32**, 838–848. doi:10.1111/cobi.13076

Laurila A, Karttunen S, Merilä J (2002) Adaptive phenotypic plasticity and genetics of larval life histories in two Rana temporaria populations. *Evolution* **56**, 617–627. doi:10.1111/j.0014-3820.2002.tb01371.x

Lenarcic AB, Svenson KL, Churchill GA, Valdar W (2012) A general Bayesian approach to analyzing diallel crosses of inbred strains. *Genetics* **190**, 413–435. doi:10.1534/genetics.111.132563

Lynch M, Walsh B (1998) *Genetics and Analysis of Quantitative Traits.* Sinauer, Sunderland MA, USA.

Macdonald SL, Llewelyn J, Moritz C, Phillips BL (2017) Peripheral isolates as sources of adaptive diversity under climate change. *Frontiers in Ecology and Evolution* **5**, 88. doi:10.3389/fevo.2017.00088

Merilä J, Hendry AP (2014) Climate change, adaptation, and phenotypic plasticity: the problem and the evidence. *Evolutionary Applications* **7**, 1–14. doi:10.1111/eva.12137

Merilä J, Söderman F, O'hara R, Räsänen K, Laurila A (2004) Local adaptation and genetics of acid-stress tolerance in the moor frog, *Rana arvalis. Conservation Genetics* **5**, 513–527. doi:10.1023/B:COGE.0000041026.71104.0a

Neff BD, Pitcher TE (2005) Genetic quality and sexual selection: an integrated framework for good genes and compatible genes. *Molecular Ecology* **14**, 19–38. doi:10.1111/j.1365-294X.2004.02395.x

Neff BD, Garner SR, Pitcher TE (2011) Conservation and enhancement of wild fish populations: preserving genetic quality versus genetic diversity. *Canadian Journal of Fisheries and Aquatic Sciences* **68**, 1139–1154. doi:10.1139/f2011-029

Newman RA (1988) Genetic variation for larval anuran (*Scaphiopus couchii*) development time in an uncertain environment. *Evolution* **42**, 763–773.

O'Brien DM, Keogh JS, Silla AJ, Byrne PG (2019) Female choice for related males in wild red-backed toadlets (*Pseudophryne coriacea*). *Behavioral Ecology* **30**, 928–937. doi:10.1093/beheco/arz031

O'Dea RE, Noble DW, Johnson SL, Hesselson D, Nakagawa S (2016) The role of non-genetic inheritance in evolutionary rescue: epigenetic buffering, heritable bet hedging and epigenetic traps. *Environmental Epigenetics* **2**, dvv014. doi:10.1093/eep/dvv014

Pakkasmaa S, Merilä J, O'hara R (2003) Genetic and maternal effect influences on viability of common frog tadpoles under different environmental conditions. *Heredity* **91**, 117–124. doi:10.1038/sj.hdy.6800289

Pederson D (1971) The estimation of heritability and degree of dominance from a diallel cross. *Heredity* **27**, 247–264. doi:10.1038/hdy.1971.88

Persson M, Räsänen K, Laurila A, Merilä J (2007) Maternally determined adaptation to acidity in *Rana arvalis*: are laboratory and field estimates of embryonic stress tolerance congruent? *Canadian Journal of Zoology* **85**, 832–838. doi:10.1139/Z07-064

Pierce BA, Harvey JM (1987) Geographic variation in acid tolerance of Connecticut wood frogs. *Copeia* 94–103. doi:10.2307/1446042

Pierce BA, Sikand N (1985) Variation in acid tolerance of Connecticut wood frogs: genetic and maternal effects. *Canadian Journal of Zoology* **63**, 1647–1651. doi:10.1139/z85-244

Pimm SL, Dollar L, Bass OL, Jr (2006) The genetic rescue of the Florida panther. *Animal Conservation* **9**, 115–122. doi:10.1111/j.1469-1795.2005.00010.x

Ralls K, Ballou JD, Dudash MR, Eldridge MD, Fenster CB, Lacy RC, Sunnucks P, Frankham R (2018) Call for a paradigm shift in the genetic management of fragmented populations. *Conservation Letters* **11**, e12412. doi:10.1111/conl.12412

Reznick D, Travis J (1996) The empirical study of adaptation in natural populations. *Adaptation* 243, 289.

Rudin Bitterli TS, Mitchell NJ, Evans JP (2018) Environmental stress increases the magnitude of nonadditive genetic variation in offspring fitness in the frog *Crinia georgiana*. *American Naturalist* **192**, 461–478. doi:10.1086/699231

Rudin-Bitterli TS, Evans JP, Mitchell NJ (2020) Geographic variation in adult and embryonic desiccation tolerance in a terrestrial-breeding frog. *Evolution* **74**, 1186–1199. doi:10.1111/evo.13973

Rudin-Bitterli TS, Evans JP, Mitchell NJ (2021) Fitness consequences of targeted gene flow to counter impacts of drying climates on terrestrial-breeding frogs. *Communications Biology* **4**(1), 1–10. doi:10.1038/s42003-021-02695-w

Rugh R (1951) *The Frog; Its Reproduction and Development*. The Blakiston Company, Philadelphia PA, USA.

Sagvik J, Uller T, Olsson M (2005) Outbreeding depression in the common frog, *Rana temporaria*. *Conservation Genetics* **6**, 205–211. doi:10.1007/s10592-004-7829-3

SäNEN KR, Laurila A, Merilä J (2003a) Geographic variation in acid stress tolerance of the moor frog, *Rana arvalis*. I. Local adaptation. *Evolution* **57**, 352–362. doi:10.1111/j.0014-3820.2003.tb00269.x

SäNEN KR, Laurila A, Merilä J (2003b) Geographic variation in acid stress tolerance of the moor frog, *Rana arvalis*. II. Adaptive maternal effects. *Evolution* **57**, 363–371. doi:10.1111/j.0014-3820.2003.tb00270.x

Semlitsch RD (1993) Adaptive genetic variation in growth and development of tadpoles of the hybridogenetic *Rana esculenta* complex. *Evolution* **47**, 1805–1818. doi:10.1111/j.1558-5646.1993.tb01271.x

Sgrò CM, Lowe AJ, Hoffmann AA (2011) Building evolutionary resilience for conserving biodiversity under climate change. *Evolutionary Applications* **4**, 326–337. doi:10.1111/j.1752-4571.2010.00157.x

Sherman CD, Wapstra E, Uller T, Olsson M (2008) Male and female effects on fertilization success and offspring viability in the Peron's tree frog, *Litoria peronii*. *Austral Ecology* **33**, 348–352. doi:10.1111/j.1442-9993.2007.01823.x

Silla AJ (2011) Effect of priming injections of luteinizing hormone-releasing hormone on spermiation and ovulation in Günther's Toadlet, *Pseudophryne guentheri*. *Reproductive Biology and Endocrinology* **9**, 68. doi:10.1186/1477-7827-9-68

Silla AJ (2013) Artificial fertilisation in a terrestrial toadlet (*Pseudophryne guentheri*): effect of medium osmolality, sperm concentration and gamete storage. *Reproduction, Fertility and Development* **25**, 1134–1141. doi:10.1071/RD12223

Silla AJ, Byrne PG (2019) The role of reproductive technologies in amphibian conservation breeding programs. *Annual Review of Animal Biosciences* **7**, 499–519. doi:10.1146/annurev-animal-020518-115056

Silla AJ, Byrne PG (2021) Hormone-induced ovulation and artificial fertilisation in four terrestrial-breeding anurans. *Reproduction, Fertility and Development* **33**(9), 615–618. doi:10.1071/RD20243

Silla AJ, Mcfadden M, Byrne PG (2018) Hormone-induced spawning of the critically endangered northern corroboree frog, *Pseudophryne pengilleyi*. *Reproduction, Fertility and Development* **30**, 1352–1358. doi:10.1071/RD18011

Silla AJ, Mcfadden MS, Byrne PG (2019) Hormone-induced sperm-release in the critically endangered Booroolong frog (*Litoria booroolongensis*): effects of gonadotropin-releasing hormone and human chorionic gonadotropin. *Conservation Physiology* **7**, coy080. doi:10.1093/conphys/coy080

Sumida M, Kotaki M, Islam MM, Djong TH, Igawa T, Kondo Y, Matsui M, Khonsue W, Nishioka M (2007) Evolutionary relationships and reproductive isolating mechanisms in the rice frog (*Fejervarya limnocharis*) species complex from Sri Lanka, Thailand, Taiwan and Japan, inferred from mtDNA gene sequences, allozymes, and crossing experiments. *Zoological Science* **24**, 547–562. doi:10.2108/zsj.24.547

Tallmon DA, Luikart G, Waples RS (2004) The alluring simplicity and complex reality of genetic rescue. *Trends in Ecology & Evolution* **19**, 489–496. doi:10.1016/j.tree.2004.07.003

Tedeschi JN, Kennington W, Tomkins JL, Berry O, Whiting S, Meekan MG, Mitchell NJ (2016) Heritable variation in heat shock gene expression: a potential mechanism for adaptation to thermal stress in embryos of sea turtles. *Proceedings. Biological Sciences* **283**, 20152320. doi:10.1098/rspb.2015.2320

Travis J, Emerson SB, Blouin M (1987) A quantitative-genetic analysis of larval life-history traits in *Hyla crucifer*. *Evolution* **41**, 145–156.

Turani B, Aliko V, Jona D (2015) In vitro fertilization and maturation of Balkan water frog (*Pelophylax kurtmuelleri*, Gayda, 1940)—a case study in reproductive amphibian biotechnology. *International Journal of Ecosystems and Ecology Science* **5**, 557–560.

Turani B, Aliko V, Faggio C (2020) Allurin and egg jelly coat impact on in-vitro fertilization success of endangered Albanian water frog, *Pelophylax shqipericus*. *Natural Product Research* **34**, 830–837. doi:10.1080/14786419.2018.1508147

Uller T, Sagvik J, Olsson M (2006) Crosses between frog populations reveal genetic divergence in larval life history at short geographical distance. *Biological Journal of the Linnean*

Society. Linnean Society of London **89**, 189–195. doi:10.1111/j. 1095-8312.2006.00673.x

Upton R, Clulow S, Calatayud NE, Colyvas K, Seeto RG, Wong LA, Mahony MJ, Clulow J (2021) Generation of reproductively mature offspring from the endangered green and golden bell frog *Litoria aurea* using cryopreserved spermatozoa. *Reproduction, Fertility and Development* **33**, 562–572. doi:10.1071/RD20296

Watts AG, Schlichting P, Billerman S, Jesmer B, Micheletti S, Fortin M-J, Funk C, Hapeman P, Muths EL, Murphy MA (2015) How spatio-temporal habitat connectivity affects amphibian genetic structure. *Frontiers in Genetics* **6**, 275. doi:10.3389/fgene.2015.00275

Weeks AR, Sgro CM, Young AG, Frankham R, Mitchell NJ, Miller KA, Byrne M, Coates DJ, Eldridge MD, Sunnucks P (2011) Assessing the benefits and risks of translocations in changing environments: a genetic perspective. *Evolutionary Applications* **4**, 709–725. doi:10.1111/j.1752-4571.2011. 00192.x

Weeks AR, Heinze D, Perrin L, Stoklosa J, Hoffmann AA, Van Rooyen A, Kelly T, Mansergh I (2017) Genetic rescue increases fitness and aids rapid recovery of an endangered marsupial population. *Nature Communications* **8**, 1071. doi:10.1038/s41467-017-01182-3

Whitlock R, Stewart GB, Goodman SJ, Piertney SB, Butlin RK, Pullin AS, Burke T (2013) A systematic review of phenotypic responses to between-population outbreeding. *Environmental Evidence* **2**, 13. doi:10.1186/2047-2382-2-13

9 Cryopreservation of amphibian genomes: targeting the Holy Grail, cryopreservation of maternal-haploid and embryonic-diploid genomes

John Clulow, Rose Upton, and Simon Clulow

INTRODUCTION

The irreversible loss of amphibian biodiversity has been well characterised and the main drivers identified (Stuart *et al.* 2004; Skerratt *et al.* 2007; Vredenburg *et al.* 2010; Bishop *et al.* 2012; Skerratt *et al.* 2016; Bower *et al.* 2017; Scheele *et al.* 2019) (Chapter 2). Recent estimates put the number of extinctions as high as 200 (Alroy 2015) with perhaps 90 of those due to chytridiomycosis alone (Scheele *et al.* 2019) and few strategies to prevent further chytrid-driven extinctions (Scheele *et al.* 2014; Clulow *et al.* 2018). As many as 40% of amphibian species may be threatened with extinction (Bishop *et al.* 2012). The pattern of amphibian decline has been particularly challenging to address, and mitigate with timely and effective conservation strategies. Common features of amphibian species' decline include (1) rapid, unanticipated declines in populations that were not previously declining or being monitored, often in remote or undisturbed habitats; and (2) populations subjected to unrelenting destruction of habitat from economic pressures and expansion of human populations. The next phase of amphibian decline driven by climatic change is already underway (Rohr and Raffel 2010; Shoo *et al.* 2011).

The nature and speed of the global amphibian decline is placing enormous pressure on the resources available for their conservation. One estimate (Bishop *et al.* 2012) suggested there were only sufficient resources to undertake captive breeding for ~50 of 360 species requiring assistance globally. There is an obvious need for additional conservation tools and strategies to increase the efficiency and effectiveness of conservation efforts for amphibians (Howell *et al.* 2021a). Biobanking of cryopreserved sperm, eggs, embryos, and other reproductive and somatic tissues (Kouba and Vance 2009; Lermen *et al.* 2009; Browne *et al.* 2011; Kouba *et al.* 2013; Browne *et al.* 2019), and the recovery of live animals from those cells and tissues using reproductive technologies (Kouba *et al.* 2009; Clulow *et al.* 2014; Clulow *et al.* 2019; Silla and Byrne 2019), could be a valuable conservation tool (Della Togna *et al.* 2020; Howell *et al.* 2021a,b) if optimised and implemented widely across conservation programs. In an ideal world, the optimum approach to biobanking would involve cryostorage and recovery of whole embryos as a single step (Clulow and Clulow 2016). This would be a simple and effective means of protecting the majority of populations and species of amphibians at risk of extinction, particularly for externally fertilising species (e.g. most anurans). The loss of many species might have been avoided had the capacity to cryopreserve amphibian embryos been available, and rolled out extensively from the 1980s when amphibian declines were first becoming apparent. Unfortunately, technical challenges associated with the cryopreservation of aquatic vertebrates' eggs and embryos due to their large size and high yolk content have limited that option to date (Clulow and Clulow 2016). Considerable success has been achieved with cryopreservation of sperm and with reproductive technologies such as gamete-collection and *in vitro* fertilisation (IVF; also referred to as artificial fertilisation, AF) in amphibians, but the absence of technologies for the cryostorage of the genomes of maternal, haploid oocytes and ova and diploid embryos may have limited the adoption of biobanking in practice. This chapter focuses on new technologies that could be applied to overcome the technical block to biobanking maternal and diploid genomes through the cryopreservation of oocytes and embryos.

CRYOPRESERVATION OF AMPHIBIAN SPERM

Advances in cryopreservation of amphibian sperm have been reviewed extensively in recent years (Kouba and Vance 2009; Kouba *et al.* 2009; 2013; Clulow *et al.* 2014; Clulow and Clulow 2016; Browne *et al.* 2019; Clulow *et al.* 2019) and will not be recapitulated here. Readers are referred to these reviews for detailed information; however, a brief overview of the history and value to conservation is given.

Amphibian sperm were among the earliest cells to be investigated for storage at sub-zero temperatures, even pre-dating the now apocryphal discovery of glycerol as a sperm cryoprotectant by Polge *et al.* (1949). Cryopreservation of amphibian sperm was first reported in 1938 when the recovery of motile sperm was reported using only sucrose as a cryoprotectant (Luyet and Hodapp 1938). Rostand (1946, 1952) also demonstrated the value of glycerol as a cryoprotectant, when he preserved motility of several anuran and urodelan species' sperm for up to 40 days at –4°C. Despite early success, there were no strong incentives to study the storage of amphibian sperm during the intervening years until the 1990s. As a result, with rare exceptions there were few publications in those years, such as the ground-breaking study by Barton and Gutman (1972), who identified dimethyl sulphoxide and ethylene glycol as favourable cryoprotectants and discovered the beneficial role of low cooling rates during cryopreservation on the recovery of amphibian sperm after cryopreservation. In contrast, there were major advances with sperm-cryopreservation protocols

for key aquacultural (Tiersch *et al.* 2007; Cabrita *et al.* 2008), agricultural (Barbas and Mascarenhas 2009; Walters *et al.* 2009), and laboratory species, as well as for humans (Sherman 1980; Walters *et al.* 2009) in the intervening years from the 1950s. Avian (Blesbois 2007) and mammalian (Walters *et al.* 2009) species predominated in these advances, and the imperatives were largely commercial, and focused on humans' and animals' health.

The concept of a sperm bank for cryopreserved human sperm has been around since at least the 1800s (Mantegazza 1866). However, it wasn't until the late 1990s, about the time of the discovery of chytridiomycosis as the primary causal agent for amphibian declines (Berger *et al.* 1998; Longcore *et al.* 1999), that the value of biobanks for amphibians was recognised (Clulow *et al.* 1999). Following this seminal paper on the need for biobanking amphibian gametes, further studies of the cryopreservation of amphibian sperm began to report stronger recovery of motility than had earlier been reported (Beesley *et al.* 1998; Browne *et al.* 1998; Mugnano *et al.* 1998) and subsequent successful generation of embryos from frozen sperm through IVF was achieved (Browne *et al.* 1998). Since then, protocols for the recovery of live, motile sperm from more than 40 species have been reported (Clulow and Clulow 2016; Browne *et al.* 2019; Clulow *et al.* 2019). Reports of the recovery of motile sperm after cryopreservation outpaced reports of fertilisation. However, the generation of live progeny has been reported for a growing proportion of species in which sperm have been cryopreserved (Mansour *et al.* 2009; Mansour *et al.* 2010; Shishova *et al.* 2011; Uteshev *et al.* 2013; Proaño and Pérez 2017; Upton *et al.* 2018; Arregui *et al.* 2020; Upton *et al.* 2021). These include F1 generations produced from cryopreserved sperm, which have then produced an F2 generation by natural spawning in *Xenopus laevis*, indicating the capacity of cryopreserved sperm to generate sexually mature, fertile individuals capable of producing viable progeny across generations (Pearl *et al.* 2017) (see also Chapter 12 for F2 generations in other species). For an example of F1 offspring produced by utilising frozen–thawed sperm, see Plates 7 and 8.

Studies of sperm protocols typically report recovery of live, motile sperm at variable, but often high rates, in various species, although the fertilising capacity of cryopreserved sperm has been investigated in fewer studies (Kouba and Vance 2009; Kouba *et al.* 2009, 2013; Clulow *et al.* 2014; Clulow and Clulow 2016; Browne *et al.* 2019; Clulow *et al.* 2019; Silla and Byrne 2019). The protocols developed have employed a range of penetrating cryoprotectants (especially dimethylsulphoxide and dimethylformamide), often at high concentrations (> 10% v/v, suggesting a high level of tolerance by amphibian sperm to these substances), and a range of non-penetrating cryoprotectants, especially sucrose and trehalose with or without additional components such as egg yolk, bovine serum albumin, and foetal calf serum. Viable cooling rates during cryopreservation have varied, but low to moderate cooling rates are most common (< 5–10°C min^{-1}), with a smaller number of protocols reporting very high cooling rates (Clulow and Clulow 2016; Browne et al. 2019). Most protocols have relied on high osmolality of diluents to inactivate sperm during cryopreservation, including recent advances in protocols to cryopreserve spermic urine (Shishova et al. 2011; Uteshev et al. 2013), which are collected initially in hypotonic urine, but the diluent osmolality is increased before freezing. Nevertheless, for some species, high osmolality of cryodiluents may be detrimental (R. Upton, S. Clulow, M. Mahony and J. Clulow, unpublished data), an area that requires further investigation.

Further refinements and improvements in protocols for cryopreservation of sperm can be expected across amphibian taxa in the future. It is likely that intracytoplasmic sperm injection (ICSI), reported for *Xenopus laevis* (Amaya and Kroll 1999; Ishibashi et al. 2007), including one report with frozen sperm nuclei (Miyamoto et al. 2015) will become a tool for the generation of live young from cryopreserved sperm with poor post-thaw recovery (Clulow et al. 1999; Clulow et al. 2014; Clulow and Clulow 2016), in addition to IVF protocols utilising free-swimming spermatozoa (Wolf and Hedrick 1971; McKinnell et al. 1976; Hollinger and Corton 1980; Waggener and Carroll 1998; Edwards et al. 2004; Kouba et al. 2009). Using androgenesis, ICSI may even have a role in recovering live young from populations and species for which there are no live females (Gillespie and Armstrong 1981; Clulow et al. 2014; Clulow et al. 2019).

There are many potential applications of cryostorage in conservation and management of threatened amphibians, including genetic rescue by the retention and recovery of lost heterozygosity and wild-type genetic diversity (Della Togna et al. 2020; Howell et al. 2021a) and even potential recovery of lost strains and species through genetic backcrossing. Application and further development of protocols for cryopreservation of the sperm of *Xenopus laevis* and *Silurana tropicalis* (Sargent and

Mohun 2005; Morrow et al. 2017; Pearl et al. 2017) will result in increasing numbers of transgenic and other strains being stored in biomedical biobanks without needing to maintain live colonies. This avenue of biobanking amphibian sperm (Hagedorn et al. 2019) is likely to produce spin-offs in technology and efficiency that will benefit and promote the use of biobanking strategies in conservation.

CHALLENGES TO THE SUB-ZERO STORAGE OF THE GENOMES OF MATERNAL HAPLOID OOCYTES AND DIPLOID EMBRYOS

The missing piece in the amphibian cryopreservation toolbox is a validated, reproducible methodology for cryopreservation of the maternal haploid and embryonic diploid genomes, (i.e. oocytes and embryos). Had the technology existed, it is certain that storage of cryopreserved early embryos would have been a major conservation tool deployed in recent decades to deal with amphibian declines and to prevent extinctions. Unfortunately, the size, structure, and biochemistry of amphibian embryos (as with fish) has so far put this important target beyond technical reach, although recent advances in technology (vitrification and laser warming) and stem-cell biology offer prospects of a way forward (Clulow and Clulow 2016).

Mature oocytes, ova, and early embryos of aquatic species are large structures (when considered as spheres). It has not been possible to directly apply cryopreservation protocols developed for mammalian embryos or oocytes to aquatic species (Clulow and Clulow 2016). Cryopreservation of mammalian oocytes and early embryos (Rall and Fahy 1985a,b; Leibo and Loskutoff 1993; Kuwayama et al. 2005; Mazur et al. 2008) has succeeded because of smaller diameters and volumes, larger surface-area-to-volume ratios, greater water and cryoprotectant permeability, and lower yolk content than those of the large, telolecithal oocytes, ova, and embryos of amphibians and fish. Proportionally, the yolk content of amphibian oocytes is even higher than those of fish, exacerbating the potential negative impacts of yolk content on cryopreservation (Wallace 1963, 1985; Hagedorn et al. 1997b; Lawson et al. 2013; Guenther et al. 2006). This problem is reviewed in more detail by Clulow et al. (2014, 2019), and only a brief overview is provided here. Mammalian oocytes and embryos that have been successfully cryopreserved fall within the range of 75–200 μm diameters (D) (mouse, bovine, human) (Clulow *et al.* 2019) compared to zebrafish (D = 800 μm) (Khosla *et al.* 2017; Liu *et al.* 2020), *Xenopus* (D = 1200–1300 μm) (Guenther *et al.* 2006; Kashiwagi *et al.* 2010)) and a wide range of diameters of other amphibian ova (Tables 9.1 and 9.2). Below we discuss some potential technical solutions that may, in the future, provide robust tools for storing mature oocytes and early embryos.

There are only a few reports of direct attempts to investigate or attempt cryopreservation of amphibian oocytes, follicles, or embryos using slow cooling (Guenther *et al.* 2006; Kleinhans *et al.* 2006; Mazur and Kleinhans 2008) or vitrification (Clulow *et al.* 2014; Derakhshan *et al.* 2017). Slow cooling resulted in formation of intracellular ice at relatively high temperatures in

the range of –8 to –10°C in *Xenopus* oocytes (Guenther *et al.* 2006; Kleinhans *et al.* 2006), well above the glass transition temperature (T_g) required to prevent formation of lethal intracellular ice crystals during cooling (Khosla *et al.* 2019). This temperature range of –8 to –10°C was close to the temperature of extracellular ice formation during cooling (Guenther *et al.* 2006; Kleinhans *et al.* 2006). Formation of ice at this temperature, in combination with the initial formation of ice at the periphery of the oocyte during cooling, suggests that the nucleation of ice in those slowly cooled oocytes was initiated external to the oocyte, before it reached lower intracellular temperatures that might result in less damaging intracellular ice formation (IIF) (closer to the temperature of

Table 9.1:   Sizes of the ova of a representative sample of anurans (frogs and toads) across several families.

Within families, species are ranked in order of increasing size of ova.

Family	Species	Diameter of ova (mm)	Reference
Alytidae	*Alytes obstetricans*	3.1	Crespo (1981)
Aromobatidae	*Allobates femoralis*	2.0	Polder (1976)
	Mannophryne trinitatis	3.5	Kenny (1969)
Arthroleptidae	*Arthroleptis wahlbergi*	2.5	Wager (1965)
Ascaphidae	*Ascaphus truei*	4.0	Noble and Putnam (1931)
Bombinatoridae	*Bombina bombina*	2.0	Berger and Michałowski (1963)
	Bombina variegata	3.0	del Pino *et al.* (2007)
Brevicipitidae	*Breviceps adspersus*	4.5	Wager (1965)
Bufonidae	*Nimbaphrynoides occidentalis*	0.5 to 0.6	Lamotte *et al.* (1964); Sandberger-Loua *et al.* (2017)
	Duttaphrynus melanostictus	1.0	Jørgensen *et al.* (1986)
	Anaxyrus quercicus	1.1	Volpe and Dobie (1959)
	Incilius valliceps	1.2	Limbaugh and Volpe (1957)
	Anaxyrus americanus	1.5	Licht (2003)
	Atelopus zeteki	1.8	General and de Patilla (2006)
	Mertensophryne micranotis	1.8	Grandison (1980)
	Anaxyrus canorus	2.0	Karlstrom (1962)
	Altiphrynoides malcomi	2.6	Grandison (1978)
	Altiphrynoides osgoodi	2.8	Grandison (1978)
	Nectophryne afra	3.5	Scheel (1970)
Ceratobatrachidae	*Platymantis guentheri*	3.0	Alcala (1962)
	Platymantis dorsalis	3.3	Alcala (1962)
	Platymantis hazelae	3.4	Alcala (1962)
Craugastoridae	*Pristimantis martiae*	3.3	Crump (1974)
	Pristimantis pseudoacuminatus	3.5	Crump (1974)
	Craugastor augusti	4.2	Jameson (1950)
Dendrobatidae	*Epipedobates machalilla*	1.6	Elinson and del Pino (2012)
	Ameerega silverstonei	2.0	Myers and Daly (1979)
	Epipedobates anthonyi	2.0	del Pino *et al.* (2007)

Family	Species	Diameter of ova (mm)	Reference
	Epipedobates tricolor	2.0	del Pino *et al.* (2007)
	Phyllobates terribilis	2.5	Myers *et al.* (1978)
	Phyllobates vittatus	2.5	Silverstone (1976)
	Ameerega ingeri	3.0	del Pino *et al.* (2007)
	Dendrobates auratus	3.5	del Pino *et al.* (2007)
Dicroglossidae	*Fejervarya cancrivora*	1.3	Alcala (1962)
	Limnonectes finchi	2.0	Alcala (1962)
	Quasipaa spinosa	3.4	Dubois (1975)
Eleutherodactylidae	*Eleutherodactylus planirostris*	2.0	Goin (1947)
	Eleutherodactylus jasperi	3.3	Drewry and Jones (1976)
	Eleutherodactylus coqui	3.5	del Pino *et al.* (2007)
Heleophrynidae	*Heleophryne purcelli*	3.5	Wager (1965)
Hemiphractidae	*Gastrotheca riobambae*	3.0	del Pino and Escobar (1981)
	Flectonotus pygmaeus	4.4	Duellman and Maness (1980)
	Gastrotheca cornuta	12.0	Duellman (1970)
Hylidae	*Pseudacris ocularis*	1.0	Gosner and Rossman (1960)
	Pseudacris crucifer	1.1	Gosner and Rossman (1960)
	Dryophytes avivoca	1.8	Volpe *et al.* (1961)
	Dryophytes cinereus	1.2	Garton and Brandon (1975)
	Smilisca cyanosticta	1.2	Pyburn (1966)
	Smilisca baudinii	1.3	Duellman and Trueb (1966)
	Pseudacris streckeri	1.4	Bragg (1942)
	Pseudacris brachyphona	1.6	Green (1938)
	Trachycephalus typhonius	1.6	Pyburn (1967)
	Boana rosenbergi	2.0	Kluge (1981)
	Dendropsophus berthalutzae	3.0	Bokermann (1963)
Hyperoliidae	*Hyperolius pusillus*	1.0	Wager (1965)
	Kassina senegalensis	1.2	Wager (1965)
	Hyperolius marmoratus	1.3	Wager (1965)
	Hyperolius tuberilinguis	1.5	Wager (1965)
	Kassina maculata	1.5	Wager (1965)
	Afrixalus fornasini	2.0	Wager (1965)
Leiopelmatidae	*Leiopelma archeyi*	4.5	Stephenson (1951)
Leptodactylidae	*Engystomops randi*	1.1	del Pino *et al.* (2007)
	Engystomops coloradorum	1.3	del Pino *et al.* (2007)
	Physalaemus cuvieri	1.9	Bokermann (1962)
Limnodynastidae	*Limnodynastes peronii*	1.5	Moore (1961)
	Limnodynastes lignarius	1.9	Tyler *et al.* (1979)
	Philoria sphagnicolus	3.3	Moore (1961)
	Philoria frosti	3.9	Littlejohn (1963)
Mantellidae	*Guibemantis liber*	1.4	Blommers-Schlösser (1975b)
Microhylidae	*Kaloula conjuncta*	1.1	Alcala (1962)
	Phrynomantis bifasciatus	1.3	Wager (1965)

(Continued)

Table 9.1:   *Continued*

Family	Species	Diameter of ova (mm)	Reference
	Anodonthyla boulengeri	2.0	Blommers-Schlösser (1975a)
	Myersiella microps	3.0	Izecksohn *et al.* (1971)
	Plethodontohyla notosticta	3.0	Blommers-Schlösser (1975a)
	Cophyla grandis	4.0	Blommers-Schlösser (1975a)
Myobatrachidae	*Crinia signifera*	1.5	Anstis (2013), p. 577
	Crinia pseudinsignifera	1.7	Main (1957)
	Crinia insignifera	1.8	Main (1957)
	Crinia georgiana	2.4	Doughty and Roberts (2003)
	Geocrinia rosea	2.4	Main (1957)
	Pseudophryne pengilleyi	3.4	Anstis (2013), p. 660
	Pseudophryne corroboree	3.5	Anstis (2013), p. 642
Odontophrynidae	*Macrogenioglottus alipioi*	1.5	Abravaya and Jackson (1978)
Pelobatidae	*Pelobates fuscus*	1.5	Berger and Michałowski (1963)
Pelodryadidae	*Litoria fallax*	1.0	R. Upton, unpublished data
	Litoria verreauxii	1.2	Anstis (1976)
	Litoria caerulea	1.3	R. Upton, unpublished data
Phrynobatrachidae	*Phrynobatrachus natalensis*	1.0	Wager (1965)
Phyllomedusidae	*Agalychnis callidryas*	2.3	Pyburn (1963)
	Phyllomedusa trinitatis	3.3	Kenny (1966, 1968)
Pipidae	*Hymenochirus boettgeri*	0.8	Rabb and Rabb (1963)
	Silurana tropicalis	0.8	Kashiwagi *et al.* (2010)
	Xenopus laevis	1.2 - 1.3	Kashiwagi *et al.* (2010); Guenther *et al.* (2006)
	Pipa pipa	6.0	Rabb and Snedigar (1960)
Ptychadenidae	*Ptychadena oxyrhynchus*	1.3	Wager (1965)
Pyxicephalidae	*Amietia fuscigula*	1.5	Wager (1965)
	Tomopterna delalandii	1.5	Wager (1965)
	Pyxicephalus adspersus	2.0	Wager (1965)
	Anhydrophryne rattrayi	2.6	Wager (1965)
	Anhydrophryne hewitti	3.0	Wager (1965)
Ranidae	*Lithobates catesbeianus*	1.0	Kabutomori *et al.* (2018)
	Lithobates clamitans	1.7	Licht (2003)
	Rana japonica	1.8	Okada (1966)
	Lithobates areolata	1.8	Volpe (1957)
	Lithobates sylvaticus	1.9	Stebbins (1951)
	Sanguirana everetti	2.2	Alcala (1962)
	Lithobates septentrionalis	2.3	Hedeen (1972)
	Rana pretiosa	2.4	Turner (1958)
	Lithobates magnaocularis	2.6	Alcala (1962)
	Rana aurora	3.0	Storm (1960)
Rhacophoridae	*Chiromantis petersii*	1.0	Coe (1974)
	Polypedates leucomystax	1.9	Alcala (1962)
	Rhacophorus pardalis	3.0	Alcala (1962)

Table 9.2: Sizes of the ova of representative samples from a range of families of urodeles (newts and salamanders).

Within families, species are ranked in order of increasing size of ova.

Family	Species	Diameter of ova (mm)	Reference
Ambystomatidae	*Ambystoma mexicanum*	2.0	Brandon (1972)
	Ambystoma dumerilii	2.2	Brandon (1972)
	Ambystoma opacum	2.7	Bishop (1941)
	Ambystoma maculatum	2.8	Bishop (1941)
	Dicamptodon ensatus	6.0	Nussbaum (1969)
Cryptobranchidae	*Cryptobranchus alleganiensis*	6.0	Bishop (1941)
Hynobiidae	*Hynobius nebulosus*	2.0	Thorn (1963)
Ichthyophiidae	*Ichthyophis glutinosus*	42.0	Breckenridge *et al.* (1979)
Plethodontidae	*Desmognathus aeneus*	1.9	Harrison (1967)
	Eurycea multiplicata	2.5	Ireland (1976)
	Hemidactyllum scutatum	2.8	Bishop (1941)
	Desmognathus marmoratus	3.0	Martof (1962)
	Desmognathus ochrophaeus	3.0	Organ (1961)
	Plethodon cinereus	3.0	Sayler (1966)
	Bolitoglossa rostrata	3.5	Houck (1977)
	Ensatina eschscholtzi	3.9	Stebbins (1954)
	Plethodon albagula	4.3	Highton (1962)
	Aneides aeneus	4.5	Gordon (1952)
	Bolitoglossa subpalmata	5.0	Vial (1968)
	Ensatina eschscholtzii	6.5	Elinson and del Pino (2012)
	Phaeognathus hubrichti	7.0	Brandon (1972)
	Aneides lugubris	7.4	Stebbins (1951)
	Batrachoseps wrighti	10.0	Stebbins (1949)
Proteidae	*Necturus maculosus*	5.5	Bishop (1941)
Rhyacotritonidae	*Rhyacotriton olympicus*	3.2	Nussbaum and Tait (1977)
Salamandridae	*Lissotriton vulgaris*	1.5	Bell and Lawton (1975)
	Notophthalmus viridescens	1.5	Bishop (1941)
Sirenidae	*Siren intermedia*	3.0	Noble and Marshall (1932)

homogeneous ice formation in the presence of high concentrations of cryoprotectants). In contrast, IIF in the oocytes of mice occurs at −41°C (Mazur *et al.* 2005; Mazur and Kleinhans 2008), closer to the temperature at which less damaging homogeneous formation of ice can occur. It is likely that the fraction of unfrozen water in the external medium is affected by cell size for reasons that are not understood, meaning that in the presence of large cells, such as frogs' oocytes, extracellular water freezes at higher temperatures, in turn inducing intracellular nucleation of ice and IIF at a higher temperature than in frogs' oocytes (Mazur and Kleinhans 2008). The difference in IIF temperatures between mice and *Xenopus* is one reason slow cooling is unlikely to avoid lethal IIF in amphibian oocytes under most, if not all, conditions, even if the concentrations of cryoprotectants are high and towards the upper limits where cytotoxicity becomes apparent (Plate 9; Clulow *et al.* 2019).

Nevertheless, the slow cooling of early stage follicles of *Rhinella marina* did show the recovery of some viable thecal cells attached to stage I and II follicles, possibly indicating recovery of some function in smaller, early stage follicles after cryopreservation (Wooi *et al.* 2018; Clulow *et al.* 2019) given that thecal cells are connected to oocytes by intercellular junctions. However, vitrification of Gosner Stage 20 larvae of *Limnodynastes peronii* (Clulow *et al.* 2014), followed by convective warming, was spectacularly unsuccessful, resulting in extensive lysis of post-thaw structures, with the disruption of yolk being particularly obvious. Nevertheless, vitrification of blastulae of *Bufotes viridis* did initially preserve embryonic structure in some blastulae, but all embryos failed by the neurula stage with obvious disruption of the neurula's surface (Derakhshan *et al.* 2017). It is not possible under the experimental conditions in those studies to differentiate the effects of cooling (vitrification) from possible IIF associated with convective (slow) warming during thawing. A rapid reverse-temperature transition in the range between the glass temperature (T_g) and the melting temperature (T_m) during thawing is required to avoid the formation of ice crystals (Plate 9). It is not easy to distinguish between damage during vitrification from damage during thawing under most experimental designs because both steps have to occur during cooling and warming of the oocytes in order to measure responses to cryopreservation. More sophisticated experimental designs are required in which cooling rates and warming rates can be varied separately (Mazur and Seki 2011).

Despite these reports, responses to cooling and thawing have not been adequately characterised in amphibian oocytes and embryos to allow definitive conclusions concerning responses to slow cooling or vitrification. More work is needed in this area. However, more extensive studies with fish embryos over many years (Harvey 1983; Harvey *et al.* 1983; Hagedorn *et al.* 1996, 1997a,b, 1998, 2004; Zhang and Rawson 1996; Isayeva *et al.* 2004; Zhang *et al.* 2005; Lin *et al.* 2009; Connolly *et al.* 2016) before the advent of laser warming (discussed below) also were unable to demonstrate recovery of viable embryos post-thaw. Taken together, the studies across amphibians and fish support the view that the large, yolky (telolecithal) eggs (Wallace 1985) of both taxa are generally refractory to cryopreservation with vitrification when convective warming is relied upon to thaw frozen structures. The significance of oocyte/embryo size to the prospects for achieving successful cryopreservation by vitrification is considered below.

VITRIFICATION AND LASER WARMING AS ONE SOLUTION TO THE CHALLENGE OF CRYOPRESERVATION OF OOCYTES AND EMBRYOS

There is a general consensus based on experimental evidence that the rate of warming during thawing is more critical than the rate of cooling during vitrification for successful cryopreservation of relatively large structures such as oocytes and embryos (Mazur and Seki 2011; Khosla *et al.* 2017). Laser warming of oocytes and embryos is a recent innovation in cryobiology that has overcome the problem of lethal IIF when tested on a limited number of aquatic species with much larger embryos than those of mammals, including zebrafish (Khosla *et al.* 2017, 2020) and coral larvae (Daly *et al.* 2018). Limited permeability of membranes affecting uptake of cryoprotectants and removal of water from cytoplasmic compartments, injury from chilling, and high yolk content (discussed above) may exacerbate problems during cryopreservation. However, there is good experimental evidence from work on zebrafish that rapid warming of vitrified embryos with a short, but intense, pulse of laser in the millisecond range allows the recovery of intact, viable embryos in contrast to the same vitrified embryos warmed by conventional convective transfer of heat (Khosla *et al.* 2017, 2020). Controlled pulses of laser can generate intracellular warming rates $\geq 10 \times 10^{6}$°C/min (in the presence of gold nanorods ~110 nm × 18 nm; peak absorbance at 1064 nm wavelength) (Khosla *et al.* 2017, 2018, 2019; Liu *et al.* 2020). These warming rates cause biological materials with a diameter less than 2000 µm (Plate 10) to transition from T_g to T_m so rapidly that no ice crystallisation occurs as the ice transitions from solid glass to liquid water. It is necessary to inject gold nanorods (GNRs) into the embryo so that the distances that heat is transferred from the GNRs to any point inside the cell are kept very short in order to achieve these maximum warming rates (Khosla *et al.* 2017, 2020).

Higher concentrations of cryoprotectant lower the temperatures of heterogeneous (T_{het}) and homogeneous ice-crystal formation and reduce the temperature range in which crystallisation of ice can occur before T_g is reached (Plate 9). This applies whether specimens are warming or cooling. In both cases, higher concentrations of cryoprotectant reduce the temperature range over which ice can form crystals. This reduces the range between T_g and T_m, which is beneficial for cryopreservation. However, higher concentrations of cryoprotectant typically are toxic to cells (Derakhshan *et al.* 2017; Clulow *et al.* 2019) and concentrations above 6 M (Plate 9) are unachievable in practice due to the extreme toxicity of cryoprotectants. Consequently, there is a trade-off between the beneficial effects of raising the concentration of cryoprotectant and the need to increase rates of cooling and warming to cope with the greater temperature range over which the crystallisation of ice can occur.

The use of laser-warming to thaw embryos was pioneered by Mazur and his colleagues who used black India ink in the cryodiluent medium to absorb laser pulses and achieve high warming rates for the oocytes of mice (Jin *et al.* 2014; Jin and Mazur 2015). The more recent use of gold nanorods (GNRs) (Khosla *et al.* 2018, 2019; Liu *et al.* 2020) has allowed much higher rates of warming that resulted in the first reports of recovery of viable embryos of aquatic species, including zebrafish (Khosla *et al.* 2017) (D = 800 µm) and coral (Daly *et al.* 2018) (D = 450 µm). Cryopreserved zebrafish embryos have gone on to complete development, achieve sexual maturity, and successfully produce an F1 generation by natural spawning, thereby completing the life cycle of the species (Khosla *et al.* 2020).

The laser-warming technique holds great promise for the successful cryopreservation of oocytes, ova, and embryos of amphibians. However, there are theoretical limits to the predicted size of cells and embryos (considered as spheres) that can be cryopreserved by vitrification and laser warming based on the rates of heating that are achievable in structures of different depths (diameters) (Khosla *et al.* 2019). These predicted limits are shown for structures heated by convective warming (no laser warming) and for structures heated by laser warming using GNRs to absorb the laser pulse (Plate 10). The theoretical limit for freezing-size increases depending on whether the GNRs are inside the cells and embryos since the distance over which heat is transferred to the centre of the structure is greatly reduced when GNRs are present internally, as opposed to only externally. The rate of warming decreases exponentially with increasing distance over which heat is transferred. The theoretical limit for structures in which GNRs are present internally is ~2000 µm (2 mm) (Plate 10).

To date, the largest aquatic embryos recovered after cryopreservation are those of zebrafish (~800 µm). There is a large range of sizes of ova across fish and amphibians. The ova of many species of amphibians (Tables 9.1, 9.2 and 9.3) are within the predicted maximum size of 2000 µm for which the critical warming rate (CWR) can be achieved by laser warming, and so may be prospects for biobanking oocytes or embryos through cryopreservation. Among a range of anurans sampled (Table 9.3), ~50% of species have D values of ≤ 2.0 mm, and ~10% ≤ 1.0 mm. The group of species with the smaller diameters may be even more favourable for cryopreservation and recovery by laser warming. The anuran with the smallest ova for which we have data is the Nimba toad (*Nimbaphrynoides occidentalis*) with oval diameters of only 0.5 to 0.6 mm, and recorded as low in content of yolk (Sandberger-Loua *et al.* 2017) (Table 9.1). These small diameters may be highly favourable for cryopreservation with laser warming, although, unusual among anurans, this species is viviparous and engages in internal fertilisation and matrotrophic development within the oviducts. This reproductive strategy makes the use of cryopreservation of embryos more complicated than for most anurans because of the internal retention of the ova during development, but not impossible. The distribution of sizes of ova is less favourable in the urodeles (Caudata; salamanders and newts; Table 9.2) with more than 80% of species having ova > 2.0 mm and 50% > 3.0 mm (Table 9.3), suggesting that fewer species in this taxon might be appropriate for the banking of embryos. It is noteworthy that *Xenopus laevis* (ovum D = 1.2 – 1.3 mm) and *Silurana tropicalis* (ovum D = 0.8 mm) (Table 9.1) are both major laboratory species in molecular and developmental biology and are generating many transgenic lines that would benefit from storage via cryopreservation of embryos; these two species fall within the predicted range for successful laser warming.

In summary, the recent spectacular advances in the cryopreservation of the embryos of fish and corals using laser warming offers hope that cryopreservation of oocytes and embryos may be a viable biobanking and conservation tool for at least a proportion of amphibians. Alternative approaches will likely be necessary for many species with larger eggs and embryos. Among these, isolation of early embryonic cells derived from the dissociated embryos may offer a source of diploid cells for cryopreservation as shown by Lawson *et al.* (2013) for *Limnodynastes peronii*. These would need to be reconstituted by techniques such as nuclear transfer or generation of chimeras by the injection of cryopreserved blastomeres into developing blastulae of the host (reviewed by Clulow *et al.* 2014; Clulow and Clulow 2016). Another approach to the storage of diploid genomes could be the use of

Table 9.3: Summary of the diameters of ova within a range of species from anuran and urodelan families (from Tables 9.1 and 9.2).

Compare to Table 1 in Khosla *et al.* (2019) for thresholds of the diameters of ova predicted to cryopreserve successfully using vitrification and laser warming: ≤ 1.0 mm, high probability; > 1 ≤ 2.0 mm predicted theoretically; > 2.0 mm low probability.

Diameters of ova (mm)	No. and % of species	≤ 1.0	> 1 ≤ 2.0	> 2.0 ≤ 3.0	> 3.0	Total
Anura	No. of species	10	55	22	26	113
	% of species	9%	49%	19%	23%	
Caudata	No. of species	0	5	9	14	28
	% of species	0%	18%	32%	50%	

primordial germ cells (PGCs) and spermatogonial stem cells (SSCs) as diploid genomes for cryostorage (discussed below).

PRIMORDIAL GERM CELLS AND SPERMATAGONIAL STEM CELLS AS VEHICLES FOR STORAGE OF DIPLOID GENOMES OF AMPHIBIANS

The previous section recognised probable limitations for the cryopreservation of intact oocytes and embryos in amphibians. An alternative approach that has prospects for wide application across amphibians is the storage of diploid genomes as cryopreserved PGCs and SSCs. This approach could have the added benefit of capturing and recovering the mitochondrial genome in the case of stored PGCs, when these would later develop into oocytes after transplantation into the developing host larvae. Sperm mitochondria are lost because the paternally derived mitochondria are not retained in the developing embryo and consequently the paternal mitochondrial lineage is lost. As with laser warming, development of these technologies are more advanced in fish but the results are encouraging for amphibians.

Cryopreservation and transplantation of primordial germ cells

Primordial germ cells are the earliest manifestation of the reproductive germ cells during development (Ogielska 2009). In anurans, they are derived from a discrete set of four cells in the vegetal pole of the blastula, continue to develop as part of the endoderm, but become clearly identifiable by size and morphology when they migrate to the germinal ridge (gonadal progenitor tissue) at about Gosner Stage 24 (Ogielska 2009). This developmental stage is the point at which anuran larvae commence feeding and become free-swimming (Gosner 1960). Primordial germ

cells are incorporated into the undifferentiated, developing gonad at about Gosner Stage 26 (early formation of the limb buds) (Ogielska 2009).

PGCs have been transplanted between blastulae and between larvae in anuran amphibians before Gosner Stage 24, and subsequently incorporated into the developing gonad (Blackler 1962; Blackler and Gecking 1972; Blitz *et al.* 2016). The resulting progeny were shown to be fertile and produced viable offspring containing the complete genome of the donor individual without any apparent contribution from the host's genome, as also reported for zebrafish (Higaki *et al.* 2010).

Transplantation has been performed both at the blastula stage (Blitz *et al.* 2016) and about the tail-bud/early larval stage (Blackler 1962), including between subspecies (Blackler 1962) and even separate species (Blackler and Gecking 1972) with viable progeny produced from the transplanted PGCs. There are no reports of amphibian PGC transplantations using cryopreserved PGCs, but this could be a potential mechanism for storing and retrieving diploid genomes if compatible host embryos or larvae are available (from the same, or even separate, species). There is at least one report of the cryopreservation of early amphibian embryonic cells (Lawson *et al.* 2013) that are precursors to PGCs; thus, the cryopreservation of amphibian PGCs is likely to be feasible.

Some encouragement for this approach comes from studies of the cryopreservation and transplantation of the PGCs of fish, whereby viable offspring have been produced (Lin *et al.* 1992; Higaki *et al.* 2010; Higaki *et al.* 2013). Nevertheless, establishing biobanking procedures for PGCs will be technically demanding and will require the following steps (Plate 11):

1. ability to correctly source the site of generation of PGCs within developing embryos/larvae
2. cryopreservation of tissue-explants containing PGCs

3. transplantation of tissue-explants containing PGCs post-thaw into recipient hosts' embryos/larvae
4. subsequent survival and development of viable larvae of the host to sexual maturation, followed by the production of viable progeny from the donor gametes produced in the host.

Utilising PGCs as the stored diploid cells would result in the generation both of male and female gonadal phenotypes and gametes, depending on the genotypic sex of the donor embryo from which the PGC explant was derived. This is because the phenotypic sex of the developing gonad is determined by the genetic sex of the donor PGCs (Slanchev *et al.* 2005; Siegfried and Nüsslein-Volhard 2008). Hence the procedure allows for the generation of both sperm and eggs from the stored PGCs, although the sex of the individual from which an explant was taken would remain unknown at the time of excision from the donor embryo or blastula. An additional benefit of utilising PGCs for the generation of gametes would be the opportunity to store and retrieve the mitochondrial genome because genetically female PGCs would direct the development of ovaries and produce oocytes following transplantation. Mitochondria are inherited maternally, so any offspring produced would be carrying the maternal mitochondria.

Cryopreservation and transplantation of spermatogonial stem cells

The storage and recovery of PGCs will be a technically challenging procedure even though the feasibility and prospects of success are high. Consequently, it may be useful to have a technically simpler approach to recovering viable pluripotent diploid cells utilising spermatogonial stem cells (SSCs), as outlined in Plate 12. This would involve only one embryonic manipulation during cryopreservation and the reconstitution of the embryo. The production of viable progeny from host testes populated by donor SSCs has been reported for several species of fish (Seki *et al.* 2017; Hagedorn *et al.* 2018; Marinović *et al.* 2018; Franěk *et al.* 2019; Marinović *et al.* 2019; Mayer 2019). The SSCs can be sourced from dissociated testicular tissue (either whole testes or, in theory, small biopsies from valuable individual males of threatened species), and in a single procedure injected into developing blastulas or larvae that are at a stage of incorporating migrating germ cells into the developing testis in the germinal ridge. Cryopreservation of SSCs has been successful in fish (Lee and Yoshizaki 2016; Lee *et al.* 2016; Seki *et al.* 2017; Marinović *et al.* 2018), and, because of the small size and unremarkable structure of these cells, is likely to be relatively straightforward for amphibian SSCs using vitrification or slow-cooling protocols. As well, injection of SSCs into the developing larvae should not be technically challenging as they only require deposition into the abdominal cavity. The primary effort in optimising such procedures is likely to involve calibrating the timing of the SSC injections to coincide with the stage of the host's larval development in which germline cells are being incorporated into the developing gonad. If successful, the procedure would result in the individual host, upon development to sexual maturity, producing sperm entirely comprising the genome of the donor individual. This means that when the host individual breeds, at least some sperm that fertilise eggs will contain the genome of the donor individual (not the host that is now breeding), thus producing progeny as if the deceased donor individual was itself breeding.

One additional problem is that gonads will be chimeras, producing both host and donor sperm cells in the adults' gonads, unless the host germline is eliminated from the developing gonad. This has been achieved in fish by utilising injections of morpholino oligonucleotides into the developing host's embryo and targeting the *dead end* gene (Yoshizaki *et al.* 2016; Li *et al.* 2017). The use of busulphan as a sterilising agent to knock down germ cells before transplantation would also be possible (Shinomiya *et al.* 2001), as would the use of triploid hosts that would not produce fertile gametes (Lee *et al.* 2016).

CONCLUSIONS

Technologies for cryopreserving amphibian genomes from both sexes are making steady advances towards the goal of efficiently storing genomes in conservation biobanks so that they can be used with assisted reproductive technologies in conservation breeding programs. Effective biobanking integrated into broader conservation programs has the potential to eliminate the risk of extinction for most species, and to provide effective tools for managing the genetic diversity of threatened populations, such as performing genetic rescue or even reviving extinct species. Technologies for the cryopreservation of sperm have achieved significant progress, resulting in the production of viable, fertile progeny. However, progress in cryopreserving and recovering maternal haploid genomes and diploid genomes has been limited by technical problems associated with cryopreserving the large oocytes and embryos typical of amphibians. This chapter has outlined several potential technologies that have strong prospects for overcoming these technical challenges, and realising the full potential of being able to biobank the amphibian genome and recover live progeny as a tool to conserve the Amphibia.

ACKNOWLEDGMENTS

The authors acknowledge funding support from the Australian Research Council, the University of Newcastle and WWF (JC); Australian Government Post-Graduate Awards (RU); and University of Canberra (SC). Plates 11 and 12 were created with BioRender.com

REFERENCES

Abravaya JP, Jackson JF (1978) Reproduction in *Macrogenioglottus alipioi* Carvalho (Anura, Leptodactylidae). *Contributions in Science, Natural History Museum of Los Angeles County* **298**, 1–9. doi:10.5962/p.241010

Alcala AC (1962) Breeding behavior and early development of frogs of Negros, Philippine Islands. *Copeia* **1962**, 679–726. doi:10.2307/1440671

Alroy J (2015) Current extinction rates of reptiles and amphibians. *Proceedings of the National Academy of Sciences of the United States of America* **112**, 13003–13008. doi:10.1073/pnas.1508681112

Amaya E, Kroll KL (1999) A method for generating transgenic frog embryos. *Methods in Molecular Biology (Clifton, N.J.)* **97**, 393–414. doi:10.1385/1-59259-270-8:393

Anstis M (1976) Breeding biology and larval development of *Litoria verreauxi* (Anura: Hylidae). *Transactions of the Royal Society of South Australia* **100**, 193–202.

Anstis M (2013) *Tadpoles and Frogs of Australia*. New Holland, Chatswood.

Arregui L, Bóveda P, Gosálvez J, Kouba AJ (2020) Effect of seasonality on hormonally induced sperm in *Epidalea calamita* (Amphibia, Anura, Bufonidae) and its refrigerated and cryopreservated storage. *Aquaculture* **529**, 735677. doi:10.1016/j.aquaculture.2020.735677

Barbas JP, Mascarenhas RD (2009) Cryopreservation of domestic animal sperm cells. *Cell and Tissue Banking* **10**, 49–62. doi:10.1007/s10561-008-9081-4

Barton HL, Gutman SL (1972) Low temperature preservation of toad spermatozoa (genus *Bufo*). *The Texas Journal of Science* **23**, 363–370.

Beesley SG, Costanzo JP, Lee RE (1998) Cryopreservation of spermatozoa from freeze-tolerant and -intolerant anurans. *Cryobiology* **37**, 155–162. doi:10.1006/cryo.1998.2119

Bell G, Lawton J (1975) The ecology of the eggs and larvae of the smooth newt (*Triturus vulgaris*). *Journal of Animal Ecology* **44**(2), 393–423. doi:10.2307/3604

Berger L, Michałowski J (1963) *Klucze do oznaczania kręgowców Polski, cz. 2. Płazy – Amphibia*. Institute of Systematic Zoology, Polish Academy of Sciences, Krakow, Poland.

Berger L, Speare R, Daszak P, Green DE, Cunningham AA, Goggin CL, Slocombe R, Ragan MA, Hyatt AD, Mcdonald KR, *et al.* (1998) Chytridiomycosis causes amphibian mortality associated with population declines in the rain forests of Australia and Central America. *Proceedings of the National Academy of Sciences of the United States of America* **95**, 9031–9036. doi:10.1073/pnas.95.15.9031

Bishop SC (1941) The salamanders of New York. *Museum Bulletin – New York State Museum* **324**. University of the State of New York, Albany, New York NY, USA.

Bishop PJ, Angulo A, Lewis JP, Moore RD, Rabb GB, Garcia Moreno JG (2012) The amphibian extinction crisis – what will it take to put the action into the Amphibian Conservation Action Plan? *Surveys and Perspectives Integrating Environment and Society* **5**, 2, <http://journals.openedition.org/sapiens/1406>.

Blackler A (1962) Transfer of primordial germ-cells between two subspecies of *Xenopus laevis*. *Development* **10**, 641–651. doi:10.1242/dev.10.4.641

Blackler AW, Gecking CA (1972) Transmission of sex cells of one species through the body of a second species in the genus *Xenopus*. II. Interspecific matings. *Developmental Biology* **27**, 385–394. doi:10.1016/0012-1606(72)90177-7

Blesbois E (2007) Current status in avian semen cryopreservation. *World's Poultry Science Journal* **63**, 213–222. doi:10.1017/S0043933907001419

Blitz IL, Fish MB, Cho KWY (2016) Leapfrogging: primordial germ cell transplantation permits recovery of CRISPR/Cas9-induced mutations in essential genes. *Development* **143**, 2868–2875. doi:10.1242/dev.138057

Blommers-Schlösser R (1975a) Observations on the larval development of some Malagasy frogs, with notes on their ecology and biology (Anura: Dyscophinae, Scaphiophryninae and Cophylinae). *Beaufortia* **24**, 7–26.

Blommers-Schlösser R (1975b) A unique case of mating behaviour in a Malagasy tree frog, *Gephyromantis liber* (Peracca, 1893), with observations on the larval development (Amphibia, Ranidae). *Beaufortia* **23**, 15–25.

Bokermann W (1962) Observações biológicas sobre Physalaemus cuvieri Fitz, 1826 (Amphibia, Salientia). *Revista Brasileira de Biologia* **22**, 391–399.

Bokermann W (1963) Girinos de anfíbios brasileiros 1. (Amphibia – Salientia). *Anais da Academia Brasileira de Ciências* **35**, 465–474.

Bower DS, Lips KR, Schwarzkopf L, Georges A, Clulow S (2017) Amphibians on the brink. *Science* **357**, 454–455. doi:10.1126/science.aao0500

Bragg AN (1942) Observations on the ecology and natural history of Anura. X. The breeding habits of *Pseudacris streckeri* Wright and Wright in Oklahoma including a description of the eggs and tadpoles. The *Wasmann Collector* **5**, 47–62.

Brandon RA (1972) Hybridization between the Mexican salamanders *Ambystoma dumerilii* and *Ambystoma mexicanum* under laboratory conditions. *Herpetologica* **28**, 199–207.

Breckenridge W, Wr B, Jayasinghe S (1979) Observations on the eggs and larvae of *Ichthyophis glutinosus*. *Ceylon Journal of Science (Biological Sciences)* **13**, 187–204.

Browne RK, Clulow J, Mahony MJ, Clark AK (1998) Successful recovery of motility and fertility of cryopreserved cane toad (*Bufo marinus*) sperm. *Cryobiology* **37**, 339–345. doi:10.1006/cryo.1998.2129

Browne RK, Li H, Robertson H, Uteshev VK, Shishova NV, Mcginnity D, Nofs S, Figiel CR, Mansour N, Lloyd R, *et al.* (2011) Reptile and amphibian conservation through gene banking and other reproduction technologies. *Russian Journal of Herpetology* **18**, 165–174.

Browne RK, Silla AJ, Upton R, Della-Togna G, Marcec-Greaves R, Shishova NV, Uteshev VK, Proaño B, Pérez OD, Mansour N, *et al.* (2019) Sperm collection and storage for the sustainable management of amphibian biodiversity. *Theriogenology* **133**, 187–200. doi:10.1016/j.theriogenology.2019.03.035

Cabrita E, Robles V, Herráez P (2008) *Methods in Reproductive Aquaculture: Marine and Freshwater species.* CRC Press, Boca Raton FL, USA.

Clulow J, Clulow S (2016) Cryopreservation and other assisted reproductive technologies for the conservation of threatened amphibians and reptiles: bringing the ARTs up to speed. *Reproduction, Fertility and Development* **28**, 1116–1132. doi:10.1071/RD15466

Clulow J, Mahony MJ, Browne RK, Pomering M, Clark AK (1999) Applications of assisted reproductive technologies (ART) to endangered anuran amphibians. In *Declines and Disappearances of Australian Frogs.* (Ed. A Campbell) pp. 219–225. Environment Australia, Canberra.

Clulow J, Trudeau V, Kouba A (2014) Amphibian declines in the twenty-first century: why we need assisted reproductive technologies. In *Reproductive Sciences in Animal Conservation.* (Eds WV Holt, JL Brown and P Comizzoli) pp. 275–316. Springer, New York NY, USA.

Clulow S, Gould J, James H, Stockwell M, Clulow J, Mahony M (2018) Elevated salinity blocks pathogen transmission and improves host survival from the global amphibian chytrid pandemic: implications for translocations. *Journal of Applied Ecology* **55**, 830–840. doi:10.1111/1365-2664.13030

Clulow J, Upton R, Trudeau V, Clulow S (2019) Amphibian assisted reproductive technologies: moving from technology to application. In *Reproductive Sciences in Animal Conservation.* 2nd edn. (Eds P Comizzoli, JL Brown and WV Holt) pp. 413–463. Springer, Cham, Switzerland.

Coe M (1974) Observations on the ecology and breeding biology of the genus *Chiromantis* (Amphibia: Rhacophoridae). *Journal of Zoology* **172**, 13–34. doi:10.1111/j.1469-7998.1974.tb04091.x

Connolly MH, Paredes E, Mazur P, Hagedorn M (2016) Toward the cryopreservation of Zebrafish embryos: tolerance to osmotic dehydration. *Cryobiology* **73**, 439. doi:10.1016/j.cryobiol.2016.09.153

Crespo EG (1981) Contribuição para o conhecimento da biologia dos Alytes ibéricos, *Alytes obstetricans boscai lataste*, 1879 e *Alytes cisternasii boscá*, 1879 (Amphibia, Salientia). Universidade de Lisboa, Tese, Faculdade de Ciência, Lisbon, Portugal.

Crump ML (1974) Reproductive strategies in a tropical anuran community. *Miscellaneous Publications of the Museum of Natural History, University of Kansas* **61**, 1–68.

Daly J, Zuchowicz N, Nuñez Lendo CI, Khosla K, Lager C, Henley EM, Bischof J, Kleinhans FW, Lin C, Peters EC, Hagedorn M (2018) Successful cryopreservation of coral larvae using vitrification and laser warming. *Scientific Reports* **8**, 15714. doi:10.1038/s41598-018-34035-0

del Pino EM, Escobar B (1981) Embryonic stages of *Gastrotheca riobambae* (Fowler) during maternal incubation and comparison of development with that of other egg-brooding hylid frogs. *Journal of Morphology* **167**, 277–295. doi:10.1002/jmor.1051670303

del Pino EM, Venegas-Ferrín M, Romero-Carvajal A, Montenegro-Larrea P, Sáenz-Ponce N, Moya IM, Alarcón I, Sudou N, Yamamoto S, Taira M (2007) A comparative analysis of frog early development. *Proceedings of the National Academy of Sciences of the United States of America* **104**, 11882–11888. doi:10.1073/pnas.0705092104

Della Togna G, Howell LG, Clulow J, Langhorne CJ, Marcec-Greaves R, Calatayud NE (2020) Evaluating amphibian biobanking and reproduction for captive breeding programs according to the Amphibian Conservation Action Plan objectives. *Theriogenology* **150**, 412–431. doi:10.1016/j.theriogenology.2020.02.024

Derakhshan Z, Nokhbatolfoghahai M, Zahiri S (2017) Cryopreservation of *Bufotes viridis* embryos by vitrification. *Cryobiology* **75**, 60–67. doi:10.1016/j.cryobiol.2017.02.003

Doughty P, Roberts JD (2003) Plasticity in age and size at metamorphosis of *Crinia georgiana* tadpoles: responses to variation in food levels and deteriorating conditions during development. *Australian Journal of Zoology* **51**, 271–284. doi:10.1071/ZO02075

Drewry GE, Jones KL (1976) A new ovoviviparous frog, *Eleutherodactylus jasperi* (Amphibia, Anura, Leptodactylidae), from Puerto Rico. *Journal of Herpetology* **10**, 161–165. doi:10.2307/1562976

Dubois A (1975) Un nouveau sous-genre (Paa) et trois nouvelles espèces du genre *Rana*. Remarques sur la phylogénie des Ranidés (Amphibiens, Anoures). *Bulletin du Muséum National d'Histoire Naturelle* **3**, 1093–1115.

Duellman WE (1970) *The Hylid Frogs of Middle America.* Volume 1. Museum of Natural History, University of Kansas, Lawrence KS, USA.

Duellman WE, Maness SJ (1980) The reproductive behavior of some hylid marsupial frogs. *Journal of Herpetology* **14**, 213–222. doi:10.2307/1563542

Duellman WE, Trueb L (1966) Neotropical hylid frogs, genus *Smilisca*. *University of Kansas Publications. Museum of Natural History* **17**, 281–375.

Edwards DL, Mahony MJ, Clulow J (2004) Effect of sperm concentration, medium osmolality and oocyte storage on artificial fertilisation success in a myobatrachid frog (*Limnodynastes tasmaniensis*). *Reproduction, Fertility and Development* **16**, 347–354. doi:10.1071/RD02079

Elinson RP, del Pino EM (2012) Developmental diversity of amphibians. *Wiley Interdisciplinary Reviews. Developmental Biology* **1**, 345–369. doi:10.1002/wdev.23

Franěk R, Marinović Z, Lujić J, Urbányi B, Fučíková M, Kašpar V, Pšenička M, Horváth Á (2019) Cryopreservation and transplantation of common carp spermatogonia. *PLoS One* **14**, e0205481. doi:10.1371/journal.pone.0205481

Garton JS, Brandon RA (1975) Reproductive ecology of the green treefrog, *Hyla cinerea*, in southern Illinois (Anura: Hylidae). *Herpetologica* **31**, 150–161.

General E, de Patilla EDC (2006) Reproductive ecology of *Atelopus zeteki* and comparisons to other members of the genus. *Herpetological Review* **37**, 284–288.

Gillespie LL, Armstrong JB (1981) Suppression of first cleavage in the Mexican axolotl (*Ambystoma mexicanum*) by heat shock or hydrostatic pressure. *The Journal of Experimental Zoology* **218**, 441–445. doi:10.1002/jez.1402180316

Goin CJ (1947) Studies on the life history of *Eleutherodactylus ricordii planirostris* (Cope) in Florida. *University of Florida Studies, Biological Sciences Series* **4**, 1–66.

Gordon RE (1952) A contribution to the life history and ecology of the plethodontid salamander *Aneides aeneus* (Cope and Packard). *American Midland Naturalist* **47**, 666–701. doi:10.2307/2422035

Gosner KL (1960) A simplified table for staging anuran embryos and larvae with notes on identification. *Herpetologica* **16**, 183–190.

Gosner KL, Rossman DA (1960) Eggs and larval development of the treefrogs *Hyla crucifer* and *Hyla ocularis*. *Herpetologica* **16**, 225–232.

Grandison AG (1978) The occurrence of *Nectophrynoides* (Anura Bufonidae) in Ethiopia. A new concept of the genus with a description of a new species. *Monitore Zoologico Italiano Supplemento* **11**, 119–172. doi:10.1080/03749444.197 8.10736579

Grandison AG (1980) Aspects of breeding morphology in *Mertensophryne micranotis* (Anura: Bufonidae): secondary sexual characters, eggs and tadpoles. *Bulletin of the British Museum, Natural History. Zoology* **39**, 299–304. doi:10.5962/p.171550

Green NB (1938) The breeding habits of *Pseudacris brachyphona* (Cope) with a description of the eggs and tadpoles. *Copeia* **1938**, 79–82. doi:10.2307/1435695

Guenther JF, Seki S, Kleinhans FW, Edashige K, Roberts DM, Mazur P (2006) Extra- and intra-cellular ice formation in Stage I and II *Xenopus laevis* oocytes. *Cryobiology* **52**, 401–416. doi:10.1016/j.cryobiol.2006.02.002

Hagedorn M, Hsu EW, Pilatus U, Wildt DE, Rall WR, Blackband SJ (1996) Magnetic resonance microscopy and spectroscopy reveal kinetics of cryoprotectant permeation in a multicompartmental biological system. *Proceedings of the National Academy of Sciences of the United States of America* **93**, 7454–7459. doi:10.1073/pnas.93.15.7454

Hagedorn M, Hsu E, Kleinhans FW, Wildt DE (1997a) New approaches for studying the permeability of fish embryos: toward successful cryopreservation. *Cryobiology* **34**, 335–347. doi:10.1006/cryo.1997.2014

Hagedorn M, Kleinhans FW, Wildt DE, Rall WF (1997b) Chill sensitivity and cryoprotectant permeability of dechorionated zebrafish embryos, *Brachydanio rerio*. *Cryobiology* **34**, 251–263. doi:10.1006/cryo.1997.2002

Hagedorn M, Kleinhans FW, Artemov D, Pilatus U (1998) Characterization of a major permeability barrier in the zebrafish embryo. *Biology of Reproduction* **59**, 1240–1250. doi:10.1095/biolreprod59.5.1240

Hagedorn M, Peterson A, Mazur P, Kleinhans FW (2004) High ice nucleation temperature of zebrafish embryos: slow-freezing is not an option. *Cryobiology* **49**, 181–189. doi:10.1016/j.cryobiol.2004.07.001

Hagedorn M, Varga Z, Walter RB, Tiersch TR (2019) Workshop report: cryopreservation of aquatic biomedical models. *Cryobiology* **86**, 120–129. doi:10.1016/j.cryobiol.2018.10.264

Hagedorn MM, Daly JP, Carter VL, Cole KS, Jaafar Z, Lager CV, Parenti LR (2018) Cryopreservation of fish spermatogonial cells: the future of natural history collections. *Scientific Reports* **8**, 6149. doi:10.1038/s41598-018-24269-3

Harrison JR (1967) Observations on the life history, ecology and distribution of *Desmognathus aeneus aeneus* Brown and Bishop. *American Midland Naturalist* **77**, 356–370. doi:10.2307/2423347

Harvey B (1983) Cooling of embryonic cells, isolated blastoderms, and intact embryos of the zebra fish *Brachydanio rerio* to −196°C. *Cryobiology* **20**, 440–447. doi:10.1016/0011-2240(83)90034-2

Harvey B, Kelley RN, Ashwood-Smith MJ (1983) Permeability of intact and dechorionated zebra fish embryos to glycerol and dimethyl sulfoxide. *Cryobiology* **20**, 432–439. doi:10.1016/0011-2240(83)90033-0

Hedeen SE (1972) Postmetamorphic growth and reproduction of the mink frog, *Rana septentrionalis* Baird. *Copeia* **1972**, 169–175. doi:10.2307/1442794

Higaki S, Eto Y, Kawakami Y, Yamaha E, Kagawa N, Kuwayama M, Nagano M, Katagiri S, Takahashi Y (2010) Production of fertile zebrafish (*Danio rerio*) possessing germ cells (gametes) originated from primordial germ cells recovered from vitrified embryos. *Reproduction (Cambridge, England)* **139**, 733–740. doi:10.1530/REP-09-0549

Higaki S, Kawakami Y, Eto Y, Yamaha E, Nagano M, Katagiri S, Takada T, Takahashi Y (2013) Cryopreservation of zebrafish (*Danio rerio*) primordial germ cells by vitrification of yolk-intact and yolk-depleted embryos using various cryoprotectant solutions. *Cryobiology* **67**, 374–382. doi:10.1016/j.cryobiol.2013.10.006

Highton R (1962) Geographic variation in the life history of the slimy salamander. *Copeia* **1962**, 597–613. doi:10.2307/1441185

Hollinger TG, Corton GL (1980) Artificial fertilization of gametes from the South African clawed frog, *Xenopus laevis*. *Gamete Research* **3**, 45–57. doi:10.1002/mrd.1120030106

Houck LD (1977) Reproductive biology of a neotropical salamander, *Bolitoglossa rostrata*. *Copeia* **1977**, 70–83. doi:10.2307/1443507

Howell LG, Frankham R, Rodger JC, Witt RR, Clulow S, Upton RM, Clulow J (2021a) Integrating biobanking minimises inbreeding and produces significant cost benefits for a threatened frog captive breeding programme. *Conservation Letters* **14**, e12776. doi:10.1111/conl.12776

Howell LG, Mawson PR, Frankham R, Rodger JC, Upton RM, Witt RR, Calatayud NE, Clulow S, Clulow J (2021b) Integrating biobanking could produce significant cost benefits and minimise inbreeding for Australian amphibian captive breeding programs. *Reproduction, Fertility and Development* **33**, 573–587. doi:10.1071/RD21058

Ireland PH (1976) Reproduction and larval development of the gray-bellied salamander *Eurycea multiplicata grisegaster*. *Herpetologica* **32**, 233–238.

Isayeva A, Zhang T, Rawson DM (2004) Studies on chilling sensitivity of zebrafish (*Danio rerio*) oocytes. *Cryobiology* **49**, 114–122. doi:10.1016/j.cryobiol.2004.05.005

Ishibashi S, Kroll KL, Amaya E (2007) Generation of transgenic Xenopus laevis: III. Sperm nuclear transplantation. *Cold Spring Harbor Protocols* **2007**, doi:10.1101/pdb.prot4840.

Izecksohn E, Jim J, Albuquerque S, Mendonça W (1971) Observações sobre o desenvolvimento e os hábitos de *Myersiella subnigra* (Miranda-Ribeiro) (Amphibia, Anura, Microhylidae). *Arquivos do Museu Nacional do Rio de Janeiro* **54**, 69–73.

Jameson DL (1950) The development of *Eleutherodactylus latrans*. *Copeia* **1950**, 44–46. doi:10.2307/1437581

Jin B, Kleinhans FW, Mazur P (2014) Survivals of mouse oocytes approach 100% after vitrification in 3-fold diluted media and ultra-rapid warming by an IR laser pulse. *Cryobiology* **68**, 419–430. doi:10.1016/j.cryobiol.2014.03.005

Jin B, Mazur P (2015) High survival of mouse oocytes/embryos after vitrification without permeating cryoprotectants followed by ultra-rapid warming with an IR laser pulse. *Scientific Reports* **5**, 9271. doi:10.1038/srep09271

Jørgensen CB, Shakuntala K, Vijayakumar S (1986) Body size, reproduction and growth in a tropical toad, *Bufo melanostictus*, with a comparison of ovarian cycles in tropical and temperate zone anurans. *Oikos* **46**, 379–389. doi:10.2307/3565838

Kabutomori J, Beloto-Silva O, Geyer RR, Musa-Aziz R (2018) *Lithobates catesbeianus* (American Bullfrog) oocytes: a novel heterologous expression system for aquaporins. *Biology Open* **7**, bio031880. doi:10.1242/bio.031880

Karlstrom EL (1962) *The Toad Genus* Bufo *in the Sierra Nevada of California: Ecological and Systematic Relationships*. University of California Press, Berkeley CA, USA.

Kashiwagi K, Kashiwagi A, Kurabayashi A, Hanada H, Nakajima K, Okada M, Takase M, Yaoita Y (2010) *Xenopus tropicalis*: an ideal experimental animal in amphibia. *Experimental Animals* **59**, 395–405. doi:10.1538/expanim.59.395

Kenny JS (1966) Nest building in *Phyllomedusa trinitatis* Mertens. *Caribbean Journal of Science* **6**, 15–22.

Kenny JS (1968) Early development and larval natural history of *Phyllomedusa trinitatis* Mertens. *Caribbean Journal of Science* **8**, 35–45.

Kenny JS (1969) The amphibia of Trinidad. *Studies on the Fauna of Curaçao and other Caribbean Islands* **29**, 1–78.

Khosla K, Wang Y, Hagedorn M, Qin Z, Bischof J (2017) Gold nanorod induced warming of embryos from the cryogenic state enhances viability. *ACS Nano* **11**, 7869–7878. doi:10.1021/acsnano.7b02216

Khosla K, Zhan L, Bhati A, Carley-Clopton A, Hagedorn M, Bischof J (2018) Physical limits of laser gold nanowarming. *Cryobiology* **85**, 161. doi:10.1016/j.cryobiol.2018.10.161

Khosla K, Zhan L, Bhati A, Carley-Clopton A, Hagedorn M, Bischof J (2019) Characterization of laser gold nanowarming: a platform for millimeter-scale cryopreservation. *Langmuir* **35**(23), 7364–7375. doi:10.1021/acs.langmuir.8b03011

Khosla K, Kangas J, Liu Y, Zhan L, Daly J, Hagedorn M, Bischof J (2020) Cryopreservation and laser nanowarming of zebrafish embryos followed by hatching and spawning. *Advanced Biosystems* **4**, 2000138. doi:10.1002/adbi.202000138

Kleinhans FW, Guenther JF, Roberts DM, Mazur P (2006) Analysis of intracellular ice nucleation in *Xenopus* oocytes by differential scanning calorimetry. *Cryobiology* **52**, 128–138. doi:10.1016/j.cryobiol.2005.10.008

Kluge AG (1981) The life history, social organization, and parental behavior of *Hyla rosenbergi* Boulenger, a nest-building gladiator frog. *Miscellaneous Publications of the Museum of Zoology of the University of Michigan* **160**, 1–170.

Kouba AJ, Vance CK (2009) Applied reproductive technologies and genetic resource banking for amphibian conservation. *Reproduction, Fertility and Development* **21**, 719–737. doi:10.1071/RD09038

Kouba AJ, Vance CK, Willis EL (2009) Artificial fertilization for amphibian conservation: current knowledge and future considerations. *Theriogenology* **71**, 214–227. doi:10.1016/j.theriogenology.2008.09.055

Kouba AJ, Lloyd RE, Houck ML, Silla AJ, Calatayud N, Trudeau VL, Clulow J, Molinia F, Langhorne C, Vance C, *et al.* (2013) Emerging trends for biobanking amphibian genetic resources: the hope, reality and challenges for the next decade. *Biological Conservation* **164**, 10–21. doi:10.1016/j.biocon.2013.03.010

Kuwayama M, Vajta G, Kato O, Leibo SP (2005) Highly efficient vitrification method for cryopreservation of human oocytes. *Reproductive Biomedicine Online* **11**, 300–308. doi:10.1016/S1472-6483(10)60837-1

Lamotte M, Rey P, Vogeli M (1964) Recherches sur l'ovaire de *Nectophrynoides occidentalis*, Batracien anoure vivipare. *Archives d'Anatomie Microscopique et de Morphologie Expérimentale* **53**, 179–224.

Lawson B, Clulow S, Mahony MJ, Clulow J (2013) Towards gene banking amphibian maternal germ lines: short-term incubation, cryoprotectant tolerance and cryopreservation of embryonic cells of the frog, *Limnodynastes peronii*. *PLoS One* **8**, e60760. doi:10.1371/journal.pone.0060760

Lee S, Iwasaki Y, Yoshizaki G (2016) Long-term (5 years) cryopreserved spermatogonia have high capacity to generate functional gametes via interspecies transplantation in salmonids. *Cryobiology* **73**, 286–290. doi:10.1016/j.cryobiol.2016.08.001

Lee S, Yoshizaki G (2016) Successful cryopreservation of spermatogonia in critically endangered Manchurian trout

(*Brachymystax lenok*). *Cryobiology* **72**, 165–168. doi:10.1016/j.cryobiol.2016.01.004

Leibo S, Loskutoff N (1993) Cryobiology of in vitro-derived bovine embryos. *Theriogenology* **39**, 81–94. doi:10.1016/0093-691X(93)90025-Z

Lermen D, Blömeke B, Browne R, Clarke ANN, Dyce PW, Fixemer T, Fuhr GR, Holt WV, Jewgenow K, Lloyd RE, *et al.* (2009) Cryobanking of viable biomaterials: implementation of new strategies for conservation purposes. *Molecular Ecology* **18**, 1030–1033. doi:10.1111/j.1365-294X.2008.04062.x

Li Q, Fujii W, Naito K, Yoshizaki G (2017) Application of dead end-knockout zebrafish as recipients of germ cell transplantation. *Molecular Reproduction and Development* **84**, 1100–1111. doi:10.1002/mrd.22870

Licht LE (2003) Shedding light on ultraviolet radiation and amphibian embryos. *Bioscience* **53**, 551–561. doi:10.1641/0006-3568(2003)053[0551:SLOURA]2.0.CO;2

Limbaugh BA, Volpe EP (1957) Early development of the Gulf Coast toad, *Bufo valliceps* Wiegmann. *American Museum Novitates* **1842**, 1–32.

Lin S, Long W, Chen J, Hopkins N (1992) Production of germline chimeras in zebrafish by cell transplants from genetically pigmented to albino embryos. *Proceedings of the National Academy of Sciences of the United States of America* **89**, 4519–4523. doi:10.1073/pnas.89.10.4519

Lin C, Zhang T, Rawson DM (2009) Cryopreservation of zebrafish (*Danio rerio*) blastomeres by controlled slow cooling. *Cryo Letters* **30**, 132–141.

Littlejohn M (1963) The breeding biology of the Baw Baw frog *Philoria frosti* Spencer. *Proceedings of the Linnean Society of New South Wales* **88**, 273–276.

Liu Y, Kangas J, Wang Y, Khosla K, Pasek-Allen J, Saunders A, Oldenburg S, Bischof J (2020) Photothermal conversion of gold nanoparticles for uniform pulsed laser warming of vitrified biomaterials. *Nanoscale* **12**, 12346–12356. doi:10.1039/D0NR01614D

Longcore JE, Pessier AP, Nichols DK (1999) *Batrachochytrium dendrobatidis* gen. et sp. nov., a chytrid pathogenic to amphibians. *Mycologia* **91**, 219–227. doi:10.2307/3761366

Luyet BJ, Hodapp EL (1938) Revival of frog's spermatozoa vitrified in liquid air. *Proceedings of the Society for Experimental Biology and Medicine* **39**, 433–434. doi:10.3181/00379727-39-10229P

Main A (1957) Studies in Australian Amphibia. 1. The genus *Crinia* Tschudi in South-western Australia and some species from south-eastern Australia. *Australian Journal of Zoology* **5**, 30–55. doi:10.1071/ZO9570030

Mansour N, Lahnsteiner F, Patzner RA (2009) Optimization of the cryopreservation of African clawed frog (*Xenopus laevis*) sperm. *Theriogenology* **72**, 1221–1228. doi:10.1016/j.theriogenology.2009.07.013

Mansour N, Lahnsteiner F, Patzner RA (2010) Motility and cryopreservation of spermatozoa of European common frog, *Rana temporaria*. *Theriogenology* **74**, 724–732. doi:10.1016/j.theriogenology.2010.03.025

Mantegazza P (1866) Fisiologia sullo sperma umano. *Rendiconti* **3**, 183–186.

Marinović Z, Li Q, Lujić J, Iwasaki Y, Csenki Z, Urbányi B, Horváth Á, Yoshizaki G (2018) Testis cryopreservation and spermatogonia transplantation as a tool for zebrafish line reconstitution. *Cryobiology* **85**, 145–146. doi:10.1016/j.cryobiol.2018.10.104

Marinović Z, Li Q, Lujić J, Iwasaki Y, Csenki Z, Urbányi B, Yoshizaki G, Horváth Á (2019) Preservation of zebrafish genetic resources through testis cryopreservation and spermatogonia transplantation. *Scientific Reports* **9**, 13861. doi:10.1038/s41598-019-50169-1

Martof BS (1962) Some observations on the role of olfaction among salientian amphibia. *Physiological Zoology* **35**, 270–272. doi:10.1086/physzool.35.3.30152812

Mayer I (2019) The role of reproductive sciences in the preservation and breeding of commercial and threatened teleost fishes. In *Reproductive Sciences in Animal Conservation*. (Eds P Comizzoli, GP Brown and WV Holt) pp. 187–224. Cham, Switzerland: Springer.

Mazur P, Kleinhans F (2008) Relationship between intracellular ice formation in oocytes of the mouse and Xenopus and the physical state of the external medium – a revisit. *Cryobiology* **56**, 22–27. doi:10.1016/j.cryobiol.2007.10.002

Mazur P, Seki S (2011) Survival of mouse oocytes after being cooled in a vitrification solution to −196°C at 95 to 70,000°C/min and warmed at 610 to 118,000°C/min: a new paradigm for cryopreservation by vitrification. *Cryobiology* **62**, 1–7. doi:10.1016/j.cryobiol.2010.10.159

Mazur P, Seki S, Pinn IL, Kleinhans F, Edashige K (2005) Extra- and intracellular ice formation in mouse oocytes. *Cryobiology* **51**, 29–53. doi:10.1016/j.cryobiol.2005.04.008

Mazur P, Leibo SP, Seidel GE (2008) Cryopreservation of the germplasm of animals used in biological and medical research: importance, impact, status, and future directions. *Biology of Reproduction* **78**, 2–12. doi:10.1095/biolreprod.107.064113

McKinnell RG, Picciano DJ, Krieg RE (1976) Fertilization and development of frog eggs after repeated spermiation induced by human chorionic gonadotropin. *Laboratory Animal Science* **26**, 932–935.

Miyamoto K, Simpson D, Gurdon JB (2015) Manipulation and in vitro maturation of *Xenopus laevis* oocytes, followed by intracytoplasmic sperm injection, to study embryonic development. *Journal of Visualized Experiments* **96**, e52496. doi:10.3791/52496

Moore JA (1961) The frogs of eastern New South Wales. *Bulletin of the American Museum of Natural History* **121**, 149–386.

Morrow S, Gosálvez J, López-Fernández C, Arroyo F, Holt WV, Guille MJ (2017) Effects of freezing and activation on membrane quality and DNA damage in *Xenopus tropicalis* and *Xenopus laevis* spermatozoa. *Reproduction, Fertility and Development* **29**, 1556–1566. doi:10.1071/RD16190

Mugnano JA, Costanzo JP, Beesley SG, Lee RE (1998) Evaluation of glycerol and dimethyl sulfoxide for the cryopreservation of spermatozoa from the wood frog (*Rana sylvatica*). *Cryo Letters* **19**, 249–254.

Myers CW, Daly JW (1979) A name for the poison frog of Cordillera Azul, eastern Peru, with notes on its biology and skin toxins (Dendrobatidae). *American Museum Novitates* **2674**, 1–24.

Myers CW, Daly JW, Malkin B (1978) A dangerously toxic new frog (*Phyllobates*) used by Emberá Indians of western Colombia, with discussion of blowgun fabrication and dart poisoning. *Bulletin of the American Museum of Natural History* **161**, 307–366.

Noble G, Putnam PG (1931) Observations on the life history of *Ascaphus truei* Stejneger. *Copeia* **1931**, 97–101. doi:10.2307/1437329

Noble GK, Marshall BC (1932) The validity of *Siren intermedia* LeConte, with observations on its life history. *American Museum Novitates* **532**, 1–17.

Nussbaum RA (1969) Nests and eggs of the Pacific giant salamander, *Dicamptodon ensatus* (Eschscholtz). *Herpetologica* **25**, 257–262.

Nussbaum RA, Tait CK (1977) Aspects of the life history and ecology of the Olympic salamander, *Rhyacotriton olympicus* (Gaige). *American Midland Naturalist* **98**, 176–199. doi:10.2307/2424723

Ogielska M (2009) The undifferentiated amphibian gonad. In *Reproduction of Amphibians*. (Eds M Ogielska and J Bartmanska) pp. 1–34. Science Publishers, Enfield NH, USA.

Okada Y (1966) *Fauna Japonica: Anura (Amphibia)*. Biogeographical Society of Japan, Tokyo.

Organ JA (1961) Studies of the local distribution, life history, and population dynamics of the salamander genus *Desmognathus* in Virginia. *Ecological Monographs* **31**, 189–220. doi:10.2307/1950754

Pearl E, Morrow S, Noble A, Lerebours A, Horb M, Guille M (2017) An optimized method for cryogenic storage of *Xenopus* sperm to maximise the effectiveness of research using genetically altered frogs. *Theriogenology* **92**, 149–155. doi:10.1016/j.theriogenology.2017.01.007

Polder W (1976) *Dendrobates, Phyllobates* en *Colostethus. Het Aquarium* **46**, 260–266.

Polge C, Smith AU, Parkes AS (1949) Revival of spermatozoa after vitrification and dehydration at low temperatures. *Nature* **164**, 666. doi:10.1038/164666a0

Proaño B, Pérez OD (2017) In vitro fertilizations with cryopreserved sperm of *Rhinella marina* (Anura: Bufonidae) in Ecuador. *Amphibian & Reptile Conservation* **11**, 1–6.

Pyburn WF (1963) Observations on the life history of the treefrog, *Phyllomedusa callidryas* (Cope). *The Texas Journal of Science* **15**, 155–170.

Pyburn WF (1966) Breeding activity, larvae and relationship of the treefrog *Hyla phaeota cyanosticta. The Southwestern Naturalist* **11**, 1–18. doi:10.2307/3669176

Pyburn WF (1967) Breeding and larval development of the hylid frog *Phrynohyas spilomma* in southern Veracruz, Mexico. *Herpetologica* **23**, 184–194.

Rabb GB, Rabb MS (1963) On the behavior and breeding biology of the African pipid frog: Hymenochirus boettgeri. *Zeitschrift für Tierpsychologie* **20**, 215–241. doi:10.1111/j.1439-0310.1963.tb01151.x

Rabb GB, Snedigar R (1960) Observations on breeding and development of the Surinam toad, *Pipa pipa. Copeia* **1960**, 40–44. doi:10.2307/1439843

Rall W, Fahy G (1985a) Vitrification: a new approach to embryo cryopreservation. *Theriogenology* **23**, 220. doi:10.1016/0093-691X(85)90126-8

Rall WF, Fahy GM (1985b) Ice-free cryopreservation of mouse embryos at −196°C by vitrification. *Nature* **313**, 573. doi:10.1038/313573a0

Rohr JR, Raffel TR (2010) Linking global climate and temperature variability to widespread amphibian declines putatively caused by disease. *Proceedings of the National Academy of Sciences of the United States of America* **107**, 8269–8274. doi:10.1073/pnas.0912883107

Rostand J (1946) Glycérine et resistance du sperme aux basses températures. *Comptes Rendus de l'Académie des Sciences (Paris)* **222**, 1524–1525.

Rostand J (1952) Sur le refroidissement des cellules spermatiques en présénce de glycérine. *Comptes Rendus Hebdomadaires des Séances de l'Académie des Sciences* **234**, 2310–2312.

Sandberger-Loua L, Müller H, Rödel M-O (2017) A review of the reproductive biology of the only known matrotrophic viviparous anuran, the West African Nimba toad, *Nimbaphrynoides occidentalis. Zoosystematics and Evolution* **93**, 105–133. doi:10.3897/zse.93.10489

Sargent MG, Mohun TJ (2005) Cryopreservation of sperm of *Xenopus laevis* and *Xenopus tropicalis. Genesis (New York, N.Y.)* **41**, 41–46. doi:10.1002/gene.20092

Sayler A (1966) The reproductive ecology of the red-backed salamander, *Plethodon cinereus*, in Maryland. *Copeia* **1966**, 183–193. doi:10.2307/1441125

Scheel J (1970) Notes on the biology of the African tree-toad, *Nectophryne afra* Buchholz & Peters, 1875 (Bufonidae, Anura) from Fernando Poo. *Revue de Zoologie et de Botanique Africaines* **81**, 225–236.

Scheele BC, Hunter DA, Grogan LF, Berger L, Kolby JE, Mcfadden MS, Marantelli G, Skerratt LF, Driscoll DA (2014) Interventions for reducing extinction risk in chytridiomycosis-threatened amphibians. *Conservation Biology* **28**, 1195–1205. doi:10.1111/cobi.12322

Scheele BC, Pasmans F, Skerratt LF, Berger L, Martel A, Beukema W, Acevedo AA, Burrowes PA, Carvalho T, Catenazzi A (2019) Amphibian fungal panzootic causes catastrophic and ongoing loss of biodiversity. *Science* **363**, 1459–1463. doi:10.1126/science.aav0379

Seki S, Kusano K, Lee S, Iwasaki Y, Yagisawa M, Ishida M, Hiratsuka T, Sasado T, Naruse K, Yoshizaki G (2017) Production of the medaka derived from vitrified whole testes by germ cell transplantation. *Scientific Reports* **7**, 43185. doi:10.1038/srep43185

Sherman JK (1980) Historical synopsis of human semen cryobanking. In *Human Artificial Insemination and Semen Preservation*. (Eds G David and WS Price) pp. 95–105. Plenum Press, New York NY, USA.

Shinomiya A, Hamaguchi S, Shibata N (2001) Sexual differentiation of germ cell deficient gonads in the medaka, *Oryzias latipes. The Journal of Experimental Zoology* **290**, 402–410. doi:10.1002/jez.1081

Shishova NR, Uteshev VK, Kaurova SA, Browne RK, Gakhova EN (2011) Cryopreservation of hormonally induced sperm for the conservation of threatened amphibians with *Rana temporaria* as a model research species. *Theriogenology* **75**, 220–232. doi:10.1016/j.theriogenology.2010.08.008

Shoo LP, Olson DH, Mcmenamin SK, Murray KA, Van Sluys M, Donnelly MA, Stratford D, Terhivuo J, Merino-Viteri A, Herbert SM, *et al.* (2011) Engineering a future for amphibians under climate change. *Journal of Applied Ecology* **48**, 487–492. doi:10.1111/j.1365-2664.2010.01942.x

Siegfried KR, Nüsslein-Volhard C (2008) Germ line control of female sex determination in zebrafish. *Developmental Biology* **324**, 277–287. doi:10.1016/j.ydbio.2008.09.025

Silla AJ, Byrne PG (2019) The role of reproductive technologies in amphibian conservation breeding programs. *Annual Review of Animal Biosciences* **7**, 499–519. doi:10.1146/annurev-animal-020518-115056

Silverstone PA (1976) A revision of the poison-arrow frogs of the genus *Phyllobates* Bibron in Sagra (Family Dendrobatidae). Revisión de las ranas venenosas del género Phyllobates Bibron en Sagra (Familia Dendrobatidae). *Natural History* **27**, 1–53.

Skerratt L, Berger L, Speare R, Cashins S, Mcdonald K, Phillott A, Hines H, Kenyon N (2007) Spread of chytridiomycosis has caused the rapid global decline and extinction of frogs. *EcoHealth* **4**, 125–134. doi:10.1007/s10393-007-0093-5

Skerratt LF, Berger L, Clemann N, Hunter DA, Marantelli G, Newell DA, Philips A, Mcfadden M, Hines HB, Scheele BC (2016) Priorities for management of chytridiomycosis in Australia: saving frogs from extinction. *Wildlife Research* **43**, 105–120. doi:10.1071/WR15071

Slanchev K, Stebler J, De La Cueva-Méndez G, Raz E (2005) Development without germ cells: the role of the germ line in zebrafish sex differentiation. *Proceedings of the National Academy of Sciences of the United States of America* **102**, 4074–4079. doi:10.1073/pnas.0407475102

Stebbins RC (1949) Observations on laying, development, and hatching of the eggs of *Batrachoseps wrighti*. *Copeia* **1949**, 161–168. doi:10.2307/1438980

Stebbins RC (1951) *Amphibians of Western North America.*: University of California Press, Berkeley CA, USA.

Stebbins RC (1954) Natural history of the salamanders of the plethodontid genus *Ensatina*. *University of California Publications in Zoology* **54**, 47–123.

Stephenson N (1951) Observations on the development of the amphicoelous frogs, *Leiopelma* and *Ascaphus*. *Zoological Journal of the Linnean Society* **42**, 18–28. doi:10.1111/j.1096-3642.1951.tb01851.x

Storm RM (1960) Notes on the breeding biology of the red-legged frog (*Rana aurora aurora*). *Herpetologica* **16**, 251–259.

Stuart SN, Chanson JS, Cox NA, Young BE, Rodrigues ASL, Fischman DL, Waller RW (2004) Status and trends of amphibian declines and extinctions worldwide. *Science* **306**, 1783–1786. doi:10.1126/science.1103538

Thorn R (1963) Contribution à l'étude d'une salamandre japonaise l'Hynobius nebulosus (Schlegel). Comportement et reproduction en captivité. *Archives Institute Grand-Ducal Luxembourg, Section des Sciences Naturelles Physiques et Mathematiques* **29**, 201–215.

Tiersch TR, Yang H, Jenkins JA, Dong Q (2007) Sperm cryopreservation in fish and shellfish. *Society of Reproduction and Fertility Supplement* **65**, 493–508.

Turner FB (1958) Life-history of the western spotted frog in Yellowstone National Park. *Herpetologica* **14**, 96–100.

Tyler MJ, Martin A, Davies M (1979) Biology and systematics of a new limnodynastine genus (Anura: Leptodactylidae) from North-Western Australia. *Australian Journal of Zoology* **27**, 135–150. doi:10.1071/ZO9790135

Upton R, Clulow S, Mahony MJ, Clulow J (2018) Generation of a sexually mature individual of the Eastern dwarf tree frog, *Litoria fallax*, from cryopreserved testicular macerates: proof of capacity of cryopreserved sperm derived offspring to complete development. *Conservation Physiology* **6**, coy043. doi:10.1093/conphys/coy043

Upton R, Clulow S, Calatayud NE, Colyvas K, Seeto RG, Wong LA, Mahony MJ, Clulow J (2021) Generation of reproductively mature offspring from the endangered green and golden bell frog *Litoria aurea* using cryopreserved spermatozoa. *Reproduction, Fertility and Development* **33**, 562–572. doi:10.1071/RD20296

Uteshev VK, Shishova N, Kaurova S, Manokhin A, Gakhova E (2013) Collection and cryopreservation of hormonally induced sperm of pool frog (*Pelophylax lessonae*). *Russian Journal of Herpetology* **20**, 105–109.

Vial JL (1968) The ecology of the tropical salamander, *Bolitoglossa subpalmata*, in Costa Rica. *Revista de Biología Tropical* **15**, 13–115.

Volpe E (1957) The early development of *Rana capito sevosa*. *Tulane Studies in Zoology* **5**, 207–225.

Volpe EP, Dobie JL (1959) The larva of the oak toad, *Bufo quercicus* Holbrook. *Tulane Studies in Zoology* **7**, 145–152.

Volpe EP, Wilkens MA, Dobie JL (1961) Embryonic and larval development of *Hyla avivoca*. *Copeia* **1961**, 340–349. doi:10.2307/1439811

Vredenburg VT, Knapp RA, Tunstall TS, Briggs CJ (2010) Dynamics of an emerging disease drive large-scale amphibian population extinctions. *Proceedings of the National Academy of Sciences of the United States of America* **107**, 9689–9694. doi:10.1073/pnas.0914111107

Wager VA (1965) *The Frogs of South Africa*. Purnell and Sons, Cape Town, South Africa.

Waggener WL, Carroll EJ (1998) A method for hormonal induction of sperm release in anurans (eight species) and in vitro fertilization in *Lepidobatrachus* species. *Development, Growth & Differentiation* **40**, 19–25. doi:10.1046/j.1440-169X.1998.t01-5-00003.x

Wallace RA (1963) Studies on amphibian yolk III. A resolution of yolk platelet components. *Biochimica et Biophysica Acta* **74**, 495–504. doi:10.1016/0006-3002(63)91392-1

Wallace RA (1985) Vitellogenesis and oocyte growth in non-mammalian vertebrates. In *Oogenesis*. (Ed. LW Browder) pp. 127–177. Springer, Boston MA, USA.

Walters EM, Benson JD, Woods EJ, Critser JK (2009) The history of sperm cryopreservation. In *Sperm Banking: Theory and Practice.* (Eds AA Pacey and MJ Tomlinson) pp. 2–10. Cambridge University Press, Cambridge, UK.

Wolf DP, Hedrick JL (1971) A molecular approach to fertilization: II. Viability and artificial fertilization of *Xenopus laevis* gametes. *Developmental Biology* **25**, 348–359. doi:10.1016/0012-1606(71)90036-4

Wooi K, Mahony M, Shaw J, Clulow J (2018) Cryopreservation of oocytes and follicular cells of the cane toad, *Rhinella marina*. *Cryobiology* **85**, 142–143. doi:10.1016/j.cryobiol.2018.10.095

Yoshizaki G, Takashiba K, Shimamori S, Fujinuma K, Shikina S, Okutsu T, Kume S, Hayashi M (2016) Production of germ cell-deficient salmonids by dead end gene knockdown, and their use as recipients for germ cell transplantation. *Molecular Reproduction and Development* **83**, 298–311. doi:10.1002/mrd.22625

Zhang T, Rawson DM (1996) Feasibility studies on vitrification of intact zebrafish (*Brachydanio rerio*) embryos. *Cryobiology* **33**, 1–13. doi:10.1006/cryo.1996.0001

Zhang T, Isayeva A, Adams SL, Rawson DM (2005) Studies on membrane permeability of zebrafish (*Danio rerio*) oocytes in the presence of different cryoprotectants. *Cryobiology* **50**, 285–293. doi:10.1016/j.cryobiol.2005.02.007

10 Culturing and biobanking of amphibian cell lines for conservation applications

Julie Strand, Barbara Fraser, Marlys L. Houck, and Simon Clulow

INTRODUCTION

The global amphibian-extinction crisis has been well documented and is now widely considered the worst modern loss of biodiversity among all vertebrate classes (Bower *et al.* 2017; Stuart *et al.* 2004). Causes are multifaceted, including emergent diseases, pollution, loss of habitat, and introduction of invasive species (see Chapter 2). Combined with the impacts from other difficult-to-manage threats such as climatic change (Cohen *et al.* 2019), hundreds more amphibian species are likely to be impacted in the near future with very few options for mitigation in the wild within the timeframe required to halt further declines and extinctions (Clulow *et al.* 2018; Bower *et al.* 2019, 2020).

One approach that can assist in preventing and reversing catastrophic loss of biodiversity from unstoppable or unpredictable threatening processes is the biobanking of genetic material in combination with reproductive technologies (Holt *et al.* 2003, 2014; Ryder and Onuma 2018; Clulow *et al.* 2019). Biobanking, sometimes referred to as genome resource banking or genome storage, involves storing embryos, gametes, or differentiated cells in a way that makes them retrievable, often involving cryopreservation (Lermen *et al.* 2009; Kouba *et al.* 2013). Reproductive technologies provide the tools required to retrieve and restore the stored genomes through techniques such as artificial insemination, *in vitro* fertilisation (IVF; also referred to as artificial fertilisation) or advanced

techniques such as cloning (Pukazhenthi and Wildt 2004; Pukazhenthi *et al.* 2006; Wildt *et al.* 2010; Clulow and Clulow 2016; Howard *et al.* 2016). Reproductive technologies have been used for decades in various taxa, especially agricultural, domestic, and experimental animals, mainly for systematic breeding programs, but in later years the focus has shifted to conservation strategies (Browne *et al.* 2011; Hildebrandt *et al.* 2012; Albl *et al.* 2016; Groeneveld *et al.* 2016; Korody *et al.* 2017; Zimkus *et al.* 2018; Browne *et al.* 2019; Strand *et al.* 2020). The benefits of biobanking and associated reproductive technologies are well known, including providing an insurance against species' extinctions, improving genetic and economic outcomes in conservation breeding programs, providing an avenue for genetic rescue of wild populations through reintroduction, or even the revival of extinct species (Clulow and Clulow 2016; Howell *et al.* 2021a,b). The gold standard for this approach is to bank embryos in a suspended state that can simply be reanimated when required, allowing embryonic, larval, and juvenile development to continue through to sexually mature individuals of both sexes. Storing gametes is also a common target, as these cells are haploid and typically easier to collect, store, and retrieve (see Chapter 9). Unfortunately, in some taxa it is not always possible to obtain or store either embryos or gametes from one or both sexes. This is generally the case with fishes and amphibians, as they typically have large eggs with a high volume of yolk, resulting in oocytes and embryos that are difficult to freeze (Clulow *et al.* 2012; Lawson *et al.* 2013; Clulow and Clulow 2016; Clulow *et al.* 2018, 2019). Storing diploid cells can be the only way to capture both the male and female genetic lines in these instances; this can pose unique challenges for reviving individual animals as it usually involves the use of more advanced reproductive technologies, especially with differentiated cells (Lawson *et al.* 2013; Clulow and Clulow 2016). Regardless, storing diploid cells is an important goal for biobanking as they are abundant, easy to obtain, and capture genome-wide diversity from both sexes.

A culture of animal cells is arguably one of the most useful technologies to be developed in concert with biobanking diploid cells. Dating back to the beginning of the 20th century (Davis 2011; Freshney 2016), culture of animal cell lines has been shown to provide a potentially unlimited source of high quality biological material, providing an expandable resource of genetic material with extensive application and usability (reviewed by Freshney 2016). The vast range of potential uses and applications range from replacing live animals in virological, pharmacological, and toxicological studies to tissue-engineering and *ex situ* conservation efforts (Yuan *et al.* 2015; Freshney 2016). Recently, the emergence of artificial fertilisation technologies that were developed from culturing embryos (Edwards 1996) demonstrated the potential of cell culture to be applied to conservation. Other potential uses for conservation and research include genome sequencing, genome expression, and cytogenetic studies (identification of species, hybridisation, chromosome analysis, etc.), including both molecular and comparative research (Houck *et al.* 2017). More advanced uses in the future could include *in vitro* studies of amphibian diseases, including chytridiomycoses caused by *Batrachochytrium dendrobatidis* (Bd) and *Batrachochytrium salamandrivorans* (Bsal) without the need for infecting whole animals, by culturing the skin cells that these fungi infect.

Attempts to preserve amphibian eggs and embryos have had limited success (Chapter 9; Clulow and Clulow 2016). However, Shinya Yamanaka's (2020) seminal work on reprogramming cell lines into pluripotent stem cells that have the ability to differentiate into any type of cell, including gametes, presents an alternative that has exciting possibilities for amphibian conservation (Clulow *et al.* 2019). Additionally, early work by Briggs and King (1957) successfully used somatic cell nuclear transfer to replace the oocyte nuclei with the nuclei of blastula cells, which then progressed through development to the tadpole stage. This work led to ongoing research, using species such as *Xenopus*, in which somatic cell nuclear transfer has been effected and induction of totipotency of somatic cell nuclei has been successful (Gurdon and Wilmut 2011). Successful somatic cell nuclear transfer may expand the current usability of advanced reproductive technologies by the inclusion of female gametes and thereby improve potential future application to conservation.

Initially, cell culture was focused on mammalian cells because of their practical application to the medical industry, whereas research on lower-order vertebrates and invertebrates has been very limited (Davis 2011; Freshney 2016). In recent years, non-model animals including reptiles, amphibians, and invertebrates have gained more focus due to pest control, outbreaks of diseases, and a desire to develop new and greater conservation efforts (Clulow and Clulow 2016; Campbell *et al.* 2020, 2021). This has led to some detailed protocols and approaches, but many of these remain focused on methods developed from cultures of mammalian cells (Masters 2003; Freshney 2016). Protocols for non-mammalian vertebrates often are developed by experimentation with variations of the methods used successfully for mammalian species, which has been the case for amphibians. The development of successful methods for the culture of amphibian cells lags well behind that of other vertebrates and is widely considered to more difficult than for any other vertebrate class. There is thus an imperative to commit more research and resources to the field of amphibian cell culture, especially to advance its application to conservation. An important first step for this is to bring together the growing body of literature on both successful and unsuccessful methods to synthesise and elucidate future research directions for the field.

This chapter attempts to help pave that path by providing a complete and up-to-date overview of published studies in amphibian cell culture that have been performed since 1959. We focus primarily on physiological and chemical conditions set for the individual established cell lines. We do not discuss studies focusing on the regeneration of limbs, lenses, and nerves, or studies focusing on cancer development using amphibian cell lines. To our knowledge, this is the first review with a focus on the establishment of primary amphibian cell lines.

Amphibian cell lines – missed opportunities and challenges ahead?

Of the 74 research papers reviewed (Table 10.1), only one published study established cell lines for the purposes of biobanking, with the cells reportedly remaining viable for up to 10 years (Okumoto 2001). However, we are aware of a small number of institutions, such as San Diego Frozen Zoo*, that have also successfully biobanked amphibian cell lines (Chemnick *et al.* 2009; Zimkus *et al.* 2018). The Frozen Zoo currently contains the largest known collection of amphibian cell lines worldwide with 95 lines from 83 individuals representing 21 species currently banked (Zimkus *et al.* 2018). Smaller collections of amphibian cell lines also exist at the American Type Culture Collection, but institutions and facilities focusing on biobanking viable cell lines are limited (Zimkus *et al.* 2018). Collectively, our data suggest that less than 0.1% of the over 8000 described amphibian species (*AmphibiaWeb* 2021) have cell lines that are currently stored in biobanks worldwide (Table 10.1). This represents a huge, missed opportunity for amphibian conservation and research and highlights a need for greater focus in this area.

Further evidence highlighting the lack of successful culture of amphibian cells around the world comes from a study into facilities involved in cryopreserving viable cell lines from terrestrial vertebrates (mammals, birds, reptiles, and amphibians) throughout Africa, Asia, and North America (Ryder and Onuma 2018). The study focused on species from the IUCN categories 'Critically Endangered' and 'Extinct in the Wild' and reported only one amphibian species from these two categories to have a cryopreserved line of cells derived from the critically endangered Malagasy frog, the golden mantella (*Mantella aurantiaca*) (Ryder and Onuma 2018). The IUCN lists 1160 amphibian species as threatened with extinction (classified as either Endangered, Critically Endangered, Extinct, or Extinct in the Wild) as of 2019 (IUCN 2021), illustrating the lack of attention in this area and the need for more institutions and facilities to participate in the establishment of amphibian cell lines to preserve their genetic and biological diversity before it is no longer possible to do so.

It is also important to note that research conducted on amphibian cell lines to date is limited to a small number of families and species. Within the order Caudata (salamanders and newts), only seven species from three families have been reported to have been the subject of research on cell culture, while similar studies within the order Anura (frogs and toads) have been reported for 20 species from eight families (Table 10.1). Among the latter, common model species such as *Rana catesbeiana*, *Rana pipiens* and *Xenopus laevis* dominate the research as model organisms (Table 10.1). The remainder of this chapter focuses on reviewing the various methods attempted to date for culturing and storing amphibian cell lines (presented in Table 10.1). It is our aim to discuss both successful and unsuccessful protocols to help identify areas that are currently hampering progress in the field, as well as to identify promising leads for future research and application.

CULTURING AND BIOBANKING AMPHIBIAN CELL LINES

Types of cell cultures

Primary cell cultures are established directly from intact or dissociated tissues and are considered 'primary' until they are subcultured (passaged); thereafter they are termed a 'cell line'. Primary cultures can give rise to cell lines that are no longer viable after several subcultures (finite cell lines) or cell lines that grow indefinitely (continuous cell lines, also known as immortal cell lines) (Davis 2011; Freshney 2016). A basic flow chart of the techniques for creating and maintaining various types of amphibian cell lines is provided in Figure 10.1. Finite and continuous cell lines display different advantages and disadvantages depending on the end use (Figure 10.1). Cell cultures must be correctly maintained and nourished, otherwise, as they are repeatedly passaged, they can change morphology, phenotype, and genotype – as some mutations may be more favourable for survival and growth within the *in vitro* environment provided, thus irrevocably changing the cells (Davis 2011; Freshney 2016). This can mean that the cell line is no longer a good model for research or suitable for biobanking. The advantage of primary cell lines over multiply passaged immortalised ones is that they tend to retain authentic genotype and phenotype. This makes primary cell lines the preferred target for many applications to conservation, including biobanking. However, the limited lifespan and slower growth rate of primary cell lines can impact their potential use in research.

Pre-culture considerations

Numerous factors including collection of samples, aseptic techniques, and laboratory and cell-culture equipment should be considered before initiating a cell culture.

Table 10.1: A complete overview of studies performed since 1959 on the culturing of amphibian cells.

Species	Tissue type	Explant (E), enzyme digestion (D), disaggregation medium (DM), or mechanical dissociation (M)	Incubation temp. °C	Media (overview only)	Antimicrobial	Cell lines established	Karyotyped	Reference
Caudata								
Andrias davidianus	Thymus, spleen, kidney	E	25	TC199, FBS	Pen/Strep	Yes	2n = 50	Yuan *et al.* (2015)
Ambystoma mexicanum	Liver	E	25	MEM, ddH₂O, FBS, HEPES	Pen/Strep/F	Primary culture	No	Clothier *et al.* (1982)
Ambystoma mexicanum	Thymus	E	26	L-15, HEPES, NaHCO₃, 2-M	Pen/Strep	Primary culture	No	Koniski and Cohen (1992)
Ambystoma tigrinum	Embryo	D	N/A	SS, BPA, HSG, HS. Several media types listed – see reference for more details		Yes	No	Jones and Elsdale (1963)
Pleurodeles waltl	Embryo	D	N/A	SS, BPA, HSG, HS. Several media types listed – see reference for more details		Yes	No	Jones and Elsdale (1963)
Salamandra salamandra	Liver	E	25	MEM, ddH₂O, FBS, HEPES	Pen/Strep/F	Primary culture	No	Clothier *et al.* (1982)
Triturus alpestri	Embryo	D	N/A	SS, BPA, HSG, HS. Several media types listed – see reference for more details		Yes	No	Jones and Elsdale (1963)
Triturus cristatus	Limp, tongue, toeclip	E	28	a-MEM, FBS, several supplements tested – see reference for more details	Pen/stresp/glut/Nor	Yes	Yes	Strand *et al.* (2021)
Triturus cristatus carnifex	Liver	E	25	MEM, ddH₂O, FBS, HEPES	Pen/Strep/F	Primary culture	No	Clothier *et al.* (1982)
Anura								
Discoglossus pictus	Embryo	D	N/A	Several media types listed – see reference for more details		Yes	No	Jones and Elsdale (1963)
Bombina maxima	Tadpole	D	25	DMEM/F12, FBS		Yes	2n = 28	Xiang *et al.* (2012)

(Continued)

Table 10.1: *Continued*

Species	Tissue type	Explant (E), enzyme digestion (D), disaggregation medium (DM), or mechanical dissociation (M)	Incubation temp. °C	Media (overview only)	Antimicrobial	Cell lines established	Karyotyped	Reference
Bombina orientalis	Embryo	D	22 and 25	L-15, ddH$_2$O, FBS		Yes	$2n = 24$	Ellinger et al. (1983)
Bufo americanus	Embryo	N/A	25	MEM, ddH$_2$O, FBS, HEPES	Pen/Strep/Myc	N/A	N/A	Freed et al. (1969)
Rhinella marina	Bladder	D	28	Ham's F-12: L-15 (7:3), NaCl, NaHCO$_3$, BEE, TS, CS, DEX, Spe, Spermi, Put, BS	Pen/strep	Yes	No	Handler et al. (1979)
Bufo vulgaris	Embryo	D	N/A	SS, BPA, HSG, HS. Several media types listed – see reference for more details		Yes	No	Jones and Elsdale (1963)
Hyla arborea japonica	Kidney and lung	E	N/A	EMEM, CS, LH	Pen/Strep/AB	Yes	$2n = 24$	Seto (1964); Stephenson and Stephenson (1970)
Hyla arborea japonica	N/A	N/A	N/A	N/A		N/A	N/A	Matsuda (1963)
Litoria caerulea	Heart, lung liver	D	31 and 37	EMEM, CS, LH	Pen/Strep/AB	Yes	$2n = 26$	Stephenson and Stephenson (1970)
Litoria infrafrenata	Adult frog toe and tadpole macerates	E	21	D-MEM, FBS, GS, ITS	Pen/Strep	Yes	$2n = 24$	Mollard (2018)
Litoria phyllochroa	Heart, lung liver	D	31 and 37	EMEM, CS, LH	Pen/Strep/AB	Yes	$2n = 26$	Stephenson and Stephenson (1970)
Leptodactylus ocellatus	Spleen	Cellular suspension	25	NCTC 109, HBSS, AS, CEE	Pen/Strep/Myc	Primary culture	$2n = 22$	Robinson and Stephenson (1967)
Limnodynastes peronii	Liver, lung, testis and heart	E	15, 20, 25 30	NCTC 109, HBSS, AS, CEE	Pen/Strep/Myc	Yes	$2n = 24$	Robinson and Stephenson (1967)

Species	Tissue		Temp	Media	Antibiotics			Reference
Rana catesbeiana	Tadpole	E, D	27 and 37	N/A	Pen/Strep/Myc	Yes	No	Gross and Lapiere (1962)
Rana catesbeiana	Tongue	N/A	N/A	NaCl-free EBSS, WEU, FBS, EMEM – see reference for antibiotic modifications		Yes	N/A	Wolf and Quimbry (1964)
Rana catesbeiana	Tadpole	D	37	N/A		Primary culture	No	Eisen and Gross (1965)
Rana catesbeiana	Liver	E	30	NaCl-free EBSS, WEU, FBS, EMEM. See reference for antibiotic modifications	Pen/Strep/Myc	Primary culture	No	Bennett et al. (1969)
Rana catesbeiana	Liver	D	25	M199, ddH$_2$O (1:1). Three other media were used – see reference for modifications	Pen/Strep, Neo/F/Myc	Primary culture	No	Blatt et al. (1969)
Rana catesbeiana	Tadpole	D	25	SS, FBS, M199 (10:1:2)	Pen/strep	Primary culture	No	Ide (1973)
Rana catesbeiana	Tadpole	D	25	L-15, ddH$_2$O FBS (5:3:1.7)	Pen/Strep	Primary culture	No	Ide (1974)
Rana catesbeiana	Liver	D	24	WM, NaCl, HEPES	G	Primary culture	No	Stanchfield and Yager (1978)
Rana catesbeiana	Liver	D	N/A	WM, 752M, 3S		Primary culture	No	Stanchfield and Yager (1980)
Rana catesbeiana	Tadpole	E	22	M199, ddH$_2$O, FBS	G/F	Primary culture	No	Ketola-Pirie and Atkinson (1983)
Rana catesbeiana	Tadpole	D	25	L-15, FBS. To dilute the conditioned medium for melanophore culture mixed filtered medium: ddH$_2$O: FBS (5.5:3:1.5)	Pen/Strep	Yes	2n = 26	Kondo and Ide (1983)
Rana catesbeiana	Liver	D	22	MF10, Glu, BSA, TPB, NaHCO$_3$, HEPES	Pen/Strep/Dex	Primary culture	No	Kubokawa and Ishii (1987)
Rana catesbeiana	Tadpole	E, DM	23	Several media types listed – see reference for more details		Yes	No	Niki and Yoshizato (1984)
Rana catesbeiana	Tadpole	D	23	Several media types listed – see reference for more details		Yes	No	Nishikawa et al. (1990)
Rana catesbeiana	Tadpole	D	22	CTS medium: RPMI-1640, HEPES, NaHCO$_3$, CTS	Pen/Strep	Yes	No	Nishikawa and Yoshizato (1986)

(Continued)

Table 10.1: *Continued*

Species	Tissue type	Explant (E), enzyme digestion (D), disaggregation medium (DM), or mechanical dissociation (M)	Incubation temp. °C	Media (overview only)	Antimicrobial	Cell lines established	Karyotyped	Reference
Rana catesbeiana	Tadpole	E	N/A	MEM, Glut, GS	Strep	Primary culture	No	Ketola-Pirie and Atkinson (1988)
Rana catesbeiana	Tadpole	D	23	RPMI 1640, 7% CTS, 10 mM NaHCO$_3$, 20 mM HEPES, Pen/Strep		Yes	No	Nishikawa *et al.* (1989)
Rana catesbeiana	Tadpole	E	22	MEM, FBS or TH		Primary culture	No	Ketola-Pirie and Atkinson (1990)
Rana catesbeiana	Skin	D	18	88.5% amphibian culture medium (Wolf and Quimby's – see above), HEPES	AA/GS	Primary culture	No	García-Díaz (1991)
Rana esculenta	Liver	E	25	MEM, ddH$_2$O, FBS, HEPES	Pen/Strep/F	Primary culture	No	Clothier *et al.* (1982)
Rana nigromaculata	Skin	D	25	L-15:ddH$_2$O:FBS (6:3:1)	Pen/Strep/F	Yes	Modal number varied	Okumoto (2001)
Rana pipiens	Embryo	D	24–27	Modifications of both Holtfreter's solution and Niu–Twitty solution; see article for precise composition of saline medium	No antibiotics are mentioned	Primary culture	No	Barth and Barth (1959)
Rana pipiens	Kidney	D	25	Media compositions – see reference	Pen	Primary culture	No	Auclair (1961)
Rana pipiens	Kidney	D	18	EMEM, WEU, CS	See reference for antibiotic solution	Primary culture	No	Shah (1962)
Rana pipiens	Embryo	D	N/A	Several media types listed – see reference for more details		Yes	No	Jones and Elsdale (1963)
Rana pipiens	Liver	D	N/A	Several media types listed – see reference for more details		Primary culture	No	Ansevin (1964)

Species	Tissue		Temp	Media	Antibiotics	Primary/Yes	No	Reference
Rana pipiens	Kidney	D	25	EMEM, dH$_2$O, CS, WEU		Primary culture	No	Malamud (1967)
Rana pipiens	Embryo	D	25	L-15, FBS		Yes	X	Freed and Mezger-Freed (1970a)
Rana pipiens	Spleen	M	N/A	MC33	Pen/Strep/F	Primary culture	No	Carver and Meints (1977)
Rana pipiens	Tadpole	E	19	EMEM with ESSS		Primary culture	No	Pollack (1980)
Rana pipiens	Liver	E	25	MEM, ddH$_2$O, FBS, HEPES	Pen/Strep/F	Primary culture	No	Clothier et al. (1982)
Rana pipiens	Skin	D	18	88.5% amphibian culture medium (Wo⁻ and Quimby's – see above), HEPES	AA/GS	Primary culture	No	García-Díaz (1991)
Rana porosa brevipoda	Skin	D	25	L-15: ddH$_2$O: FBS (6:3:1)	Pen/Strep/F	Yes	Modal number varied	Okumoto (2001)
Rana sphenocephalia	Liver	E	25	MEM, ddH$_2$O, FBS, HEPES	Pen/Strep/F	Primary culture	No	Clothier et al. (1982)
Rana temporia	Embryo	D	N/A	Several medic types listed – see reference for more details		Yes	No	Jones and Elsdale (1963)
Rana temporia	Liver	E	25	MEM, ddH$_2$O, FBS, HEPES	Pen/Strep/F	Primary culture	No	Clothier et al. (1982)
Xenopus laevis	Embryo	D	N/A	Several medic types listed – see reference for more details		Yes	No	Jones and Elsdale (1963)
Xenopus laevis	Liver, spleen, kidney	E	25	MEM, ddH$_2$O, FBS, HEPES	Pen/Strep/Myc	Yes	No	Balls and Ruben (1966)
Xenopus laevis	Liver	N/A	N/A	N/A		N/A	N/A	Regan et al. (1968)
Xenopus laevis	Skin	E	N/A	N/A		Primary culture	No	McGarry and Vanable (1969)
Xenopus laevis	Liver, kidney, ovary	E	25	L-15, ddH$_2$O, FBS		Primary culture	No	Simnett and Balls (1969)
Xenopus laevis	Embryo	D	25	Several medic types listed – see reference for more details		Primary culture	No	Laskey (1970)

(Continued)

Table 10.1: *Continued*

Species	Tissue type	Explant (E), enzyme digestion (D), disaggregation medium (DM), or mechanical dissociation (M)	Incubation temp. °C	Media (overview only)	Antimicrobial	Cell lines established	Karyotyped	Reference
Xenopus laevis	Tadpole	D	25	L-15, ddH$_2$O, FBS	Pen/Strep/Myc	Yes	No	Gurdon and Laskey (1970)
Xenopus laevis	Spleen, liver, kidney, leg muscle	D	25	L-15, ddH$_2$O, FBS, CS, TPB	Pen/Strep/F	Yes	X	Arthur and Balls (1971)
Xenopus laevis	Liver	E	26	Several media types listed – see reference for more details		Yes	No	Solursh and Reiter (1972)
Xenopus laevis	Tadpole	D	28	NCTC 109, ddH$_2$O, LH, Glut, NaHCO$_3$		Yes	No	Pudney et al. (1973)
Xenopus laevis	Thymus and spleen	N/A	–	Several media types listed – see reference for more details		Primary culture	No	Weiss and Du Pasquier (1973)
Xenopus laevis	Liver	E	25	Serum-free medium (modified Wolf and Quimby's)	G/F	Primary culture	No	Green et al. (1976)
Xenopus laevis	Embryo	D	25	L-15, FBS, GS	F	Yes	No	Miller and Daniel (1977)
Xenopus laevis	Tadpole	D	26	L-15, FBS, TPB	Pen/Strep/F	Primary culture	No	Wahli and Weber (1977)
Xenopus laevis	Liver	D	25	Basic medium (Wolf and Quimby's), HEPES, Glut, PI	Pen/Strep/G	Primary culture	No	Wangh et al. (1979)
Xenopus laevis	Tadpole	N/A	25	MEM, dH2O, FBS		Yes	X (2n = 36)	Anizet et al. (1981)
Xenopus laevis	Liver	E	25	MEM, ddH2O, FBS, HEPES	Pen/Strep/F	Primary culture	No	Clothier et al. (1982)
Xenopus laevis	Embryo	N/A	20	N/A		N/A	N/A	Asashima and Grunz (1982)
Xenopus laevis	Liver	D	23	L-15, Glu, Ins	Pen/Strep	Primary culture	No	Kawahara et al. (1983)
Xenopus laevis	Liver	D	25	Several media types listed – see reference for more details		Primary culture	No	Kawahara et al. (1985)

Species	Tissue	D/E/M	Temp	Media	Antimicrobials			Reference
Xenopus laevis	Skin and embryo	D	N/A	L-15, FBS	GS	Yes	No	Ellison et al. (1985)
Xenopus laevis	Parenchymal cells	D	21	CM, Dex, Ins. TH		Primary culture	No´	Aprison et al. (1986)
Xenopus laevis	Larvae	E	25	L-15	Pen/Strep/GS/AB	Primary culture	No	Mathisen and Miller (1989)
Xenopus laevis	Stage 8 blastulae	D	25	See reference		Yes	No	Godsave and Slack (1989)
Xenopus laevis	Skin samples of tadpoles and froglets	D	25	Several media types listed – see reference for more details		Yes	No	Nishikawa et al. (1990)
Xenopus laevis	Embryo and blastulae	E	N/A	N/A		Primary culture	No	Slack et al. (1990)
Xenopus laevis	Liver	D	N/A	N/A		N/A	N/A	Baby and Hayashi (1991)
Xenopus laevis	Tadpole	E	21	Wolf and Quimby's medium, Ins	G/Strep/Glut	Primary culture	No	Tata and Baker (1991)
Xenopus laevis	Tadpole	E	20	MEM, ddH$_2$C, FBS, HEPES	Pen/Strep/F	Yes	No	Clothier et al. (1982)
Xenopus laevis	Skin	N/A	27	L-15, FBS	Pen/Strep	Yes	No	Nuttall et al. (1999)
Xenopus laevis	Limb	E	28	L-15, FBS	Pen/Strep	Yes	X ($2n = 21$)	Sinzelle et al. (2012)
Xenopus laevis	Skin	E	26	Serum-free DMEM		Primary culture	No	Groot et al. (2012)
Xenopus laevis	Liver and bone marrow	M	27	ASFM, FBS, XS	Pen/Strep/KM	Yes	No	Grayfer and Robert (2013)
Xenopus laevis	Tadpole	D	20	L-15, FCS		Yes	No	Tamura et al. (2015)
Xenopus laevis	Liver and bone marrow	N/A	N/A	N/A		Yes	N/A	Yaparla and Greyfer (2018)
Xenopus laevis	N/A	N/A	N/A	N/A		N/A	N/A	Seki et al. (1995)

Abbreviations: CS – calf serum; DMEM – Dulbecco's modified eagle medium; EBSS – Earle's balanced salt medium; EMEM – Eagle's minimal essential medium; M199 – medium 199; CM –Coon's media; WM – Waymouth's medium; 752M – 752/1 medium; ASFM – ASF media; MF10 – medium F-10; L-15; MEM – minimal essential medium; Glu – glucose; TPB – tryptose phosphate broth; 2-M – 2-mercaptoethanol; SS – Steinberg's solution; PI – porcine insulin; BPA – bovine plasma albumin; Ins – insulin; HSG – human serum globulin; HS – MC33 – MC33 medium; Glut – glutamine; BEE – beef embryo extract; TS – toad serum; TH – thyroid hormone (3,3',5-triiodothyronine, T$_3$); Dex – dexamethasone Spe – spermidine; Spermi – spermidine; Put – putrescine; BS – bovine horse serum; LH – lactalbumin hydrolysate; ITS – 10 µg/mL insulin, bovine pancreas; 5.5 µg/mL human transferrin; 5 ng/mL sodium selenite; XS – Xenopus laevis serum; AS – avian serum; CEE – chick embryo extract; GS – gentamicin insulin; WEU – whole egg ultrafiltrate. **Antimicrobials:** Pen – penicillin; Strep – streptomycin; F –fungizone; Nor – normacin; Myc – mycostatin; AB –amphotericin B; Neo – neomycin; G – gentamicin; AA – antibiotic sulfate; antimycotic; KM – kanamycin.

N/A – not available; X – missing information.

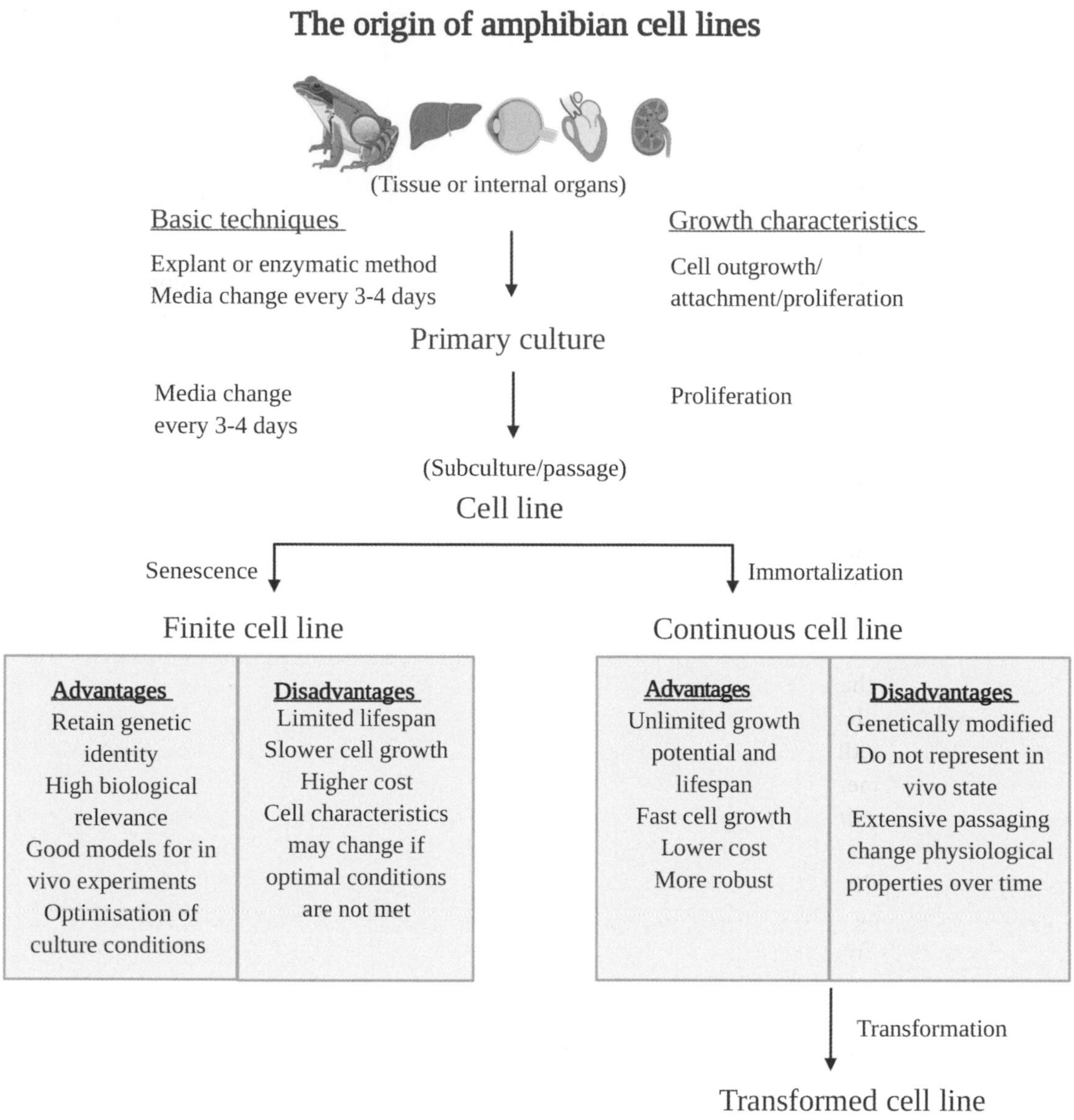

The origin of amphibian cell lines

(Tissue or internal organs)

Basic techniques

Explant or enzymatic method
Media change every 3-4 days

Growth characteristics

Cell outgrowth/
attachment/proliferation

Primary culture

Media change
every 3-4 days

Proliferation

(Subculture/passage)

Cell line

Senescence

Immortalization

Finite cell line

Advantages	Disadvantages
Retain genetic identity	Limited lifespan
High biological relevance	Slower cell growth
Good models for in vivo experiments	Higher cost
Optimisation of culture conditions	Cell characteristics may change if optimal conditions are not met

Continuous cell line

Advantages	Disadvantages
Unlimited growth potential and lifespan	Genetically modified
Fast cell growth	Do not represent in vivo state
Lower cost	Extensive passaging change physiological properties over time
More robust	

Transformation

Transformed cell line

Figure 10.1: Overview of the creation and maintenance of various types of amphibian cell lines. Figure inspired by Davis (2011). Created with BioRender.com

Country and region-specific requirements for collection, export, and ethical permission should also be considered before collecting samples, especially for threatened species (Zimkus *et al.* 2018). It is sometimes possible to use a mobile laboratory when working in the field, but typically samples are prepared for transport in a transport medium, including antibiotics, before transportation to a permanent laboratory facility. The successful culturing of amphibian cell lines is reliant on both the freshness of the sample and the correct use of aseptic techniques (Houck *et al.* 2017). Therefore, tissue samples should be processed

as soon as possible after collection under strict aseptic conditions. Freshney (2016) provided a thorough overview of laboratory design, setup, aseptic conditions, and equipment and materials required for cell culture.

Collection and preparation of samples

There are several factors to consider when planning the method of sampling, including the quality of the tissue, the type of cell required, the end use of the culture, (i.e. for research or biobanking), and the conservation status of the species. Types of amphibian tissues that have previously been utilised for tissue culture are outlined in Table 10.1. Even though the quality of the tissue is important for successfully establishing cell lines, little information regarding the success of different types of tissue or the relative success of fresh *versus* preserved tissue has been published. Recent research by Strand *et al.* (2021) reported the successful culture of three different types of tissue, including limb, tongue, and toe-clip, with tongue tissue found to be the most successful in the establishment of cell lines. Additionally, Strand *et al.* (2021) found that, even though both fresh and cryopreserved tissues were useful for establishing cell lines, fresh tissue was superior. A non-invasive technique is always preferred when working with amphibians. For example, cells from toe-clips or toe-webbing biopsies have been successfully cultured and are non lethal to the animal (Mollard 2018). Cells from whole tadpoles, kidney, foot, tongue, and eye have been cultured successfully, with cells from the foot typically the most successful, but all required pre-culture treatment with antimicrobial agents (Houck *et al.* 2017). Tissue samples from internal organs that have been collected postmortem due to injury or euthanasia have the advantage of reducing contamination with external microbes. Additionally, this non-invasive method of acquiring samples is important when working with endangered species.

Amphibian tissue cultures can be initiated either by an explant method, where millimetre-sized pieces of the tissue are cultured, or by using enzyme digestion of the tissue sample to release cells from the surrounding matrix. Among studies that were published over the past six decades, the latter was the most commonly used for the establishment of cell lines, with more than 60% of studies using variations of enzyme digestion (Table 10.1). Collagenase was found to be the most commonly used enzyme for establishing cultures, used occasionally for short-term primary cultures, but more often for the establishment of longer-term cell lines (Table 10.1). Conversely, the explant method (Figure 10.2) was sometimes used for long-term cell lines, but most often used for short-term primary cultures (Table 10.1). Protocols for explant and enzyme digestion methods have been developed by the San Diego Frozen Zoo®. The explant method is their preferred method when setting up amphibian cell lines because the cells appear to benefit from the close contact with cells and cell signalling as they emerge from the tissue providing a higher success rate. In contrast, the enzyme digestion method produces single cells distant from each other which often quickly die out (Houck *et al.* 2017).

CONDITIONS OF CELL CULTURE
Temperature
Incubation temperature for tissue culture is dependent on the normal physiological temperature of the sample's source. As amphibians are ectothermic, the temperature of their bodily tissues and expelled gametes is aligned with the external environment. Although amphibians are exposed to a wide range of temperatures, the optimal temperature for culturing amphibian cells typically ranges from 20 to 30°C and varies between families and species (Houck *et al.* 2017). Considerable work conducted by the San Diego Frozen Zoo® describes optimal temperatures ranging from 20 to 23°C for species from colder climates and 27–30°C for tropical species (Zimkus *et al.* 2018).

Choice of culture flask
Amphibian cells grow very slowly *in vitro*, with the process taking on average more than 150 days (range = 19–596 days) before the number of cells are sufficient to freeze (Zimkus *et al.* 2018). The slow *in vitro* growth rate may indicate that the current conditions employed for growth of amphibian cells are suboptimal. To promote cell growth, the use of a small culture flask or dish will enhance intercellular contact, which is generally known to have a positive effect on growth in terms of signalling and cellular differentiation (Freshney 2016). Indeed, when establishing amphibian cell cultures it is recommended to use a 12.5 cm^2 culture flask instead of the standard 25 cm^2 culture flask if the sample is 10 mm or less (Houck, unpublished data). Additionally, the use of attachment factors and extracellular matrix (ECM) components have been found to promote the adhesion of cells and to enhance their spreading and proliferation (Davis 2011; Freshney 2016). Attachment factors and ECM components are used to increase the surface charge of the substrate and thereby enhance cell-surface binding, thereby influencing the differentiation, structure, and function of the cultured cells. Popular ECM components used in cell culture include fibronectin, laminin, and collagens or Matrigel^{TM} (Davis 2011; Grefte *et al.* 2012; Freshney 2016). Matrigel is used to form a thin substrate layer or gel within the culture flask. Among its other components, Matrigel normally contains a variety of growth factors. Most of these ECM components also have been found useful to anchor organ explants (Elliget and Trump 1991; Davis 2011; Grefte *et al.* 2012; Freshney 2016). There are many different types of attachment

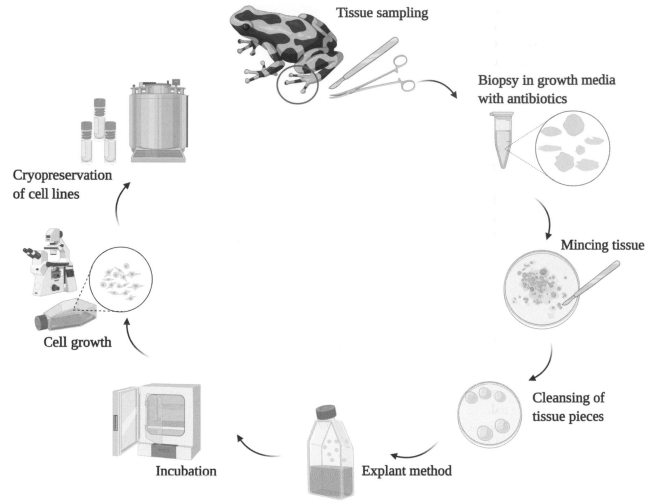

Figure 10.2: The process of non-invasive sampling and explant culturing for cryopreservation of amphibian cells. Created with BioRender.com.

factors or ECM components available, including different concentrations, and it is highly recommended that coating conditions are optimised for each species and type of cell (see Davis 2011 and Freshney 2016 for further details).

Culture medium and CO$_2$

The survival and proliferation of cells are dependent on the composition of the culture medium (Freshney 2016). The culture medium provides all the necessary nutrients, vitamins, cofactors, amino acids, and other supplements that are needed to support cellular functions and proliferation of cells (Davis 2011; Freshney 2016). Media are categorised as either natural or artificial. Natural media consist of only components necessary for growth and proliferation, whereas artificial media have several artificial components and additional nutrients such as

provided by serum (see 'Supplements', below) (Davis 2011; Verma *et al.* 2020).

Successful growth of mammalian, avian, and reptilian cells has been observed in media such as α- Minimum Essential Medium (α-MEM) or Gibco Roswell Park Memorial Institute medium (RPMI) supplemented with a buffer, antibiotics, and 10% foetal bovine serum (FBS) (Houck *et al.* 2017). There is still very little consensus about a successful culture medium specifically for amphibians. However, it is clear there is not a 'one-size-fits-all' medium that will suit all cell types from all species. In fact, a variety of media have been used to culture amphibian cells. The most widely used growth-medium employed for culturing amphibian cells to date is Leibovitz's L-15 medium, supplemented with 10% FBS, 100 U/mL penicillin, and 100 μg/mL streptomycin

(Table 10.1). L-15 medium uses a non-bicarbonate buffering system and was originally developed for use under conditions of atmospheric CO_2; that is, in an environment where CO_2 is not regulated (Davis 2011; Freshney 2016). These qualities make L-15 especially useful for sampling in the field where laboratory equipment is limited. Alternatively, supplemented α-MEM is also a preferred medium for optimal growth of cells across most vertebrate taxa including mammals, birds, reptiles, and amphibians (Houck *et al.* 2017).

Media must be buffered to maintain the cells within their normal range of pH, thereby limiting the effect on pH by waste products that are generated by the cells during culturing (Davis 2011; Freshney 2016). Bicarbonate is frequently used as a buffer due to its low toxicity and nutritional value; however, this buffer requires an enriched atmosphere of 5–10% CO_2 to maintain the appropriate pH (Davis 2011; Freshney 2016). 4-(2-hydroxyethyl)-1-piperazineethanesulfonic acid (HEPES) is another, more robust, buffer ideal for use in culture media as it does not require a CO_2-enriched environment, thus facilitating the use of incubators without regulation of CO_2. The pH level of the medium can be monitored visually by the inclusion of phenol red, which is a widely used indicator for this application (Davis 2011; Freshney 2016). The tonicity of amphibian blood has been found to be lower than that of mammalian blood; therefore osmolalities recommended for amphibian cells are ~60–65% of the osmolalities found in mammalian cell-culture media (Freed and Mezger-Freed 1970b; Balls and Worley 1973). Balls and Worley (1973) found media with osmolalities ranging from 120 to 320 mOsm/kg gave the best growth rates for explants from *Xenopus laevis*, with the optimal osmolality peaking at 190 mOsm/kg (Balls and Worley 1973*)*.

Although amphibian cell cultures appear to have the same CO_2 requirements as mammalian cells (5–6%), some studies have reported that a lower oxygen environment is beneficial when culturing amphibian cells (Houck *et al.* 2017). Due to the very limited information on oxygen levels, we strongly suggest further research to optimise these conditions, both for different species and for the level of type of cell, to ensure optimal conditions of growth.

Supplements

A deficiency or imbalance of micronutrients in the cell culture medium can have a significant negative effect on the viability and genomic stability of cultured cells (Arigony *et al.* 2013). Optimal nutrition is therefore essential for cells, particularly in longer-term culture and when biobanking is the intended end use. Artificial media are usually supplemented with sera, most commonly foetal bovine serum (FBS), also known as foetal calf serum (FCS), to ensure optimal growth and proliferation of the cells. Serum is usually added at a rate of 10%; however, the concentration used can vary from 5% up to 20% FBS (Davis 2011; Freshney 2016; Houck *et al.* 2017; Verma *et al.* 2020). There are several advantages of using sera as they provide basic nutrients, as well as an appropriate osmotic pressure within the culture. Moreover, they contain hormones such as insulin (essential for growth), proteins such as transferrin (transports iron), fibronectin (improves cell-surface attachment), minerals such as Ca^{2+}, Mg^{2+}, Fe^{2+}, K^+ and Na^+, and growth factors such as epidermal growth factor (EGF) and fibroblast growth factor (FGF), which sustain cellular proliferation and differentiation (Davis 2011; Freshney 2016; Houck *et al.* 2017; Verma *et al.* 2020). These factors can all be added in isolation along with other factors such as thyroid hormone (3,3′,5-triiodothyronine, T3) (Ketola-Pirie and Atkinson 1990).

The application of amphibian sera should also be considered when working with amphibian cell lines, as they may be more suitable for amphibian cells than are the widely used mammalian sera; however, little research has been conducted to date that evaluates their efficacy. Grayfer and Robert (2013) successfully developed primary cultures of *Xenopus laevis* liver and bone-marrow cells by the addition of *Xenopus laevis* serum to their medium, which was also supplemented with 10% FBS (Grayfer and Robert 2013). Similarly, Handler *et al.* (1979) added 1% toad serum, along with 0.5% calf serum, to their medium to culture bladder cells from *Rhinella marina* (Handler *et al.* 1979). Culture media supplemented with amphibian sera may provide a balance of nutrients more suited to provide the optimal *in vitro* environment for amphibian cells and warrants further research.

Despite the benefits of sera in promoting the growth of cells, it is important to recognise several disadvantages to using sera, including the financial cost; batch-to-batch variation in the levels of nutrients and growth factors present; and high risk of contamination with viruses, fungi, and mycoplasma (Verma *et al.* 2020). The lack of uniformity in constituents increases the possibility of giving inconsistent results. In fact, serum-free media has previously been used successfully for culturing amphibian liver cells (Stanchfield and Yager 1978, 1980; Kawahara *et al.* 1983; Nuttall *et al.* 1999), *Ambystoma mexicanum* thymus explants (Koniski and Cohen 1992), *Xenopus laevis* larvae (Mathisen and Miller 1989), tadpoles' cells (Pudney *et al.* 1973), and *Xenopus laevis* skin explants (Groot *et al.* 2012). Nevertheless, the use of sera is universally accepted as being typically required for optimal vitality and growth of most cells in culture and, while not ideal, the batch-to-batch variability and risk of contamination may be limited by additional screenings of quality and contamination.

Antimicrobials

Microbial contamination is a critical issue when working with amphibian cell lines compared to other taxa such as mammals and birds (Zimkus *et al.* 2018). The moist, semi-aquatic or fully aquatic environments that most amphibians inhabit expose them to higher loads of bacteria and fungi, which pose a challenge to the culture of amphibian cells even with the use of antibiotics and antimycotics (Zimkus *et al.* 2018). In addition, the slow growth rate and thus need for long-term cultures contribute further issues for contamination as biological contaminants can alter the phenotype and genotype of the cell lines (Segeritz and Vallier 2017).

As seen in Table 10.1, a small number of studies have succeeded in establishing primary cultures without applying antibiotics (Barth and Barth 1959; Freedt and Mezger-Freed 1970; Ellinger *et al.* 1983; Groot *et al.* 2012). While the use of antibiotics and antimycotics varies widely among studies (Table 10.2), typically they are regarded as a crucial part of collecting and culturing samples. A combination of 100 U/mL penicillin (penicillin G or benzylpenicillin) and 100 µg/mL streptomycin is the most commonly used broad-spectrum antibiotic treatment (Bianchi and Molina 1967; Handler *et al.* 1979; Koniski and Cohen 1992; Yuan *et al.* 2015), with some studies adding an antimycotic such as Fungizone[TM] (Stephenson and Stephenson 1970; Arthur and Balls 1971; Clothier *et al.* 1982; Okumoto 2001) or mycostatin (Seto 1964; Balls and Ruben 1966; Robinson and Stephenson 1967). Another broad-spectrum antibiotic, gentamicin, has also been used in various studies, either alone (Stanchfield and Yager 1978; Mollard 2018) or in combination with other antibiotics, such as Fungizone or a mixture of penicillin G, streptomycin and amphotericin B (Ketola-Pirie and Atkinson 1983; García-Díaz 1991). Of

note, all antimicrobials come with some level of cytotoxicity and their application may inhibit growth (Davis 2011; Freshney 2016). Having said this, some studies were found to have used two (Wangh *et al.* 1979), three (Mathisen and Miller 1989) and up to 10 times (Mollard 2018) the dose recommended by the supplier. The maximum tolerated dose varies depending on the species and the type of cell. Ideally, optimal antimicrobial doses should be determined during the initial phases of culturing.

To further reduce possible contamination by environmental microbes, it has been suggested to keep wild-caught animals in captivity for several weeks (Zimkus *et al.* 2018). Additionally, when working with explants of amphibian tissue, aseptic conditions are extremely important and it has been shown that bacterial and fungal contamination can be significantly reduced by adding a 30-s wash-step with gentamicin (50 mg/mL) to the aseptic protocol during pre-culture preparation (Strand *et al.* in press).

Maintenance and passaging of cultures

The culture media for cells must be periodically supplemented to meet the physiological demands for nutrients and antibiotics. Antibiotics are only stable in culture media for a limited time, ~3–5 days depending on the antibiotic and the temperature at which the culture is held (Davis 2011; Freshney 2016). To ensure adequate growth for most animal cells, it is recommended that media are changed every third day. A periodic change of media minimises the build-up of harmful waste products, which is normally detected through a drop in pH (Freshney 2016). For amphibian cells it is sometimes beneficial to only change half of the media and then top up with fresh media, thereby retaining a greater volume of the original media that contains beneficial components secreted by the cells as they grow (Houck, unpublished data).

There is currently no standard protocol for when to passage amphibian cells, but we recommend ~85–90% confluency (J. Strand, unpublished data), which is slightly higher than that used for mammalian cells, for which it is normally recommended to passage at 70–90% confluency, depending on the type of cell (Freshney 2016; Segeritz and Vallier 2017). The main reason for this is the sensitivity of the cells, as passaging them can have a negative impact on adhesion to new surfaces and on survival (J. Strand, unpublished data).

Table 10.2: Antibiotics and antimycotics used in culturing animal cells.

Antibiotics	Gram (+)	Gram (–)	Yeast	Fungi	Mycoplasma	Working concentration (listed by supplier)
Penicillin-G	X					
Gentamicin	X	X			X	0.5–50 µg/mL (Thermo Fisher Scientific)
Streptomycin	X	X				50–100 µg/mL (Thermo Fisher Scientific)
Normocin	X	X		X	X	100 µg/mL (InvivoGen)
Ampicillin	X	X				10–25 µg/mL (Thermo Fisher Scientific)
Ciprofloxacin hydrocloride	X	X			X	1 µg/mL (Serva)
Erythromycin	X	X			X	50–200 mg/L (Sigma-Aldrich)
Kanamycin	X	X			X	100 µg/mL (Thermo Fisher Scientific)
MRA (MP Biomedicals)					X	0.5 µg/mL (Fisher Scientific)
Neomycin	X	X				50 µg/ml (Thermo Fisher Scientific)
Polymixin B		X				100 units/mL (Thermo Fisher Scientific)
Tetracycline hydrocloride	X	X				10 mg/L (Merck)
Tylosin tartrate					X	8 mg/L (Merck)
Antimycotics						
Fungizone (amphotericin B)			X	X		0.25–2.50 µg/mL (Thermo Fisher Scientific)
Nystatin (or mycostatin)			X	X		50 mg/L (Sigma-Aldrich)

Data compiled from Freshney (2016) and Zimkus *et al.* (2018)

Note: X indicates the specific target of each antibiotic/antimycotic.

Cryopreservation: protocols for freezing and thawing

Today the most common cryopreservation technique for cell lines includes a cryoprotective agent (CPA), which lowers the glass transition temperature (Elliott *et al.* 2017) and subjects the cells to a cooling process at a rate of 1°C/min down to –80°C for storage. Controlled freezing can be achieved using programmable freezers or insulated boxes. The most common CPA used for amphibian cells is dimethylsulfoxide (DMSO), but others include glycerol, dimethylacetamide, and/or non-penetrating CPAs, such as sugars, egg yolk, serum protein, and milk (Davis 2011; Clulow and Clulow 2016; Freshney 2016). Cooling and thawing rates are often species-specific and cell-specific and must be optimised to prevent intracellular formation of ice, but also rapid enough to minimise the time of exposure to CPA solutions, which can be cytotoxic (Davis 2011; Clulow and Clulow 2016; Freshney 2016). Generally, cell lines should be thawed rapidly (Davis 2011). Mammalian cells normally are thawed in a 37°C water bath (Davis 2011; Houck *et al.* 2017), whereas 30°C has been

successful for amphibian cells (Strand *et al.* 2021). The need for a quick-thawing process is also important to reduce CPA cytotoxicity on sensitive cells. To limit the cytotoxic impact, CPAs can be removed immediately following thawing by centrifugation and either extracting and replacing the medium, or by diluting the CPA by adding a larger volume of fresh medium to the solution (Davis 2011).

To date, only four published studies have described freezing and thawing protocols for amphibian cell lines, highlighting the need for further research (Table 10.1). Yuan *et al.* (2015) suspended trypsinised cells in 3 ml TC199 medium + 20% FBS with 10% DMSO before transferring the cell suspension to cryovials kept in a foam box and quickly transferring them for storage at –80°C (Yuan *et al.* 2015). The cells were successfully recovered post thaw in a 37°C water bath (Yuan *et al.* 2015). Similarly, Xiang *et al.* (2012) re-suspended cells in a freezing medium consisting of 10% DMSO. Moreover, Xiang *et al.* (2012) used 50% FBS and 40% distilled water and cooled the cells over 20 h using a programmable cryopreservation system before storing samples in liquid nitrogen (LN_2). Frozen cryovials were subsequently thawed in a 30°C water bath; after thawing, more than 80% were viable and displayed good patterns of proliferation (Xiang *et al.* 2012). Mollard (2018) used culture medium containing 10% DMSO as a freezing medium in cryovials that were frozen overnight at –80°C before storage in LN_2. Samples were subsequently placed on ice and thawed by rubbing the cryovials between finger and thumb until the ice crystals were dissolved. Post-thaw viability was estimated to be within the range of 30–60% (Mollard 2018). The freezing and thawing protocols used by Strand *et al.* (2021) are similar to the standard procedures recommended by the San Diego Frozen Zoo®, where cells are re-suspended in complete αMEM with 10% DMSO (Houck *et al.* 2017). The cryovials are then placed in a CoolCell container and stored at –80°C for 4–24 h before being stored in LN_2 (Houck *et al.* 2017). Cells are removed from the LN_2 and immediately thawed in a 30°C water bath. Finally, the thawed cells are re-suspended in 4 mL of complete αMEM and incubated at a temperature based on the range in temperature in the species' natural habitat. Real-time measurements did not show any changes in post-thaw viability in terms of proliferation or adhesion (Strand *et al.* 2021).

CONCLUSIONS

The unprecedented global amphibian crisis has necessitated interventionist approaches for conserving large groups of species from extinction due to threatening processes that cannot easily be mitigated within a short time. Biobanking combined with other complementary reproductive technologies, such as artificial fertilisation, are a way to help prevent further catastrophic loss of biodiversity. Moreover, more advanced reproductive technologies, such as somatic cell nuclear transfer and the possibility of reprogramming cell lines into iPSC (induced pluripotent stem cells), have the potential to expand the usability of biobanked amphibian cell cultures and, additionally, female gametes could also be included to improve the success of conserving amphibians.

This chapter provided a complete overview of the current information available on culturing and storing amphibian cell lines, which is a critical first step towards improving existing techniques and technologies. Several challenges exist when culturing amphibian cell lines, in particular, one of the main reasons many amphibian cell lines fail is the high incidence of fungal and bacterial contamination, which often results in initially successful cell lines failing rapidly. Thus, we suggest that improving aseptic techniques from beginning to end, as well as further research optimising protocols for the use of antimicrobials, are likely to be one of the keys to increasing success for amphibian cell lines. An additional challenge is the large variation in the conditions required for successful cell culture among amphibian species and types of cells, resulting in the need to optimise cell culture and biobanking conditions between taxa and types of samples. This includes refining multiple variables such as temperature, media, CO_2 levels, supplements, antibiotics, and freezing/thawing protocols. The information compiled within this chapter provides a comprehensive go-to overview of cell lines and specific protocols attempted on the 28 amphibian species studied to date. This chapter should serve as a tool to narrow the 'dos and don'ts' for amphibian cell culture and biobanking of cell lines based on current knowledge. However, further research is required to continue to improve techniques for culturing and storing amphibian cells, including studies focusing on the preparation of samples; choice and dose of antimicrobials; temperature of cultures; the components of media; attachment factors such as addition of amphibian sera; conditioned media; and additional growth factors to improve the outcomes. The broad range of factors requiring further investigation demonstrates how far cell culture of amphibians is behind that of other vertebrates such as mammals, birds, and reptiles. By combining all the essential information crucial for establishing successful amphibian cell lines in one location within this chapter, we are hopeful that others will utilise this knowledge to help move the field forward. Finally, we suggest that improved collaboration and exchange of knowledge between institutions attempting amphibian cell culture will not only elevate the general level of information and expertise but will also increase the availability of material available for safeguarding and sustaining both *in situ* and *ex situ* conservation efforts.

REFERENCES

Albl B, Haesner S, Braun-Reichhart C, Streckel E, Renner S, Seeliger F, Wolf E, Wanke R, Blutke A (2016) Tissue sampling guides for porcine biomedical models. *Toxicologic Pathology* **44**, 414–420. doi:10.1177/0192623316631023

AmphibiaWeb (2021) University of California, Berkeley CA, USA, <https://amphibiaweb.org>.

Anizet MP, Huwe B, Pays A, Picard JJ (1981) Characterization of a new cell line, XL2, obtained from *Xenopus laevis* and determination of optimal culture conditions. *In Vitro* **17**, 267–274. doi:10.1007/BF02618137

Ansevin KD (1964) Aggregative and histoformative performance of adult frog liver cells *in vitro*. *The Journal of Experimental Zoology* **155**, 371–380. doi:10.1002/jez.1401550309

Aprison BS, Martin-Morris L, Spolski RJ, Wangh LJ (1986) Estrogen-dependent DNA synthesis in cultures of *Xenopus* liver parenchymal cells. *In Vitro Cellular & Developmental Biology* **22**, 457–464. doi:10.1007/BF02623446

Arigony ALV, de Oliveira IM, Machado M, Bordin DL, Bergter L, Prá D, Pêgas Henriques JA (2013) The influence of micronutrients in cell culture: a reflection on viability and genomic stability. *BioMed Research International* **2013**, 597282. doi:10.1155/2013/597282

Arthur E, Balls M (1971) Amphibian cells *in vitro*. Growth of *Xenopus* cells in a soft agar medium and on an agar surface. *Experimental Cell Research* **64**, 113–118. doi:10.1016/0014-4827(71)90199-6

Asashima M, Grunz H (1982) Effects of inducers on inner and outer gastrula ectoderm layers of *Xenopus laevis*. *Differentiation* **23**, 206–212. doi:10.1111/j.1432-0436.1982.tb01284.x

Auclair W (1961) Cultivation of monolayer cultures of frog renal cells. *Nature* **192**, 467–468. doi:10.1038/192467a0

Baby TG, Hayashi S (1991) Presence of ornithine decarboxylase antizyme in primary cultured hepatocytes of the frog *Xenopus laevis*. *Biochimica et Biophysica Acta* **1092**, 161. doi:10.1016/0167-4889(91)90150-V

Balls M, Ruben LN (1966) Cultivation *in vitro* of normal and neoplastic cells of *Xenopus laevis*. *Experimental Cell Research* **43**, 694–695. doi:10.1016/0014-4827(66)90048-6

Balls M, Worley RS (1973) Amphibian cells *in vitro*. *Experimental Cell Research* **76**, 333–336. doi:10.1016/0014-4827(73)90384-4

Barth LG, Barth LJ (1959) Differentiation of cells of the *Rana pipiens* gastrula in unconditioned medium. *Development* **7**, 210–222. doi:10.1242/dev.7.2.210

Bennett TP, Kriegstein H, Glenn JS (1969) Thyroxine stimulation of ornithine transcarbamylase activity and protein synthesis in tadpole (RNA *catesbeiana*) liver in organ culture. *Biochemical and Biophysical Research Communications* **34**, 412–417. doi:10.1016/0006-291X(69)90397-0

Bianchi NO, Molina JO (1967) DNA replication patterns in somatic chromosomes of *Leptodactylus ocellatus* (Amphibia, Anura). *Chromosoma* **22**, 391–400. doi:10.1007/BF00286544

Blatt LM, Kim K, Cohen PP (1969) The effect of thyroxine on ribonucleic acid synthesis by premetamorphic tadpole liver cell suspensions. *The Journal of Biological Chemistry* **244**, 4801–4807. doi:10.1016/S0021-9258(18)94274-2

Bower DS, Lips KR, Schwarzkopf L, Georges A, Clulow S (2017) Amphibians on the brink. *Science* **357**, 454–455. doi:10.1126/science.aao0500

Bower DS, Lips KR, Amepou Y, Richards S, Dahl C, Nagombi E, Supuma M, Dabek L, Alford RA, Schwarzkopf L, *et al.* (2019) Island of opportunity: can New Guinea protect amphibians from a globally emerging pathogen? *Frontiers in Ecology and the Environment* **17**, 348–354. doi:10.1002/fee.2057

Bower DS, Jennings CK, Webb RJ, Amepou Y, Schwarzkopf L, Berger L, Alford RA, Georges A, McKnight DT, Carr L, *et al.* (2020) Disease surveillance of the amphibian chytrid fungus *Batrachochytrium dendrobatidis* in Papua New Guinea. *Conservation Science and Practice* **2**, e256. doi:10.1111/csp2.256

Briggs R, King TJ (1957) Changes in the nuclei of differentiating endoderm cells as revealed by nuclear transplantation. *Journal of Morphology* **100**, 269–312.

Browne RK, Li H, Robertson H, Uteshev VK, Shishova NR, McGinnity D, Nofs S, Figiel CR, Mansour N, Lloyd RE, *et al.* (2011) Reptile and amphibian conservation through gene banking and other reproduction technologies. *Russian Journal of Herpetology* **18**, 165–174.

Browne RK, Silla AJ, Upton R, Della-Togna G, Marcec-Greaves R, Shishova NV, Uteshev VK, Proaño B, Pérez OD, Mansour N, *et al.* (2019) Sperm collection and storage for the sustainable management of amphibian biodiversity. *Theriogenology* **133**, 187–200. doi:10.1016/j.theriogenology.2019.03.035

Campbell L, Cafe SL, Upton R, Doody JS, Nixon B, Clulow J, Clulow S (2020) A model protocol for the cryopreservation and recovery of motile lizard sperm using the phosphodiesterase inhibitor caffeine. *Conservation Physiology* **8**, coaa044. doi:10.1093/conphys/coaa044

Campbell L, Clulow J, Howe B, Upton R, Doody JS, Clulow S (2021) Efficacy of short-term cold storage prior to cryopreservation of spermatozoa in a threatened lizard. *Reproduction, Fertility and Development* **33**, 555–561. doi:10.1071/RD20231

Carver FJ, Meints RH (1977) Studies of the development of frog hemopoietic tissue *in vitro*. Spleen culture assay of an erythropoietic factor in anemic frog blood. *The Journal of Experimental Zoology* **201**, 37–46. doi:10.1002/jez.1402010105

Chemnick LG, Houck ML, Ryder OA (2009) Banking of genetic resources: the Frozen Zoo° at the San Diego Zoo. In *Conservation Genetics in the Age of Genomics*. (Eds G Amato, R Desalle, HC Rosenblum and OA Ryder) pp. 124–130. Columbia University Press, New York NY, USA.

Clothier RH, Balls M, Hostry GS, Robertson NJ, Horner SA (1982) Amphibian organ culture in experimental toxicology: the effects of paracetamol and phenacetin on cultured tissues from urodele and anuran amphibians. *Toxicology* **25**, 31–40. doi:10.1016/0300-483X(82)90082-8

Clulow J, Clulow S (2016) Cryopreservation and other assisted reproductive technologies for the conservation of threatened amphibians and reptiles: bringing the ARTs up to speed. *Reproduction, Fertility and Development* **28**, 1116–1132. doi:10.1071/RD15466

Clulow J, Clulow S, Guo J, French AJ, Mahony MJ, Archer M (2012) Optimisation of an oviposition protocol employing human chorionic and pregnant mare serum gonadotropins in the barred frog *Mixophyes fasciolatus* (Myobatrachidae). *Reproductive Biology and Endocrinology* **10**, 60. doi:10.1186/1477-7827-10-60

Clulow J, Pomering M, Herbert D, Upton R, Calatayud N, Clulow S, Mahony MJ, Trudeau VL (2018) Differential success in obtaining gametes between male and female Australian temperate frogs by hormonal induction: a review. *General and Comparative Endocrinology* **265**, 141–148. doi:10.1016/j.ygcen.2018.05.032

Clulow J, Upton R, Trudeau VL, Clulow S (2019) Amphibian assisted reproductive technologies: moving from technology to application. In *Reproductive Sciences in Animal Conservation*. 2nd edn. (Eds P Comizzoli, JL Brown and WV Holt) pp. 413–463. Springer, Cham, Switzerland.

Cohen JM, Civitello DJ, Venesky MD, McMahon TA, Rohr JR (2019) An interaction between climate change and infectious disease drove widespread amphibian declines. *Global Change Biology* **25**, 927–937. doi:10.1111/gcb.14489

Davis JM (2011) *Animal Cell Culture: Essential Methods*. John Wiley & Sons, Inc., Hoboken NJ, USA.

Edwards RG (1996) The history of assisted human conception with especial reference to endocrinology. *Experimental and Clinical Endocrinology & Diabetes* **104**, 183–204. doi:10.1055/s-0029-1211443

Eisen AZ, Gross J (1965) The role of epithelium and mesenchyme in the production of a collagenolytic enzyme and a hyaluronidase in the anuran tadpole. *Developmental Biology* **12**, 408–418. doi:10.1016/0012-1606(65)90006-0

Elliget KA, Trump BF (1991) Primary cultures of normal rat kidney proximal tubule epithelial cells for studies of renal cell injury. *In Vitro Cellular & Developmental Biology* **27**, 739–748. doi:10.1007/BF02633220

Ellinger MS, Sharif S, Bemiller PM (1983) Amphibian cell culture: established fibroblastic line from embryos of the discoglossid frog, *Bombina orientalis*. *In Vitro* **19**(5) 429–434. doi:10.1007/BF02619560

Elliott GD, Wang S, Fuller BJ (2017) Cryoprotectants: a review of the actions and applications of cryoprotective solutes that modulate cell recovery from ultra-low temperatures. *Cryobiology* **76**, 74–91. doi:10.1016/j.cryobiol.2017.04.004

Ellison TR, Mathisen PM, Miller L (1985) Developmental changes in keratin patterns during epidermal maturation. *Developmental Biology* **112**, 329–337. doi:10.1016/0012-1606(85)90403-8

Freed JJ, Mezger-Freed L, Schatz SA (1969) Characteristics of cell lines from haploid and diploid anuran embryos. In *Biology of Amphibian Tumors*. (Ed. M Mizell) pp. 101–111. Springer Verlag, Berlin, Germany. doi:10.1007/978-3-642-85791-1_7

Freed JJ, Mezger-Freed L (1970a) Stable haploid cultured cell lines from frog embryos. *Proceedings of the National Academy of Sciences of the United States of America* **65**, 337–344. doi:10.1073/pnas.65.2.337

Freed JJ, Mezger-Freed L (1970b) Culture methods for anuran cells. In *Methods in Cell Physiology*. (Ed. DM Prescott) pp. 19–46. Academic Press, New York NY, USA.

Freshney IR (2016) *Culture of Animal Cells: A Manual and Basic Technique and Specialized Applications*. John Wiley & Sons, Inc., Hoboken NJ, USA.

García-Díaz F (1991) Whole-cell and single channel K$^+$ and Cl$^-$ currents in epithelial cells of frog skin. *The Journal of General Physiology* **98**, 131–161. doi:10.1085/jgp.98.1.131

Godsave SF, Slack JMW (1989) Clonal analysis of mesoderm induction in *Xenopus laevis*. *Developmental Biology* **134**, 486–490. doi:10.1016/0012-1606(89)90122-X

Grayfer L, Robert J (2013) Colony-stimulating factor-1-responsive macrophage precursors reside in the amphibian (*Xenopus laevis*) bone marrow rather than the hematopoietic subcapsular liver. *Journal of Innate Immunity* **5**, 531–542. doi:10.1159/000346928

Green CD, Tata JF, Hill M (1976) Direct induction by estradiol of vitellogenin synthesis in organ cultures of male *Xenopus laevis* liver. *Cell* **7**, 131–139. doi:10.1016/0092-8674(76)90263-4

Grefte S, Vullinghs S, Kuijpers-Jagtman AM, Torensma R, Von den Hoff JW (2012) Matrigel, but not collagen I, maintains the differentiation capacity of muscle derived cells *in vitro*. *Biomedicals Material* **7**, 055004. doi:10.1088/1748-6041/7/5/055004

Groeneveld LF, Gregusson S, Guldbrandtsen B, Hiemstra SJ, Hveem K, Kantanen J, Lohi H, Stroemstedt L, Berg P (2016) Domesticated animal biobanking: land of opportunity. *PLoS Biology* **14**, e1002523. doi:10.1371/journal.pbio.1002523

Groot H, Munuz-Carmargo C, Moscoso J, Riveros G, Salazar V, Florez FK, Mitrani E (2012) Skin micro-organs from several frog species secrete a repertoire of powerful antimicrobials in culture. *The Journal of Antibiotics* **65**, 461–467. doi:10.1038/ja.2012.50

Gross J, Lapiere CM (1962) Collagenolytic activity in amphibian tissues: a tissue culture assay. *Proceedings of the National Academy of Sciences of the United States of America* **48**, 1014–1022. doi:10.1073/pnas.48.6.1014

Gurdon JB, Laskey RA (1970) The transplantation of nuclei from single cultured cells into enucleate frogs' eggs. *Journal of Embryology and Experimental Morphology* **24**, 227. doi:10.1242/dev.24.2.227

Gurdon JB, Wilmut I (2011) Nuclear transfer to eggs and oocytes. *Cold Spring Harbor Perspectives in Biology* **3**, a002659. doi:10.1101/cshperspect.a002659

Handler JS, Steele B, Sahib MK, Wade JB, Preston AS, Lawson NL, Johnson JP (1979) Toad urinary bladder epithelial cells in culture: maintenance of epithelial structure, sodium transport, and response to hormones. *Proceedings of the National Academy of Sciences of the United States of America* **76**, 4151–4155. doi:10.1073/pnas.76.8.4151

Hildebrandt TB, Hermes R, Saragusty J, Potier R, Schwammer HM, Balfanz F, Vielgrader H, Baker B, Bartels P, Göritz F (2012) Enriching the captive elephant population genetic pool through artificial insemination with frozen-thawed semen collected in the wild. *Theriogenology* **78**, 1398–1404. doi:10.1016/j.theriogenology.2012.06.014

Holt WV, Abaigar T, Watson PF, Wildt DE (2003) Genetic resource banks for species conservation. In *Reproductive Science and Integrated Conservation*. (Eds WV Holt, AR Picard, CJ Rodger and DE Wildt) pp. 267–280. Cambridge University Press, Cambridge, UK.

Holt WV, Brown JL, Comizzoli P (2014) Reproductive science as an essential component of conservation biology. In *Reproductive Sciences in Animal Conservation*. (Eds WV Holt, JL Brown and P Comizzoli) pp. 3–14. Springer, New York NY, USA.

Houck ML, Lear TL, Charter SJ (2017) Animal cytogenetics. In *The AGT Cytogenetics Laboratory Manual*. (Eds M Arsham, M Barch and H Lawce) pp. 1055–1102. John Wiley & Sons, Inc, Hoboken NJ, USA.

Howard JG, Lynch C, Santymire RM, Marinari PE, Wildt DE (2016) Recovery of gene diversity using long-term cryopreserved spermatozoa and artificial insemination in the endangered black-footed ferret. *Animal Conservation* **19**, 102–111. doi:10.1111/acv.12229

Howell LG, Frankham R, Rodger JC, Witt RR, Clulow S, Upton RMO, Clulow J (2021a) Integrating biobanking minimizes inbreeding and produces significant cost benefits for a threatened frog captive breeding programme. *Conservation Letters* **14**, e12776. doi:10.1111/conl.12776

Howell LG, Mawson PR, Frankham R, Rodger JC, Upton RMO, Witt RR, Calatayud NE, Clulow S, Clulow J (2021b) Integrating biobanking could produce significant cost benefits and minimize inbreeding for Australian amphibian captive breeding programs. *Reproduction, Fertility and Development* **33**, 573–585. doi:10.1071/RD21058

Ide H (1973) Effects of ACTH on melanophores and iridophores isolated from bullfrog tadpoles. *General and Comparative Endocrinology* **21**, 390–397.

Ide H (1974) Proliferation of amphibian melanophores *in vitro*. *Developmental Biology* **41**, 380–384. doi:10.1016/0012-1606(74)90314-5

IUCN (2021) *IUCN Amphibian Red List Summary Statistics*, <https://www.iucnredlist.org/statistics>.

Jones KW, Elsdale TR (1963) The culture of small aggregates of amphibian embryonic cells *in vitro*. *Journal of Embryology and Experimental Morphology* **11**, 135. doi:10.1242/dev.11.1.135

Kawahara A, Sato K, Amano M (1983) Regulation of protein synthesis by estradiol 17β, dexamethasone and insulin in primary cultured *Xenopus* hepatocytes. *Experimental Cell Research* **148**, 423–436. doi:10.1016/0014-4827(83)90164-7

Kawahara A, Ishikawa S, Amano M (1985) In vitro growth of adult amphibian (*Xenopus laevis*) hepatocytes and characterization of hepatocyte-proliferating activity in homologous serum. *Experimental Cell Research* **159**, 344–352. doi:10.1016/S0014-4827(85)80008-2

Ketola-Pirie C, Atkinson BG (1983) Cold- and heat-shock induction of new gene expression in cultured amphibian cells. *Canadian Journal of Biochemistry and Cell Biology* **61**, 462–471. doi:10.1139/o83-062

Ketola-Pirie CA, Atkinson BG (1988) 3,3′,5-triiodothyronine-induced differences in water-insoluble protein synthesis in primary epidermal cell cultures from the hind limb of premetamorphic *Rana catesbeiana* tadpoles. *General and Comparative Endocrinology* **69**, 197–204. doi:10.1016/0016-6480(88)90006-8

Ketola-Pirie CA, Atkinson BG (1990) Thyroid hormone-induced differential synthesis of water-insoluble proteins in epidermal cell cultures from the hind limb of *Rana catesbeiana* tadpoles in stages XII–XV and XVI–XIX. *General and Comparative Endocrinology* **79**, 275–282. doi:10.1016/0016-6480(90)90113-Z

Kondo H, Ide H (1983) Long-term cultivation of amphibian melanophores: *in vitro* ageing and spontaneous transformation to a continuous cell line. *Experimental Cell Research* **149**, 247–256. doi:10.1016/0014-4827(83)90396-8

Koniski AD, Cohen N (1992) Reproducible proliferative responses of salamander (*Ambystoma mexicanum*) lymphocytes cultured with mitogens in serum-free medium. *Developmental and Comparative Immunology* **16**, 441–451. doi:10.1016/0145-305X(92)90028-B

Korody ML, Pivaroff C, Nguyen TD, Peterson SE, Ryder OA, Loring JF (2017) Four new induced pluripotent stem cell lines produced from northern white rhinoceros with non-integrating reprogramming factors. *BioRxiv* **202499**, 1–3. doi:10.1101/202499

Kouba AJ, Lloyd RE, Houck ML, Silla AJ, Calatayud N, Trudeau VL, Clulow J, Molinia F, Langhorne C, Vance C, Arregui L *et al.* (2013) Emerging trends for biobanking amphibian genetic resources: the hope, reality and challenges for the next decade. *Biological Conservation* **164**, 10–21. doi:10.1016/j.biocon.2013.03.010

Kubokawa K, Ishii S (1987) Receptors for native gonadotropins in amphibian liver. *General and Comparative Endocrinology* **68**, 260–270. doi:10.1016/0016-6480(87)90037-2

Laskey RA (1970) The use of antibiotics in the preparation of amphibian cell cultures from highly contaminated material. *Journal of Cell Science* **7**, 653–659. doi:10.1242/jcs.7.3.653

Lawson B, Clulow S, Mahony MJ, Clulow J (2013) Towards gene banking amphibian maternal germ lines: short-term incubation, cryoprotectant tolerance and cryopreservation of embryonic cells of the frog, *Limnodynastes peronii*. *PLoS One* **8**, e60760. doi:10.1371/journal.pone.0060760

Lermen D, Blomeke B, Browne R, Clarke A, Dyce P, Fixemer T, Fuhr GR, Holt WV, Jewgenow K, Lloyd R *et al.* (2009) Cryobanking of viable biomaterials: implementation of new strategies for conservation purposes. *Molecular Ecology* **18**, 1030–1033. doi:10.1111/j.1365-294X.2008.04062.x

Malamud D (1967) DNA synthesis and the mitotic cycle in frog kidney cells cultivated *in vitro*. *Experimental Cell Research* **45**, 277–280. doi:10.1016/0014-4827(67)90179-6

Masters J (2003) *Animal Cell Culture*. Oxford University Press, New York NY, USA.

Mathisen PM, Miller L (1989) Thyroid hormone induces constitutive keratin gene expression during *Xenopus laevis* development. *Molecular and Cellular Biology* **9**, 1823–1831.

Matsuda K (1963) Culture techniques with some amphibian tissues and a chromosome study of the tree frog, *Hyla arborea japonica*. *Zoological Magazine* **72**, 105–109.

McGarry MP, Vanable JW (1969) The role of cell division in *Xenopus laevis* skin gland development. *Developmental Biology* **20**, 291–303. doi:10.1016/0012-1606(69)90016-5

Miller L, Daniel JC (1977) Comparison of *in vivo* and *in vitro* ribosomal RNA synthesis in nucleolar mutants of *Xenopus laevis*. *In Vitro* **13**, 557–563. doi:10.1007/BF02627851

Mollard R (2018) Culture, cryobanking and passaging of karyotypically validated native Australian amphibian cells. *Cryobiology* **81**, 201–205. doi:10.1016/j.cryobiol.2018.03.004

Niki K, Yoshizato K (1984) *In vitro* regression of tadpole tail by thyroid hormone. *Development, Growth & Differentiation* **26**, 329–338. doi:10.1111/j.1440-169X.1984.00329.x

Nishikawa A, Yoshizato K (1986) Hormonal regulation of growth and life span of bullfrog tadpole tail epidermal cells cultured *in vitro*. *The Journal of Experimental Zoology* **237**, 221–230. doi:10.1002/jez.1402370208

Nishikawa A, Kaiho M, Yosttizato IA (1989) Cell death in the anuran tadpole tail: thyroid hormone induces keratinization and tail-specific growth inhibition of epidermal cells. *Developmental Biology* **131**, 337–344. doi:10.1016/S0012-1606(89)80007-7

Nishikawa A, Shimizu-Nishikawa K, Miller L (1990) Isolation, characterization and in vitro culture of larval adult epidermal cells of the frog *Xenopus laevis*. *In Vitro Cellular & Developmental Biology* **26**, 1128–1134. doi:10.1007/BF02623689

Nuttall ME, Lee JC, Murdock PR, Badger AM, Wang F, Laydon JT, Hofmann GA, Pettman GR, Lee JA, Parihar A, *et al.* (1999) Amphibian melanophore technology as a functional screen for antagonists of G-protein coupled 7-transmembrane receptors. *Journal of Biomolecular Screening* **4**, 269–277. doi:10.1177/108705719900400508

Okumoto H (2001) Establishment of three cell lines derived from frog melanophores establishment of three cell lines derived from frog melanophores. *Zoological Science* **18**, 483–496. doi:10.2108/zsj.18.483

Pollack ED (1980) Target-dependent survival of tadpole spinal cord neurites in tissue culture. *Neuroscience Letters* **16**, 269–274. doi:10.1016/0304-3940(80)90009-9

Pudney M, Varma MG, Leake CJ (1973) Establishment of a cell line (XTC-2) from the South African clawed toad, *Xenopus laevis*. *Experientia* **29**, 466–467. doi:10.1007/BF01926785

Pukazhenthi BS, Wildt DE (2004) Which reproductive technologies are most relevant to studying, managing and conserving wildlife? *Reproduction, Fertility and Development* **16**, 33–46. doi:10.1071/RD03076

Pukazhenthi BS, Comizzoli P, Travis AJ, Wildt DE (2006) Applications of emerging technologies to the study and conservation of threatened and endangered species. *Reproduction, Fertility and Development* **18**, 77–90. doi:10.1071/RD05117

Regan JD, Cook JS, Lee WH (1968) Photoreactivation of amphibian cells in culture. *Journal of Cellular Physiology* **71**, 173–176. doi:10.1002/jcp.1040710208

Robinson ES, Stephenson M (1967) A karyological study of cultured cells of *Limnodynastes peroni* (Anura: Leptodactylidae) check on tadpoles. *Cytologia* **32**, 200–207. doi:10.1508/cytologia.32.200

Ryder OA, Onuma M (2018) Viable cell culture banking for biodiversity characterization and conservation. *Annual Review of Animal Biosciences* **6**, 83–98. doi:10.1146/annurev-animal-030117-014556

Segeritz C, Vallier L (2017) Cell culture: growing cells as model systems *in vitro*. In *Basic Science Methods for Clinical Researchers*. (Eds M Jalali, FYL Saldanha and M Jalali) pp. 151–172. Academic Press, Boston MA, USA.

Seki T, Kikuyama S, Yanihara N (1995) *In vitro* development of *Xenopus* skin glands producing 5-hydroxytryptamine and caerulein. *Experientia* **51**, 1040–1044. doi:10.1007/BF01946912

Seto T (1964) The karyotype of *Hyla arborea japonica* with some remarks on heteromorphism of the sex chromosome. *Journal Faculty of Science Hokkaido University* **15**, 366–373.

Shah VC (1962) An improved technique of preparing primary cultures of isolated cells from adult frog kidney. *Experientia* **18**, 239–240. doi:10.1007/BF02148322

Simnett JD, Balls M (1969) Cell proliferation in xenopus tissues: a comparison of mitotic incidence *in vivo* and in organ culture. *Journal of Morphology* **127**, 363–372. doi:10.1002/jmor.1051270307

Sinzelle L, Thuret R, Hwang H, Herszberg B, Paillard E, Bronchain OJ, Stemple DL, Dhorne-pollet S, Pollet N (2012) Characterization of a novel *Xenopus tropicalis* cell line as a model for *in vitro* studies. *Genesis* **50**, 316–324. doi:10.1002/dvg.20822

Slack JMW, Darlington BG, Gillespie LL, Godsave SF, Isaacs HV, Paterno GD (1990) Mesoderm induction by fibroblast growth factor in early *Xenopus* development. *Philosophical Transactions of the Royal Society of London* **327**, 75–84.

Solursh M, Reiter RS (1972) Long-term cell culture of two differentiated cell types from the liver of larval and adult *Xenopus laevis*. *Zeitschrift für Zellforschung und Mikroskopische Anatomie* **128**, 457–469. doi:10.1007/BF00306982

Stanchfield JE, Yager JD (1978) An estrogen responsive primary amphibian liver cell culture system. *Experimental Cell Research* **116**, 239–252. doi:10.1016/0014-4827(78)90445-7

Stanchfield JE, Yager JD (1980) Primary induction of vitellogenin synthesis in monolayer cultures of amphibian hepatocytes. *The Journal of Cell Biology* **84**, 468–475. doi:10.1083/jcb.84.2.468

Stephenson EM, Stephenson NG (1970) Karyotypes of two Australian hylids. *Chromosoma* **50**, 38–50. doi:10.1007/BF00293908

Strand J, Thomsen H, Jensen JB, Marcussen C, Nicolajsen TB, Skriver MB, Søgaard IM, Ezaz T, Purup S, Callesen H, Pertoldi C (2020) Biobanking in amphibian and reptilian conservation and management: Opportunities and challenges. *Conservation Genetics Resources* **12**, 709–725. doi:10.1007/s12686-020-01142-y

Strand J, Callesen H, Pertoldi C, Purup S (2021) Establishing cell lines from fresh or cryopreserved tissue from the great crested newt (*Triturus cristatus*): a preliminary protocol. *Animals (Basel)* **11**, 35–46. doi:10.3390/ani11020367

Strand J, Callesen H, Pertoldi C, Purup S (in press) Amphibian cell lines: usable tissue types and differences within a species. *Amphibian & Reptile Conservation*.

Stuart SN, Chanson JS, Cox NA, Young BE, Rodrigues ASL, Fischman DL, Waller RW (2004) Status and trends of amphibian declines and extinctions worldwide. *Science* **306**, 1783–1786. doi:10.1126/science.1103538

Tamura K, Takayama S, Ishii T, Mawaribuchi S, Takamatsu N, Ito M (2015) Apoptosis and differentiation of *Xenopus* tail-derived myoblasts by thyroid hormone. *Journal of Molecular Endocrinology* **54**, 185–192. doi:10.1530/JME-14-0327

Tata JR, Baker A (1991) Prolactin inhibits both thyroid hormone-induced morphogenesis and cell death in cultured amphibian larval tissues. *Developmental Biology* **146**, 72–80. doi:10.1016/0012-1606(91)90447-B

Verma A, Verma M, Singh A (2020) Animal tissue culture principles and applications. In *Animal Biotechnology*. 2nd edn. (Eds AS Verma and A Singh) pp. 269–293. Academic Press, Boston MA, USA.

Wahli W, Weber R (1977) Factors promoting the establishment of primary cultures of liver cells from *Xenopus* larvae. *Wilhelm Roux's Archives of Developmental Biology* **182**, 347–360. doi:10.1007/BF00848385

Wangh LJ, Osborne JA, Hentschel CC, Tilly R (1979) Parenchymal cells purified from *Xenopus* liver and maintained in primary culture synthesize vitellogenin in response to estradiol-17β and serum albumin in response to dexamethasone. *Developmental Biology* **70**, 479–499. doi:10.1016/0012-1606(79)90040-X

Weiss N, Du Pasquier L (1973) Factors affecting the reactivity of amphibian lymphocytes in a miniaturized technique of the mixed lymphocyte culture. *Journal of Immunological Methods* **3**, 273–285. doi:10.1016/0022-1759(73)90023-9

Wildt DE, Comizzoli P, Pukazhenthi B, Songsasen N (2010) Lessons from biodiversity – the value of nontraditional species to advance reproductive science, conservation, and human health. *Molecular Reproduction and Development* **77**, 397–409. doi:10.1002/mrd.21137

Wolf K, Quimbry MC (1964) Amphibian cell culture: permanent cell line from the bullfrog (*Rana catesbeiana*). *Science* **144**, 1578–1580. doi:10.1126/science.144.3626.1578

Xiang Y, Xiang Y, Gao Q, Gao Q, Su W, Su W, Zeng L, Zeng L, Wang J, Wang J, *et al.* (2012) Establishment, characterization and immortalization of a fibroblast cell line from the Chinese red belly toad *Bombina maxima* skin. *Cytotechnology* **64**, 95–105. doi:10.1007/s10616-011-9399-9

Yamanaka S (2020) Pluripotent stem cell-based cell therapy – promise and challenges. *Cell Stem Cell* **27**, 523–531. doi:10.1016/j.stem.2020.09.014

Yaparla A, Greyfer L (2018) Isolation and culture of amphibian (*Xenopus laevis*) sub-capsular liver and bone marrow cells. In *Xenopus: Methods in Molecular Biology*. (Ed, K Vleminckx) pp. 275–281. Humana Press, New York NY, USA.

Yuan J, Chen Z, Huang X, Gao X, Zhang Q (2015) Establishment of three cell lines from Chinese giant salamander and their sensitivities to the wild-type and recombinant ranavirus. *Veterinary Research* **46**, 58. doi:10.1186/s13567-015-0197-9

Zimkus BM, Hassapakis CL, Houck ML (2018) Integrating current methods for the preservation of amphibian genetic resources and viable tissues to achieve best practices for species conservation. *Amphibian & Reptile Conservation* **12**, e165.

11　Linking *in situ* and *ex situ* populations of threatened amphibian species using genome resource banks

Carrie K. Kouba and Allison R. Julien

INTRODUCTION

Amphibian diversity and genetic fitness is threatened in ~43% of species today (Stuart *et al.* 2004; Allentoft and O'Brian 2010). Growing pressure from anthropogenic factors such as climatic change, disease, and destruction of habitats has increased the need for linking *in situ* and *ex situ* conservation programs for genetic stability and conserving the populations of threatened amphibian species. Although the concept of linking captive and wild populations is bidirectional, the focus of this chapter will be on gene flow from the wild into captive breeding colonies for genetic management of the *ex situ* populations. We first describe concerted efforts to conserve some of the most at-risk amphibian species by establishing conservation breeding programs that eventually aim to reintroduce progeny back into wild settings. The discussion then addresses the prominent role that genetic diversity plays in stability of *ex situ* populations and how genetic management strategies have traditionally relied on the import of new founder animals from *in situ* populations. Next, we outline the advantages and disadvantages of obtaining new founder lineages of amphibians through importation of live animals, compared to using reproductive technologies and the biobanking of gametes, followed by the practical application of reproductive technologies in field settings. Finally, three North American case studies are presented, demonstrating successful linking of

in situ and *ex situ* populations of threatened anuran species through the National Amphibian Genome Bank to bring in new founder lineages. In the case of the Puerto Rican crested toad (*Peltophryne lemur*), the cycle is completed as new mixed wild–captive progeny are reintroduced back into the wild. Concluding remarks identify other conservation breeding programs that are in an excellent position to utilise this same strategy for maintaining genetically robust assurance populations, and potential genetic enrichment of coupled wild and captive populations of threatened amphibians.

Ex situ programs support in situ amphibian conservation

The International Union for the Conservation of Nature (IUCN) describes *ex situ* conservation as implementing conditions under which individuals, or their progeny, are spatially restricted with respect to their natural habitat and patterns of movement, or are removed from many of their natural ecological processes, and are managed on some level by humans (IUCN/SSC 2014). Restricting a species to a local area, or translocating animals into a captive setting for breeding, gamete-collection, or re-release are all modes of *ex situ* modulation of a species' spatial movement and environmental influences. The most prominent strategy for *ex situ* management is the formation of assurance colonies wherein wild-caught individuals are maintained in a restricted setting, such as zoological institutions, private reserves, or rescue centres (Rorabaugh and Sredl 2014). *Ex situ* habitats within these facilities may be open structures subject to natural photoperiods and changes in temperature, such as the caged outdoor enclosures used to house Chiricahua leopard frogs (*Lithobates chiricahuensis*) at the Turner Ladder Ranch in New Mexico, USA (Figure 11.1A). Smaller enclosures (tanks) under artificially controlled climates typically are found in zoos but may also be arranged in isolated prefabricated constructs such as shipping containers (Figure 11.1B, C). These 'frog pods' have also been used to build rescue centres near wild habitat, such as at the El Valle Amphibian Conservation Center (EVACC) located in de Anton, Panama, where a population of

Panamanian golden frogs (*Atelopus zeteki*) are held. Regardless of the captive setting, *ex situ* management is based on spatial separation of animals from their natural habitat for species' stability, propagation, and biosecurity rather than through passive implementation of policy on the native landscape. Generally, all of these types of *ex situ* programs focus on breeding amphibians held in captivity for species' conservation, the stability of populations, and the eventual reintroduction of offspring when *in situ* conditions become favourable (Harding *et al.* 2016).

There has been a 57% increase in captive breeding and repopulation since the Amphibian Conservation Action Plan (ACAP) was established in 2007 (Gascon 2007; Harding *et al.* 2016). Prior to the ACAP fewer than 50% of

Figure 11.1: Housing and the arrangements of field laboratory conditions. (A) Naturalised holding pens for Chiricahua leopard frogs (*Lithobates chiricahuensis*) at Ladder Ranch, New Mexico, USA. (B) Zoological breeding facility, Fort Worth, Texas, USA. (C) Frog pod holding Panamanian golden frogs (*Atelopus zeteki*) at the El Valle Amphibian Conservation Center in de Anton, Panama. (D) Portable laboratory set-up for collection, analysis, and biobanking of sperm from the Chiricahua leopard frog. Photograph credits: Panels A, B, and D by Carrie Kouba, Mississippi State University; panel C by Vicky Poole, Fort Worth Zoo.

ex situ breeding programs intended to reintroduce captive-bred animals (Griffiths and Pavajeau 2008), and instead focused on education and outreach. For highly threatened and endangered species, or those extirpated in the wild, *ex situ* breeding for reintroduction is a targeted goal, with the hope that eventually the native habitat can be recovered enough to support re-establishment of the species. However, the threats to *in situ* population stability are often complex and require years, if not decades, to overcome (Scheele *et al.* 2014; Harding *et al.* 2016; Lewis *et al.* 2019). In the interim, captive assurance colonies serve as an active participant in the conservation of these species and must be planned for the long-term stability of the *ex situ* population both in terms of total numbers of individuals and the genetic diversity represented. For most at-risk species, active management of populations is facilitated primarily by incorporating new founder lines, and through breeding and transfer to assure that a species' population is genetically stable (Taylor *et al.* 2003). New genetic lineages are typically sourced from wild amphibian populations, usually in the form of adult animals or larvae, and are required to maintain the viability of *ex situ* populations over many generations. Alternatively, recent advances in transport and storage of gametes from amphibians provide means of increasing gene flow from animals *in situ*, but without depletion or disruption of the wild population (Koúba *et al.* 2012; Burger *et al.* 2021).

GENETIC MANAGEMENT OF *EX SITU* AMPHIBIAN POPULATIONS

Replenishment of genetic diversity of captive populations with wild lineages is one strategy used to combat the inbreeding of *ex situ* populations and reduce the genetic drift that can lead to lower reproductive fitness, greater mortality of offspring, reduced resistance to disease, and lower adaptive capacity (Ralls and Ballou 1983; Frankham 2005; McCartney-Melstad and Shaffer 2015; Smallbone *et al.* 2016). In *ex situ* populations, Araki *et al.* (2007) found that captivity imposed negative genetic effects that resulted in reduced reproductive success of subsequent generations of fish after just two breeding cycles and that fecundity decreased by 40% per captive reared generation. Moreover, species' adaptation to the captive environment may promote fixation of alleles that are deleterious in nature when captive-bred offspring are reintroduced (Lynch and O'Hely 2001; Kraaijeveld-Smit

et al. 2006; Blanchet *et al.* 2008). Studies on reproductive success and early life-history stages of the critically endangered dusky gopher frog (*Lithobates sevosus*) demonstrated that higher genetic fitness resulted in improved rates of survival of egg clutches and larval stages up to metamorphosis (Richter *et al.* 2009; Richter and Nunziata 2014). Further analysis revealed that inbreeding selection, rather than the purging of deleterious alleles, was the primary driver of developmental failure, which then allowed more genetically fit lineages to yield metamorphosed individuals with potential to sustain the species in the wild (Richter and Nunziata 2014). Besides decreasing fecundity of adults, or survivorship of fertilised eggs to metamorphosed juveniles, other signs of inbreeding depression in amphibian captive breeding programs can be seen as deformations of limbs or spine, often leading to reduced fitness and failed reintroductions (Spielman *et al.* 2004; Frankham 2005). For amphibians, inserting new genetic variation into *ex situ* populations exhibiting inbreeding depression can help salvage compromised lineages, spread new alleles or capture lost variants (Johnson *et al.* 2010). Outbreeding depression can also occur if the level of genetic divergence between the *ex situ* and *in situ* populations is too high and can also lead to negative consequences for fitness (Chapter 8).

A successful captive breeding program ensures retention of the founders' genetics that contribute to the species' genetic diversity, which is most effectively achieved by maintaining a large enough *ex situ* population to avoid inbreeding (Frankham *et al.* 2010). Inbreeding depression of genetic diversity in just five generations has been observed in wild species harbouring an effective population (N_e) of only 50 individuals, while data compiled over 30 years suggest $N_e > 100$ is required to minimise total loss of fitness to below 10% (Frankham *et al.* 2014). However, the susceptibility of captive populations to inbreeding depression may actually be higher due to artificial selection, unsuitability to the *ex situ* environment, behavioural incompatibilities, and regular population bottlenecks (Frankham 2009; Mattila *et al.* 2012). Baker (2007) found that 67% of programs in zoos under the Association of Zoos and Aquariums (AZA) maintain captive assurance colonies of less than the recommended minimum of 100 animals. Unfortunately, many conservation breeding programs cannot realistically accommodate large enough captive populations to maintain the required genetic diversity for sustainability. Cited barriers to holding threatened amphibians in zoos include lack of resources, legal obstacles to holding or obtaining individuals (particularly endangered species), biosecurity, financial challenges, insufficient expertise, and limited planning of collections (Kouba *et al.* 2013; Michaels *et al.* 2014; Brady *et al.* 2017).

Management of populations of at-risk amphibian species

The AZA manages several anuran species through structured Species Survival Plans (SSPs) assisted by the Population Management Center (PMC, Lincoln Park Zoo, Chicago, Illinois, USA), which provides recommendations

of breeding pairs based on known pedigree and mean kinship to achieve the highest genetic diversity in the *ex situ* population. Population genetic parameters for select species in North American AZA-SSP programs are provided in Table 11.1. In this assessment, founder genetic lines are generally assumed to be unrelated unless kinship testing has been conducted, or the individuals are known to be related (e.g. brought in as egg clutches) (Hogg *et al.* 2019). Usually genetic lines are traced to the original animals captured or rescued from native sites and the numbers are known absolutely from historical records maintained by each SSP. In the case of the dusky gopher frog, *Lithobates*

Table 11.1:　Population genetic parameters for *ex situ* amphibian populations in the North American Association of Zoos and Aquariums Species Survival Plans and programs for managing the conservation of populations.

Genetic summary *ex situ* anuran populations	*Atelopus zeteki* (Ahogado)	*Atelopus zeteki* (Soro)	*Atelopus varius* (Marta)	*Anaxyrus baxteri*	*Anaxyrus houstonensis*	*Lithobates sevosus*	*Peltophryne lemur*
Ex situ population/ no. institutions	556/16	594/46	254/4	509/NR	452/4	624/15	1146/30
Founders	12	20	6	8	38	38	48
Founder genome equivalents (FGE)	8.06	7.49	4.18	3.43	13.04	22.13	16.54
Gene diversity (GD %)	93.80	93.33	88.03	85.44	96.14	97.74	96.98
Population mean kinship (MK)	0.0620	0.0668	0.1197	0.1456	0.0383	0.0226	0.0302
Mean inbreeding (F)	0.0070	0.0259	0.1500	0.1202	0.0079	0.1331	0.0270
Effective population size (N_e/N)	0.0350	0.0585	0.0859	0.1262	0.0493	0.00/0.01*	0.0857
Percentage of known pedigree	93.7	81.8	20.9	92.7	82.1	88	95.8
Years to 90% gene diversity	8	18	N/A	N/A	7	3	144
Years to 10% loss of gene diversity	27	62	42	37	10	4	174
Gene diversity at 100 years	56.8	NR	66.7	61.3	0	7.4	94.5
Gene diversity at 10 generations	76.0	84.6	75.6	79.8	0	43	94.9
Mean generation time (years)	4.8	5.4	5.4	2	4	3.2	4.4
Population growth rate (5 year/ projected)	1.043/1.217	0.959/1.073	1.188/1.098	1.03/1.03	0.934/0.884	0.982/1.000	1.061/1.09
Annual required no. metamorphs added	104	187	45	243	121	100	193
Reference	Poole *et al.* (2018)	Poole *et al.* (2019)	Poole *et al.* (2020)	Hornyak *et al.* (2020)	Mayes *et al.* (2019)	Reichling *et al.* (2018)	Barber *et al.* (2018)

* N_e/N projections with founders included (without founders = 0.0000). N/A = not applicable; NR = not reported.

sevosus, the *ex situ* population originated from importation of 40 individuals in 2001 from the last two existing breeding sites and from fertilised egg masses rescued in 2003 (Richter *et al.* 2009). Importantly, no *ex situ* natural breeding of this species was observed until 2019, and sexing and breeding of the captive population in the interim was conducted using assisted reproductive technologies, specifically ultrasonography, hormone-induced release of gametes, and *in vitro* fertilisation (IVF; also referred to as artificial fertilisation) (Kouba *et al.* 2009; Graham *et al.* 2016; Bronson and Vance 2019). Association of Zoos and Aquariums (AZA)–accredited institutions are required to maintain detailed records of breeding events and lineages, which are managed by a designated studbook keeper and species coordinator to organise and facilitate annual exchanges of animals between holding institutions and *in situ* sources. In each amphibian conservation breeding program hundreds of individual lineages are tracked over multiple generations, a truly herculean effort in the curation of data and the analysis of populations. With the exception of *Atelopus varius* and *Anaxyrus baxteri*, management of the amphibian species under SSP recommendations has led to growth of each population to over 500 individuals and a projected 90% genetic diversity for at least three years forthcoming (Table 11.1). Other North American amphibian species such as the boreal toad (*Anaxyrus boreas boreas*), Chiricahua leopard frog (*Lithobates chiricahuensis*), Houston toad (*Anaxyrus houstonensis*), mountain yellow-legged frog (*Rana muscosa*), and Oregon spotted frog (*Rana pretiosa*) have established *ex situ* breeding colonies and plans for managing populations coordinated by state or federal agencies, which also work in conjunction with AZA-accredited zoos and aquaria (Pierce 2006; Hallock 2013; USFWS 2018).

Founder animals are obtained from the wild and used to establish or replenish captive assurance colonies, but the number of founders varies widely by species and institution, thereby impacting genetic diversity, with smaller numbers of founders often leading to more rapid genetic decline (Robert 2009; Hogg *et al.* 2019). To maintain a conservation breeding colony, while preventing loss of genetic diversity to avoid inbreeding depression, a minimum of 15 unrelated founder contributions and a minimum population size of 100 individuals should be targeted (Witzenberger and Hochkirch 2011). From this basis, managed breeding programs employ a pedigree-based, mean-kinship approach to select breeding pairs (Hogg

et al. 2019). Ideally, genetic studies of founder animals should be conducted before removing adults or larvae from the wild, as wild individuals may have unique, separate alleles or may be inbred already at the time of capture, particularly when the founder comes from a small *in situ* population (Vredenburg *et al.* 2007; Richter and Nunziata 2014). Additionally, recommendations for genetic analysis and cross-classified breeding experiments to identify genetic compatibility between amphibian populations have recently been made after observing reduced fertilisation rates and the viability of offspring in controlled population-mixing experiments of *Pseudophryne bibronii* (Byrne and Silla 2020). In this species, it was evident that outbreeding depression resulted in drastic consequences for fitness, highlighting the need to consider the consequences of both inbreeding and outbreeding on the genetic viability of species (see Chapter 8). In fact, investigation of the Zoological Information Management System (ZIMS) suggests 58% of amphibian species held in captivity already are likely to have genetic markers available to enable analysis (Jensen *et al.* 2020). Unfortunately, the genetics of founder animals cannot easily be assessed in real time or under field conditions to determine whether to risk translocating the animals into the *ex situ* population, leaving a period of uncertainty in the real value of a newly proposed founder's genetic contribution. Even if a founder animal is deemed to be too closely related to the current *ex situ* population's genetic manifold, the genetics may become more independent if reintroduced as a founder to future generations. Successful application of a generational leapfrog of the genetics of founder animals requires the founder individual (potentially of unknown age) to survive and remain reproductively viable for years, a situation that is frequently unattainable. In some captive breeding programs, such as those for *Atelopus* spp. and *Anaxyrus baxteri*, there is no potential for introducing new founder genetics, and genes from the oldest lines added back into the *ex situ* population at generational intervals might provide otherwise lost alleles needed for adaptive stability to natural habitats (Frankham *et al.* 2014; Albert *et al.* 2015).

Ex situ breeding colonies can be considered extremely isolated fragments of a species, but with the compounding effect that the population may no longer experience all the crucial natural environmental cycles that shape the species' adaptive diversity. In the wild, fragmented populations of amphibian species face increasing isolation and risk of genetic loss that can lead to the collapse of species at the local level. Amphibian *ex situ* populations, which are similarly isolated, face these same risks of inbreeding and loss of allelic diversity, and even more so as captive populations are often further divided across multiple institutions. Overall species' genetic stability depends on coordinating both intra-population and inter-population gene flow to maintain diversity in each subpopulation of the captive community. However, stable gene structure in small fragmented wild populations of amphibian species can be important to maintain when location-specific adaptations arise (Frankham *et al.* 2002; Albert *et al.* 2015). When the captive breeding colony targets tadpoles for reintroduction, genetic analysis of founders aligns populations that may have slightly different genetic profiles that provide needed advantages in adapting to their particular environments (Frankham 2008; Albert *et al.* 2015). Thus, it is also necessary to maintain variants in the parallel *ex situ* colony to assure suitability of animals intended for reintroduction to the specified site. A case in point is the management of the Panamanian golden frog, which is divided into three SSPs, in which each coordinates genetic flow within the location-specific subpopulations identified as Ahogado, Soro, or Marta, established in original rescues after the species had collapsed from a deadly fungal disease (chytridiomycosis) (Table 11.2; Woodhams *et al.* 2011; Lewis *et al.* 2019). The *ex situ* population of *Lithobates chiricahuensis* is also managed for breeding by housing and pairing only animals from a subpopulation based on *in situ* sites, separated by 20–200 miles and which feature distinct habitats (Rorabaugh and Sredl 2014). Although

threatened primarily by destruction of habitat and by invasive species, some native sites of *Lithobates chiricahuensis* are also contaminated with chytrid fungus; limiting crossover between spatially distinct subpopulations not only maintains the environmental adaptations needed for the given subpopulation to succeed when reintroduced, but also mitigates the transfer of disease to those distant sites and throughout the species as a whole (Christman and Jennings 2014).

LINKING *IN SITU* AND *EX SITU* POPULATIONS

Incorporating new founder genetics into an existing population of amphibians can be accomplished through three primary modes of acquisition (Figure 11.2). A long-serving practice is translocating adult individuals from *in situ* to *ex situ* settings for breeding purposes. If adults are unobtainable, then fertilised egg masses, fractions of egg masses, or groups of hatchlings are commonly brought into captivity to be raised and eventually added to the gene pool. An emerging alternative is to acquire new genetic lineages using germplasm repositories to move gametes from *in situ* amphibian populations to *ex situ* breeding programs (Swanson *et al.* 2007; Burger *et al.* 2021).

Importation of adult amphibians

Many captive assurance colonies rely on the regular importation of animals from the wild in order to 'phase out' older, non-reproductive animals and to introduce genetic variation into captive populations to prevent

Table 11.2:   Biobanked samples of sperm collected from hormonally treated amphibians that have produced living offspring.

Species *Scientific name* Common name	Biobanked sperm No. of males	Sperm concentration ($\times 10^6$)	Per cent motility of sperm, pre-freeze/post-thaw	IVF Tadpoles/ metamorphs*	Reference
Ambystoma tigrinum Tiger salamander	6	N/A	N/A	21/1*	Marcec (2016)
Anaxyrus boreas boreas Boreal toad	4	3.1	85/39	119/89	Langhorne (2016)
Anaxyrus houstonensis Houston toad	11	5.64	83/31	13/2	Kouba *et al.* (2021)
Lithobates chiricahuensis Chiricahua leopard frog	22	2.33	55/39	16/0	Kouba *et al.* (2021)
Lithobates sevosus Dusky gopher frog	9	–	90/27	290/42*	Langhorne (2016)
Peltophryne lemur Puerto Rican crested toad	8	9.35	70/27	55/46*	Burger *et al.* (2021)

* Offspring that have grown to adulthood and produced a second generation of tadpoles and metamorphs.

Strategy 1: Adult import

Advantages

- Easy to collect
- Ready to breed same year
- Genotype can be determined

Disadvantages

- Depletion of in-situ population
- Low numbers for import
- Unknown reproductive potential
- Unknown age, limited life span
- Transportation and disease risk
- Quarantine and holding costs
- Negative behavior, stress
- Legal barriers

Strategy 2: Larvae import

Advantages

- Medium difficulty of collection
- Known age of cohort
- One founder line, many siblings
- Higher survival rate than in situ

Disadvantages

- Depletion of in situ population
- Very narrow collection window
- Lineage unknown
- Single genetic line / generation
- Years to next reproduction
- Quarantine and holding costs
- Legal restrictions

Strategy 3: Gamete import

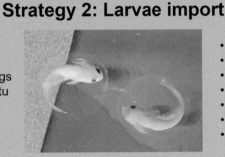

Advantages

- One time cost investment
- Many founders represented
- No animals removed from wild
- Multiple crossings / genetic line
- Multiple institutions / genetic line
- Multiple years / genetic line
- No quarantine or holding costs

Disadvantages

- Protocol development required
- Skilled personnel to execute
- Repository curation
- IVF to produce next generation
- Lower fertilization rate
- Lineage unknown

Figure 11.2: Comparison of import strategies of new founder genetics to *ex situ* breeding programs. Advantages and disadvantages address the processes necessary to incorporate a new founder genetic line into an *ex situ* population, with the end-point of creating the next generation.

inbreeding. Practically speaking, abstraction of wild individuals for captive breeding colonies is not a difficult process in itself, but the rationale and justification are more complex. The primary advantages of translocating adults include the potential of the individual to reproduce in the immediate breeding season and the ability to assess the degree of genetic heterozygosity offered to the *ex situ* population through genotyping (Figure 11.2). Disadvantages range from limiting factors such as unknown age and reproductive potential, to negative effects of an undesirable genetic profile or high risk of transmissible disease. While there are amphibian species that are long-lived, particularly among urodeles (> 15 years;

e.g. Cryptobranchidae, Ambystomatidae), other species might breed for only a few years (< 5 years, e.g. *Anaxyrus houstonensis*, *Litoria booroolongensis*), in turn creating the need to replenish captive populations more frequently than for longer-lived taxa (McCartney-Melstad and Shaffer 2015).

Mortality can occur in transport, or following arrival in quarantine, due to the stresses and risks that come from physical relocation to captive environments or unforeseen exposure to foreign pathogens (Cayuela *et al.* 2019), one of the main disadvantages of importing adults. Moreover, many amphibian species do not readily breed in captivity and bringing in more individuals from wild populations may actually have little positive effect on captive breeding (evidenced by the discrepancy between the number of founders and the effective founder equivalents in the amphibian populations managed by AZA-SSP programs) (Table 11.1). More recent concerns about the negative behavioural and genetic effects of captivity have created the opinion that animals in captive assurance colonies should only be held temporarily, and be bred and released as quickly as possible (Robert 2009; Hogg *et al.* 2019). To counteract these issues, even more animals may be taken from the wild than needed to compensate for potential losses and offset non-breeding individuals, promoting a cycle of over abstraction of valuable wild genetic lineages to accomplish meagre progress towards the goals of conservation in the management of captive populations.

Of concern is that augmentation of *ex situ* population genetics through the repeated practice of importing animals decreases genetic variation in the wild populations. There are several amphibian species with numbers that have dwindled drastically in recent years, and any removal of wild animals for captive-assurance colonies can be detrimental. For example, the dusky gopher frog (*Lithobates sevosus*) has an estimated wild breeding population of ~100–200 individuals in historic ranges around two breeding ponds (Richter *et al.* 2009), which face constant threats such as predation and disease. Further removal of members of their population to increase captive breeding can come at a high cost to the wild gene pool. Amphibians typically have high levels of genetic differentiation across isolated populations within a species due to factors such as low dispersal rates and predation (Funk *et al.* 2005; Allentoft and O'Brian 2010). As a consequence, in small wild populations of a threatened species, the most genetically diverse individuals tend to survive early life stages and become reproductively valuable adults for propagation of the species *in situ* (Richter *et al.* 2014). Removal of adult individuals can be devastating for small populations and often with no guaranteed benefit to the targeted *ex situ* breeding colony.

Disease

Wild amphibians often harbour diseases that can threaten the stability of captive colonies. Quarantine practices and testing reveal infected individuals, which are then typically withheld or culled from the breeding population, ending the genetic line and reproductive output. Spread of the chytrid fungus is of particular concern to

amphibians, and translation of animals at almost any life stage is highly regulated (Woodhams *et al.* 2011; Lips 2016). Panamanian golden frog populations have been devastated by chytridiomycosis in the environment, prompting enormous efforts to save the species through rescue and extensive captive management (Woodhams *et al.* 2011; Lewis *et al.* 2019). Applications of reproductive technologies have benefited breeding and health, aiding in maintenance of the species' genetic diversity (Della Togna *et al.* 2017; Bronson *et al.* 2021). The chytrid fungus also poses a high risk to the threatened boreal toad (*Anaxyrus boreas boreas*) in which it infects the larval stage via transmission through water, thereby threatening generations of tadpoles (Pilliod *et al.* 2010). Ranaviruses are iridoviruses that cause cutaneous ulceration, haemorrhaging, and eventual necrosis of organs and death in fish, amphibians, and some reptiles. Moreover, the infection can affect multiple amphibian life stages, killing both adult and larval animals (Lesbarrères *et al.* 2012). One of the most threatening aspects of ranaviruses is their ability to cross-contaminate different taxa through contact or environmental transmission (e.g. via water). Outbreaks of ranaviruses have been observed all over the world including in remote areas in China where it infects the Chinese giant salamander (*Andrias davidianus*). In yet another example, Houston toads (*Anaxyrus houstonensis*) are susceptible to chlamydia, and the transport and breeding of captive individuals for genetic management naturally risk spreading the disease across institutions and biocontainment facilities. Alternatively, the collection and import of gametes (e.g. sperm) from wild individuals, rather than importing the animals themselves, offers a lower risk of the transmission of disease and provides a broader approach to introducing new founder-lines to *ex situ* amphibian colonies.

Importing amphibian embryos or larvae

Since amphibian species tend to have many offspring at once, a less invasive strategy is the collection of a portion of the fertilised egg mass from *in situ* breeding sites. When clutches of fertilised eggs or developing larvae are collected from the wild, many siblings are reared as a single genetic cohort. Although these can be spread throughout the *ex situ* population, they cannot be genetically crossed with other lineages until the animals are reproductively mature (1–5 years depending on the species). Tadpoles can carry diseases as easily as adults and, because of their aquatic lifestyles and tendency to crowd together, disease can be more prolific and rigorous testing should be performed on larvae before integration into captive populations or released into wild environments. However, while relocated adults have higher mortality rates than individuals left in their native environment, the first generation offspring of translocated animals have similar survival rates to the offspring *in situ* (Cayuela *et al.* 2019).

Importing amphibians' gametes

An emerging option is to increase gene flow from *in situ* to *ex situ* populations (as well as between *ex situ* populations) through transfer of gametes rather than translocation of live animals, so captive assurance colonies can be maintained or enriched while wild populations continue to breed and remain genetically diverse. Figure 11.3 highlights two primary modes of the transport and storage of gametes to bridge *in situ* and *ex situ* amphibian populations using reproductive technologies. As viable unfertilised eggs are extremely difficult to obtain *in situ*, the more tenable option is to import sperm from wild males to fertilise eggs from captive females through *in vitro* fertilisation. Although not practical in conservation breeding programs, for decades, spermatozoa from testes macerates of sacrificed males have been cryopreserved, thawed, and used in IVF in several

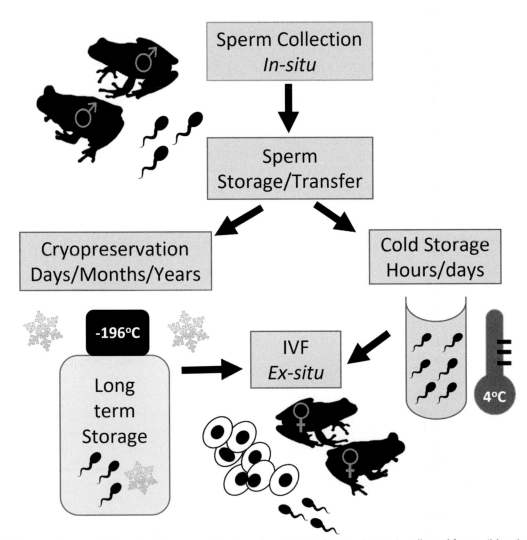

Figure 11.3: Process of gamete biobanking to connect *in situ* and *ex situ* populations. Sperm is collected from wild male founders, stored and shipped. Cold stored sperm must be used for IVF within hours or a day at the *ex situ* breeding site. Cryopreserved sperm can be stored in a germplasm repository and withdrawn as needed at any time, over multiple seasons and at any *ex situ* breeding facility for IVF.

common species (Browne *et al.* 1998; Proano and Perez 2017; Uteshev *et al.* 2019). In the field, sperm can easily be obtained as spermic urine or milt in response to exogenous hormones delivered non-invasively such as through the nares in a method introduced by Julien *et al.* (2019), by transdermal absorption (Snyder *et al.* 2012), or through minimally invasive intraperitoneal, intramuscular, or subcutaneous injection (Chapter 7). Short-term cold storage of sperm samples at 0–5°C maintains the motility and viability of spermatozoa over a few hours or days (Arregui *et al.* 2020; Langhorne *et al.* 2021), thereby permitting transport to *ex situ* breeding facilities for new genetic pairings using IVF procedures (Kouba *et al.* 2011), although the most rapid decrease in the motility and viability of spermatozoa in spermic urine occurs in the first few hours after collection. Active sperm is needed on demand for IVF when the female oviposits because the eggs have a short period before the capacity for fertilisation drops significantly. Cryobanking permits sperm to be stored indefinitely (Chapters 9 and 12), eliminating the problem of gametes being expressed asynchronously as sperm samples can be rapidly thawed and applied immediately when eggs are expressed, thus increasing coordination of the availability of gametes to maximise the success of genetic pairing (Kouba *et al.* 2012). The trade-off is that freeze–thaw processes are physically damaging to sperm cells and must be optimised to recover motility and the potential for fertilisation. Another advantage is improved animal welfare as the importing of gametes eliminates the risk of stress or disease encountered during the transport or quarantine of live animals. Furthermore, bringing in gametes from the wild creates a faster turnover rate for the production of offspring, and allows for multiple genetic crossings in the same breeding season. Genome resource banking has numerous other benefits that are reviewed elsewhere (Kouba and Vance 2009; Uteshev *et al.* 2019; see Chapters 9 and 12).

Cryobanking amphibian sperm is not a trivial process and requires skilled personnel for proper execution. When collected from living animals by hormonal induction, anuran sperm is fragile and short-lived when activated in the low osmolality (< 70 mOsmol) of the urine where it rapidly consumes cellular energy stores (Kouba *et al.* 2003, 2009). Protecting sperm cells from damage by ice crystals and dehydration requires suspending them in mixtures of cryodiluent and cryoprotectant that is then chilled in a stepwise fashion to control the rate and changes in phase of the sperm solution as it freezes.

Amphibian spermatozoa are typically frozen in 100–200 µL sealed cryostraws rather than 'open' vials, which are used in frozen-pellet methods. Proper loading of cryostraws is critical for successful biobanking because without appropriate gaps for expansion of fluid during freezing, and increased vapour pressure upon rapid thawing, the straw explodes and the sperm sample is lost entirely. Sealed cryostraws also prevent transmission of disease through the liquid nitrogen (LN_2) medium used in long-term germplasm repositories, thereby alleviating pathogenic contamination (Larmen *et al.* 2014). Successful amphibian sperm cryobanking requires both a pre-tested procedure and a skilled investigator to handle the process of freezing and thawing the sample. Another challenging aspect is the process of labelling and tracking the genetic lineages so the frozen sperm can be accessed on demand when eggs become available for IVF. Up front, cryobanking sperm may be a more intensive process than collecting and translocating animals or larvae to augment *ex situ* amphibian population genetics. However, the initial investment in time and effort to collect and store numerous new founder lines is recovered by increasing storage capacity and longevity that allows more diversification of genetic pairings each season, and over many generations, which was demonstrated by Howell *et al.* (2020) to be far more cost-efficient in the long term than maintaining live animals.

PRACTICAL ASPECTS OF BIOBANKING GAMETES *IN SITU*
Legal and professional considerations
Threatened amphibians, with or without existing conservation breeding programs, usually are listed under federal or state protection. Permits are required for access to, and acquisition of, wild animals as well as the collection of gametes. Moreover, additional authorisation is typically required from the relevant institutional animal care and use committee or animal ethics committee to ensure that all procedures comply with the requirements for the health, safety, and welfare of animals. Collaborations with federal, state, or private institutions are needed to solidify the rationale for sampling animals and for approval of permits. Representatives from institutions holding captive amphibians generally work with governmental officers and/or team members for recovery of threatened species to determine the number of animals required for genetic management, while field biologists may assist with locating and

capturing animals, thereby ensuring that animal welfare and *in situ* management policies are followed.

Working conditions in the field

Wild amphibians can be difficult to locate, especially in populations of sufficient size for collecting live animals or gametes. Many species reside in areas that are environmentally challenging to access. Populations of the Puerto Rican crested toad (*Peltophryne lemur*) are subject to severe weather, specifically hurricanes and collapsed land structures that make it difficult to locate animals safely, and in many cases management programs lack sufficient funding and skilled personnel to clear areas from damage. In alpine species, such as the boreal toad (*Anaxyrus boreas boreas*), Wyoming toad (*Anaxyrus baxteri*), corroboree frog (*Pseudophryne corroboree*), and baw baw frog (*Philoria frosti*), travel is restricted most of the year due to snow, and animals are only accessible by hiking expeditions. At the opposite extreme, the Chiricahua leopard frog resides, often on private land, in the deserts of New Mexico where overnight temperatures are below freezing but soar to > 40°C a few hours after sunrise. Another desert-dwelling species is the Sonoran tiger salamander (*Ambystoma tigrinum stebbinsi*), found in isolated sites across the deserts of northern Mexico and Arizona. Safety and access aside, the animals themselves often can only be obtained during the breeding season and at night, sometimes restricting the window of opportunity for collecting animals or gametes to only a week or two.

Equipment and supplies

Implementation of reproductive technologies in field settings requires manageable equipment and supplies. In sexually monomorphic species, or to assess follicular development in females, ultrasonography is becoming increasingly adopted and valuable to have in the field (Chapters 3 and 6). In addition, microscopic evaluation of spermatozoa is required before cryopreservation. Thus, multiple power sources with long battery lives, generators, or car batteries are critical to power the equipment required for assessing gametes. Sperm cryopreservation requires liquid nitrogen (LN_2) both for freezing and storage and can be difficult to obtain in remote locations. Moreover, LN_2 is hazardous and evaporates rapidly, limiting its availability and transport-options over long periods in the field. While these limitations can be challenging, they can be overcome with sufficient planning, field technicians, funding, and logistical support from multiple agencies.

CASE STUDIES

While sperm cryopreservation and post-thaw recovery of motility have been successfully reported for numerous common and threatened amphibian species, there are only a few threatened species in which viable offspring have been produced from frozen sperm that was collected by non-lethal means (i.e. hormone-induced sperm-release) (Table 11.2). The integration of wild genetics into captive populations through cryopreservation and transfer of gametes has only been achieved in three of these species to date (see Table 11.2), but with application to other species on the near horizon.

Puerto Rican crested toad

The most successful example of coupling *in situ* and *ex situ* conservation though biobanking is in the Puerto Rican crested toad (*Peltophryne lemur*), which is critically endangered in its native environment. In 2020, earthquakes destroyed the karst formations that form its native habitat, potentially eradicating the few remaining populations on the island. Annually, wild populations of the toad are supplemented by releases of tadpoles from coordinated *ex situ* breeding events. To manage the long-term genetics of the captive population, new adult individuals are needed from the wild to minimise inbreeding, with an influx of 10 new founder individuals proposed over the next few years (Barber *et al.* 2018). Although the current *ex situ* population size is large (*n* = 1146; Table 11.1), a significant number of the adults do not contribute genetic material to the next generation through natural breeding, thereby leading to bias in representation of founders in subsequent generations (FGE = 16.54; Table 11.1) (Sanchez-Montes *et al.* 2017).

In coordination with the importation of animals, sperm from hormonally stimulated males was collected and cryopreserved *in situ* and then transported and stored at the National Amphibian Genome Bank at Mississippi State University. After 14 months, the sperm was transported to Fort Worth Zoo, where captive females were hormonally induced to lay eggs in order to conduct IVF using the frozen sperm from wild males. Thawed sperm from males contributed to four genetically unique cohorts, producing 46 metamorphs,

24 of which remain in the captive population and have grown to adulthood, eight are deceased and the remaining 14 were released back to the wild (Table 11.2) (Burger *et al.* 2021). The addition of eight founder lineages through the cryobanking of gametes and assisted reproductive technologies reached 80% of the SSP's target of 10 new founder lines in one attempt at breeding, without removing individuals from the *in situ* population. Moreover, the IVF process permitted unrepresented females in the assurance colony to contribute to the gene pool. Thus, linkage of *in situ* and *ex situ* populations through the use of reproductive technologies has introduced genetic variation into the captive population, which can subsequently be used in future efforts to bolster genetic diversity within the wild population. Storage of sperm samples from genetically valuable males of numerous species, such as the Puerto Rican crested toad, allows for a constant replenishment of genetic variation and a hedge against inbreeding both in captive and wild populations.

Boreal toad

The boreal toad (*Anaxyrus boreas boreas*) is a threatened high-alpine species fragmented across the southern Rocky Mountains and under severe pressure from the destruction of its habitat and from the disease chytridiomycosis (Pilliod *et al.* 2010). Efforts to recover the species have been championed by the Colorado Parks and Wildlife Native Aquatic Species Restoration Facility (NASRF), which focuses on captive breeding to maintain genetic diversity within the defined regional subpopulation. Yet, this conservation breeding program faces a myriad of challenges that stem from low fecundity, a potential biannual reproductive cycle, and an ageing breeding population. Efforts to increase genetic diversity in the captive population at NASRF involved development of reproductive technologies for the species and included obtaining sperm from wild males to import into the breeding population using cold-storage methodologies and cryopreservation (Langhorne 2016; Langhorne *et al.* 2021). Release of sperm was induced using hormonal therapy and collected samples were subsequently transported by cold storage to the NASRF facility where post-thawed sperm was used for IVF procedures with captive females. Similar to the case study with Puerto Rican crested toads, thawed sperm from wild male boreal toads was used to fertilise eggs from captive females, following

hormonal induction, resulting in the production of 119 tadpoles and 89 metamorphs, of which several progressed to adulthood.

Chiricahua leopard frog

The Chiricahua leopard frog (*Lithobates chiricahuensis*) is a threatened ranid species endemic to the southwestern United States and parts of Mexico, and which breeds from March through September in small ponds, *cienagas*, and streams. As a way to increase populations, captive breeding programs have been established across several institutions including Turner Enterprise's outdoor ranarium at Ladder Ranch (semi-captive management), the Fort Worth Zoo, and the Phoenix Zoo (captive management). These institutions have spearheaded conservation efforts through the reintroduction of egg masses, tadpoles, and metamorphic froglets into native and historic locations across the southern United States. Since 1995, the Phoenix Zoo has released 26 egg masses, 15 562 tadpoles, 11 119 juveniles, and 140 adult frogs while the Fort Worth Zoo has released 1600 captive-bred tadpoles since 2015, all produced by natural pairings. While these reproductive successes are positive, natural breeding within these programs has been sporadic, with few genetic lines represented, and far below targeted reintroduction numbers needed for recovery. In addition, breeding events are heavily based upon seasonal factors such as spring rains and changes in summer temperature, limiting both the number of attempted breeding events and subsequent production of offspring. A complicating factor is that the species is fragmented into small, isolated populations with no possible migration between sites, and captive breeding efforts have focused on maintaining genetic pairings of individuals with those subpopulations to avoid potential out-breeding depression. In an effort to facilitate incorporation of new founder lines from wild animals representing each site, and without depleting the numbers of adults *in situ*, the breeding programs have begun the process of obtaining cryopreserved sperm from wild males. In 2018 and 2021, sperm samples from males representing eight different populations were collected and frozen on site in southern New Mexico (Figure 11.1 D). These frozen gametes were transported to the Fort Worth Zoo where two captive females were hormonally induced to lay eggs, which were fertilised with thawed sperm from wild males to produce offspring (Table 11.2).

In situ and *ex situ* populations of amphibians poised to be linked through genome resource banks

Several other threatened amphibian species with existing *ex situ* breeding colonies are well positioned to implement reproductive technologies and the cryopreservation of sperm to link the *in situ* and *ex situ* populations for increased genetic management, particularly through biobanking. As mentioned above, there is significant investment in procedural development and obtaining skilled personnel for *in situ* collection and freezing of gametes, and implementation of IVF procedures. Captive Houston toads (*Anaxyrus houstonensis*) have been induced to express sperm that was cryopreserved and then thawed several days later to conduct IVFs with eggs from captive females, producing 19 tadpoles of which two are juveniles progressing to adulthood (Kouba *et al.* 2021). The application of this methodology is now proven and ready for field collections of sperm from wild male Houston toads. The detailed cryopreservation process for hormonally induced sperm for all of these described species was initially developed for the critically endangered dusky gopher frog (*Lithobates sevosus*) by Langhorne (2016), who obtained 42 metamorphs from 60 tadpoles in IVF attempts with frozen–thawed sperm. It has also been demonstrated that sperm from the tiger salamander (*Ambystoma tigrinum*) can be collected after hormonal therapy, frozen, and thawed to produce viable offspring via IVF. Of concern to long-term breeding efforts is whether or not the offspring generated from frozen sperm are themselves reproductively viable. To address this question, threatened dusky gopher frogs (*Lithobates sevosus*), produced from frozen–thawed sperm and IVF, progressed to adulthood and have themselves produced sperm that was used in subsequent IVF experiments to produce an F2 generation of tadpoles (Vance *et al.* 2018; Kouba *et al.* 2021). Puerto Rican crested toads have also reached adulthood and parented offspring (tadpoles) during hormonally induced pair breeding (not IVF), and the resulting 5000 tadpoles were released into the wild (Burger *et al.* 2021). A female tiger salamander produced from cryopreserved sperm was raised to adulthood and, after hormonal therapy, produced eggs that were then fertilised and yielded larval salamanders that are approaching two years of age (Figure 11.2 centre panel). All of these breeding programs have proven that IVF using biobanked sperm can yield viable offspring, and now we are in position to implement biobanking as a means of incorporating new founder lineages into current *ex situ* breeding populations of threatened amphibians.

CONCLUSIONS

Captive assurance colonies play an integral part in the conservation and management of threatened and endangered amphibian species. Maintaining a high level of genetic diversity in captive colonies is imperative for the recovery of wild populations that are experiencing drastic declines in numbers of animals. Cryopreservation and biobanking of male gametes to link *in situ* and *ex situ* groups of threatened and endangered species have been recommended conservation strategies for decades, yet only recently has the series of reproductive techniques been implemented in a coordinated fashion to create offspring for reintroduction into the wild. Here, we show how reproductive technologies, such as hormonal therapy, IVF, and cryopreservation of sperm can be used to link *in situ* and *ex situ* populations of endangered species of amphibians, especially where at-risk captive populations exist and long-term sustainability is a challenge without the influx of new genetic variation from wild sources.

REFERENCES

Albert EM, Fernández-Beaskoetxea S, Godoy JA, Tobler U, Schmidt BR, Bosch J (2015) Genetic management of an amphibian population after a chytridiomycosis outbreak. *Conservation Genetics* **16**, 103–111. doi:10.1007/s10592-014-0644-6

Allentoft M, O'Brian J (2010) Global amphibian declines, loss of genetic diversity and fitness: a review. *Diversity (Basel)* **2**, 47–71. doi:10.3390/d2010047

Araki H, Cooper B, Blouin MS (2007) Genetic effects of captive breeding decline in the wild. *Science* **318**, 100–103. doi:10.1126/science.1145621

Arregui L, Bóveda P, Gosálvez J, Kouba AJ (2020) Effect of seasonality on hormonally induced sperm in *Epidalea calamita* (Amphibia, Anura, Bufonidae) and its refrigerated and cryopreservated storage. *Aquaculture* **529**, 735677. doi:10.1016/j.aquaculture.2020.735677

Baker A (2007) Animal ambassadors: an analysis of the effectiveness and conservation impact of *ex situ* breeding efforts. In *Catalysts for Conservation: A Directive for Zoos in the 21st Century.* (Eds A Zimmermann, M Hatchwell, L Dickie and C West) pp. 139–154 Cambridge University Press, Cambridge, UK.

Barber D, Smith D, Schad K (2018) *Puerto Rican Crested Toad (*Peltophryne lemur*). AZA Species Survival Plan˚ Designation Green Program Population Analysis & Breeding and Transfer Plan.* AZA Population Management Center, Chicago IL, USA.

Becker MH, Richards-Zawacki CL, Gratwicke B, Belden LK (2014) The effect of captivity on the cutaneous bacterial community of the critically endangered Panamanian golden frog (*Atelopus zeteki*). *Biological Conservation* **176**, 199–206. doi:10.1016/j.biocon.2014.05.029

Blanchet S, Páez DJ, Bernatchez L, Dodson JJ (2008) An integrated comparison of captive-bred and wild Atlantic salmon (*Salmo salar*): implications for supportive breeding programs. *Biological Conservation* **141**, 1989–1999. doi:10.1016/j.biocon.2008.05.014

Brady L, Young R, Goetz M, Dawson J (2017) Improving zoos' conservation potential through understanding barriers to holding globally threatened amphibians. *Biodiversity and Conservation* **26**, 2735–2749. doi:10.1007/s10531-017-1384-y

Bronson E, Vance CK (2019) Anuran reproduction. In *Fowler's Zoo Wild Animal Medicine, Current Therapy*. Volume 9. (Eds RE Miller, N Lamberski and P Calle) pp. 371–379. Elsevier Inc., St Louis MO, USA.

Bronson E, Guy EL, Murphy KJ, Barrett K, Kouba AJ, Poole VA, Kouba CK (2021) Influence of oviposition-inducing hormone on spawning and mortality in the endangered Panamanian golden frog (*Atelopus zeteki*). *BMC Zoology* **6**, 17. doi:10.1186/s40850-021-00076-8

Browne RK, Clulow J, Mahony M, Clark A (1998) Successful recovery of motility and fertility of cryopreserved cane toad (*Bufo marinus*) sperm. *Cryobiology* **37**, 339–345. doi:10.1006/cryo.1998.2129

Burger I, Julien AR, Kouba AJ, Counsell KR, Barber D, Pacheco C, Kouba CK (2021) Linking *in situ* and *ex situ* populations of the endangered Puerto Rican crested toad (*Peltophryne lemur*) through genome banking. *Conservation Science and Practice* 3(11), e525. doi:10.1111/csp2.525

Byrne PG, Silla AJ (2020) An experimental test of the genetic consequences of population augmentation in an amphibian. *Conservation Science and Practice* **2**, 194. doi:10.1111/csp2.194

Cayuela H, Gillet L, Laudelout A, Besnard A, Bonnaire E, Levionnois P, Muths E, Dufrene M, Kinet T (2019) Survival cost to relocation does not reduce population self-sustainability in an amphibian. *Ecological Applications* **29**, 1–15. doi:10.1002/eap.1909

Christman BL, Jennings RD (2018) 'Distribution of the amphibian chytrid fungus *Batrachochytrium dendrobatidis* (Bd) in New Mexico'. Report to New Mexico Department of Game and Fish, Conservation Science Division, Santa Fe NM, USA.

Della Togna G, Trudeau V, Gratwicke B, Evans M, Augustine L, Chia H, Bronikowski E, Murphy J, Comizzolii P (2017) Effects of hormonal stimulation on the concentration and quality of excreted spermatozoa in the critically endangered Panamanian golden frog (*Atelopus zeteki*). *Theriogenology* **91**, 27–35. doi:10.1016/j.theriogenology.2016.12.033

Frankham R (2005) Genetics and extinction. *Biological Conservation* **126**, 131–140. doi:10.1016/j.biocon.2005.05.002

Frankham R (2008) Genetic adaptation to captivity in species conservation programs. *Molecular Ecology* **17**, 325–333. doi:10.1111/j.1365-294X.2007.03399.x

Frankham R (2009) Genetic architecture of reproductive fitness and its consequences. In *Adaptation and Fitness in Animal Populations: Evolutionary and Breeding Perspectives on Genetic Resource Management*. (Eds J van der Werf, HU Graser, R Frankham and C Gondro) pp. 15–39. Springer, Dordrecht, Holland.

Frankham R, Briscoe DA, Ballou JD (2002) *Introduction to Conservation Genetics*. Cambridge University Press, New York NY, USA.

Frankham R, Ballou JD, Briscoe DA (2010) *Introduction to Conservation Genetics*. 2nd edn. Cambridge University Press, Cambridge, UK.

Frankham R, Bradshaw CJA, Brook BW (2014) Genetic adaptation to captivity in species conservation. *Biological Conservation* **170**, 56–63. doi:10.1016/j.biocon.2013.12.036

Funk WC, Greene AE, Corn PS, Allendorf FW (2005) High dispersal in a frog species suggests that it is vulnerable to habitat fragmentation. *Biology Letters* **1**, 13–16. doi:10.1098/rsbl.2004.0270

Gascon C, Collins JP, Moore RD, Church DR, McKay J, Mendelson III J (Eds) (2007) *Amphibian Conservation Action Plan: Proceedings IUCN/SSC Amphibian Conservations Summit 2005*. World Conservation Union (IUCN), Gland, Switzerland.

Graham KM, Kouba AJ, Langhorne CJ, Marcec RM, Willard ST (2016) Biological sex identification in the endangered dusky gopher frog (*Lithobates sevosa*): a comparison of body size measurements, secondary sex characteristics, ultrasound imaging, and urinary hormone analysis methods. *Reproductive Biology and Endocrinology* **14**, 41. doi:10.1186/s12958-016-0174-9

Griffiths RA, Pavajeau L (2008) Captive breeding, reintroduction, and the conservation of amphibians. *Conservation Biology* **22**, 852–861. doi:10.1111/j.1523-1739.2008.00967.x

Hallock L (2013) *Draft State of Washington Oregon Spotted Frog Recovery Plan*. Washington Department of Fish and Wildlife, Olympia WA, USA.

Harding G, Griffiths RA, Pavajeau L (2016) Developments in amphibian captive breeding and reintroduction programs. *Conservation Biology* **30**, 340–349. doi:10.1111/cobi.12612

Hogg CJ, Wright B, Morris KM, Lee AV, Ivy JA, Grueber CE, Belov K (2019) Founder relationships and conservation management: empirical kinships reveal the effect on breeding programmes when founders are assumed to be unrelated. *Animal Conservation* **22**, 348–361. doi:10.1111/acv.12463

Hornyak V, Armstrong S, Navarro A (2020) *Wyoming Toad (Anaxyrus baxteri). AZA Species Survival Plan* Designation Yellow Program Population Analysis & Breeding and Transfer Plan. AZA Population Management Center, Chicago IL, USA.

Howell LG, Frankham R, Rodger JC, Witt RR, Clulow S, Upton RMO, Clulow J (2020) Integrating biobanking minimizes inbreeding and produces significant cost benefits for a threatened frog captive breeding programme. *Conservation Letters* **14**, e12776.

IUCN/SSC (2014) *Guidelines on the Use of Ex Situ Management for Species Conservation*. Version 2. 0. IUCN Species Survival Commission, Gland, Switzerland

Jensen EL, McClenaghan B, Ford B, Lentini A, Kerr KCR, Russello MA (2020) Genotyping on the ark: a synthesis of genetic resources available for species in zoos. *Zoo Biology* **39**, 257–262. doi:10.1002/zoo.21539

Johnson HE, Mills S, Wehausen JD, Stephenson TR, Luikart G (2010) Translating effects of inbreeding depression on component vital rates to overall population growth in endangered bighorn sheep. *Conservation Biology* **25**(6), 1240–1249.

Julien AR, Kouba AJ, Kabelik D, Feugang JM, Willard ST, Kouba CK (2019) Nasal administration of gonadotropin releasing hormone (GnRH) elicits sperm production in Fowler's toads (*Anaxyrus fowleri*). *BMC Zoology* **4**, e3e3. doi:10.1186/s40850-019-0040-2

Kouba AJ, Vance CK (2009) Applied reproductive technologies and genetic resource banking for amphibian conservation. *Reproduction, Fertility and Development* **21**, 719–737. doi:10.1071/RD09038

Kouba AJ, Vance CK, Frommeyer MA, Roth TL (2003) Structural and functional aspects of *Bufo americanus* spermatozoa: effects of inactivation and reactivation. *Journal of Experimental Zoology. Part A, Comparative Experimental Biology* **295A**, 172–182. doi:10.1002/jez.a.10192

Kouba AJ, Vance CK, Willis EL (2009) Artificial fertilization for amphibian conservation: Current knowledge and future considerations. *Theriogenology* **71**, 214–227. doi:10.1016/j.theriogenology.2008.09.055

Kouba AJ, Willis E, Vance C, Hasenstab S, Reichling S, Krebs J, Linhoff L, Snoza M, Langhorne CJ, Germano JM (2011) Development of assisted reproductive technologies for the endangered Mississippi gopher frog (*Lithobates sevosa*) and sperm transfer for *in vitro* fertilization. *Reproduction, Fertility and Development* **24**, 170. doi:10.1071/RDv24n1Ab116

Kouba AJ, Vance CK, Calatayud N, Rowlison T, Langhorne C, Willard ST (2012) Assisted reproduction technologies (ART) for amphibians. In *Amphibian Husbandry Resource Guide*. 2nd edn. (Eds VA Poole and S Grow) pp. 60–118. Amphibian Taxon Advisory Group, American Association of Zoos and Aquariums, Silver Spring MD, USA.

Kouba AJ, Lloyd RE, Houck ML, Silla AJ, Calatayud N, Trudeau VL, Clulow J, Molinia F, Langhorne C, Vance CK, *et al.* (2013) Emerging trends for biobanking amphibian genetic resources: The hope, reality and challenges for the next decade. *Biological Conservation* **164**, 10–21. doi:10.1016/j.biocon.2013.03.010

Kouba CK, Julien AR, Burger I, Barber D, Lampert SS, Kouba AJ (2021) Cryopreserved sperm and hormone therapy produce an F1-generation of adults and their F2-generation of offspring in four amphibian species. *Reproduction, Fertility and Development* **34**(2), 248–248.

Kraaijeveld-Smit FJL, Griffiths RA, Moore RD, Beebee TJC (2006) Captive breeding and the fitness of reintroduced species: A test of the responses to predators in a threatened amphibian. *Journal of Applied Ecology* **43**, 360–365. doi:10.1111/j.1365-2664.2006.01137.x

Langhorne CJ (2016) Developing assisted reproductive technologies for endangered North American anurans. PhD thesis. Mississippi State University, USA.

Langhorne CJ, Calatayud NE, Kouba CK, Willard ST, Smith T, Ryan PL, Kouba AJ (2021) Efficacy of hormone stimulation on sperm production in an alpine amphibian (*Anaxyrus boreas boreas*) and the impact of short-term storage on sperm quality. *Zoology* **146**, e125912. doi:10.1016/j.zool.2021.125912

Larman MG, Hashimoto S, Morimoto Y, Gardner DK (2014) Cryopreservation in ART and concerns with contamination during cryobanking. *Reproductive Medicine and Biology* **13**, 107–117. doi:10.1007/s12522-014-0176-2

Lesbarrères D, Balseiro A, Brunner J, Chinchar VG, Duffus A, Kerby J, Miller DL, Robert J, Schock DM, Waltzek T, Gray MJ (2012) Ranavirus: past, present and future. *Biological Letters* **8**, 481–483. doi:10.1098/rsbl.2011.0951

Lewis CHR, Richards-Zawacki CL, Ibanez R, Luedtke J, Voyles J, Houser P, Gratwicke B (2019) Conserving Panamanian harlequin frogs by integrating captive-breeding and research programs. *Biological Conservation* **236**, 180–187. doi:10.1016/j.biocon.2019.05.029

Lips KR (2016) Overview of chytrid emergence and impacts on amphibians. *Philosophical Transactions of the Royal Society B* **371**, 20150465.

Lynch M, O'Hely M (2001) Captive breeding and the genetic fitness of natural populations. *Conservation Genetics* **2**, 363–378. doi:10.1023/A:1012550620717

Marcec R (2016) Development of assisted reproductive technologies for endangered North American salamanders. PhD thesis. Mississippi State University, USA.

Mattila ALK, Duplouy A, Kirjokangas M, Lehtonen R, Rastas P, Hanski I (2012) High genetic load in an old isolated butterfly population. *Proceedings of the National Academy of Sciences of the USA* **109**, e2496–e2505.

Mays S, Barber D, Senner P (2019) Houston Toad (*Anaxyrus houstonensis*). *AZA Species Survival Plan* Designation *Yellow Program Population Analysis & Breeding and Transfer Plan*. AZA Population Management Center, Chicago IL, USA.

McCartney-Melstad E, Shaffer HB (2015) Amphibian molecular ecology and how it has informed conservation. *Molecular Ecology* **24**, 5084–5109. doi:10.1111/mec.13391

Michaels CJ, Gini BF, Preziosi RF (2014) The importance of natural history and species-specific approaches in amphibian ex-situ conservation. *The Herpetological Journal* **24**, 135–145.

Pierce LJS (2006) *Boreal Toad (Bufo boreas boreas) Recovery Plan*. New Mexico Department of Game and Fish Conservation Sciences Division, Santa Fe NM, USA.

Pilliod DS, Hossack BR, Bahls PF, Bull EL, Corn PS, Hokit G, Maxell BA, Munger JC, Wyrick A (2010) Non-native salmonids affect amphibian occupancy at multiple spatial scales. *Diversity & Distributions* **16**, 959–974. doi:10.1111/j.1472-4642.2010.00699.x

Poole V, Barrett K, Schad K, Senner P (2018) *Panamanian Golden Frog (Ahogado) (Atelopus zeteki). AZA Species Survival Plan˚ Designation Yellow Program Population Analysis & Breeding and Transfer Plan.* AZA Population Management Center, Chicago IL, USA.

Poole V, Barrett K, Senner P (2019) *Panamanian Golden Frog (Soro) (Atelopus zeteki). AZA Species Survival Plan˚ Designation Yellow Program Population Analysis & Breeding and Transfer Plan.* AZA Population Management Center, Chicago IL, USA.

Poole V, Barrett K, Nuetzel H, Bladow R (2020) *Panamanian Golden Frog (Marta) (Atelopus varius). AZA Species Survival Plan˚ Designation Yellow Program Population Analysis & Breeding and Transfer Plan.* AZA Population Management Center, Chicago IL, USA.

Proano B, Perez OD (2017) *In vitro* fertilizations with cryopreserved sperm of *Rhinella marina* (Anura; bufonidae) in Ecuador. *Amphibian & Reptile Conservation* **11**, 1–6.

Ralls K, Ballou JD (1983) Extinction: lessons from zoos. In *Genetics and Conservation: A Reference for Managing Wild Animal and Plant Populations.* (Eds CM Schonewald-Cox, SM Chambers, B MacBryde and L Thomas) pp. 164–183. Benjamin Cummings, Menlo Park CA, USA.

Reichling S, Lance D, Bryan CG (2018) *Dusky Gopher Frog (Lithobates sevosa). AZA Species Survival Plan˚ Designation Yellow Program Population Analysis & Breeding and Transfer Plan.* AZA Population Management Center, Chicago IL, USA.

Richter SC, Nunziata SO (2014) Survival to metamorphosis is positively related to genetic variability in a critically endangered amphibian species. *Animal Conservation* **17**, 265–274. doi:10.1111/acv.12088

Richter S, Crother B, Broughton R (2009) Genetic consequences of population reduction and geographic isolation in the critically endangered frog, *Rana sevosa. Copeia* **2009**, 799–806. doi:10.1643/CH-09-070

Robert A (2009) Captive breeding genetics and reintroduction success. *Biological Conservation* **142**, 2915–2922. doi:10.1016/j.biocon.2009.07.016

Rorabaugh JC, Sredl MJ (2014) Herpetofauna of the 100-mile circle: Chiricahua leopard frog (*Lithobates chiricahuensis*). *Sonoran Herpetologist* **27**, 61–70.

Sánchez-Montes G, Ariño AH, Vizmanos JL, Wang J, Martínez-Solano I (2017) Effects of sample size and full sibs on genetic diversity characterization: a case study of three syntopic Iberian pond-breeding amphibians. *The Journal of Heredity* **108**, 535–543. doi:10.1093/jhered/esx038

Scheele BC, Hunter DA, Grogan LF, Berger L, Kolby JE, McFadden MS, Marantelli G, Skerratt LF, Driscoll DA (2014) Interventions for reducing extinction risk in chytridiomycosis-threatened amphibians. *Conservation Biology* **28**, 1195–1205. doi:10.1111/cobi.12322

Smallbone W, van Oosterhout C, Cable J (2016) The effects of inbreeding on disease susceptibility: *Gyrodactylus turnbulli* infection of guppies, *Poecilia reticulata. Experimental Parasitology* **167**, 32–37. doi:10.1016/j.exppara.2016.04.018

Snyder WE, Trudeau VL, Loskutoff NM (2012) A noninvasive, transdermal absorption approach for exogenous hormone induction of spawning in the northern cricket frog, *Acris crepitans*: a model for small endangered amphibians. *Reproduction, Fertility and Development* **25**, 232–233. doi:10.1071/RDv25n1Ab168

Spielman D, Brook B, Briscoe D, Frankham R (2004) Does inbreeding and loss of genetic diversity decrease disease resistance? *Conservation Genetics* **5**, 439–448. doi:10.1023/B:COGE.0000041030.76598.cd

Stuart SN, Chanson JS, Cox NA, Young BE, Rodrigues AS, Fischman DL, Waller RW (2004) Status and trends of amphibian declines and extinctions worldwide. *Science* **306**, 1783–1786. doi:10.1126/science.1103538

Swanson WF, Magarey GM, Herrick JR (2007) Sperm cryopreservation in endangered felids: developing linkage of *in situ-ex situ* populations. *Society of Reproduction and Fertility Supplement* **65**, 417–432.

Taylor EB, Stamford MD, Baxter JS (2003) Population subdivision in westslope cutthroat trout (*Oncorhynchus clarki lewisi*) at the northern periphery of its range: evolutionary inferences and conservation implications. *Molecular Ecology* **12**, 2609–2622. doi:10.1046/j.1365-294X.2003.01937.x

Uteshev VK, Gakhova EN, Kramarova LI, Shishova NV, Kaurova SA, Browne RK (2019) Cryobanking of amphibian genetic resources in Russia: past and future. *Russian Journal of Herpetology* **26**, 319–324. doi:10.30906/1026-2296-2019-26-6-319-324

USFWS (2018) *Recovery Plan for the Southern California Distinct Population Segment of the Mountain Yellow-Legged Frog* (Rana muscosa). US Fish and Wildlife Service, Pacific Southwest Region, Sacramento CA.

Vance CK, Julien AR, Counsell K, Marcec R, Agcanas LA, Tucker A, Kouba AJ (2018) Amphibian ART over the generations: frozen sperm offspring produce viable F2 generation. *Cryobiology* **85**, 178. doi:10.1016/j.cryobiol.2018.10.220

Vredenburg VT, Bingham R, Knapp R, Morgan JAT, Moritz C, Wake D (2007) Concordant molecular and phenotypic data delineate new taxonomy and conservation priorities for the endangered mountain yellow-legged frog. *Journal of Zoology* **271**, 361–374.

Witzenberger KA, Hochkirch A (2011) Ex situ conservation genetics: a review of molecular studies on the genetic consequences of captive breeding programmes for endangered animal species. *Biodiversity and Conservation* **20**, 1843–1861. doi:10.1007/s10531-011-0074-4

Woodhams DC, Bosch J, Briggs C, Cashins S, Davis LR, Lauer A, Muths E, Puschendorf R, Schmidt BR, Sheafor B, Voyles J (2011) Mitigating amphibian disease: strategies to maintain wild populations and control chytridiomycosis. *Frontiers in Zoology* **8**, 8. doi:10.1186/1742-9994-8-8

12 Genome resource banks as a tool for amphibian conservation

Andy J. Kouba

INTRODUCTION

Biological collections are a foundation for scientific discovery by temporally and spatially preserving ecological and biological information, oftentimes advancing science in ways unanticipated at the time of preservation. These biological collections can broadly be broken down into two categories: (1) natural history collections primarily focused on non-living specimens (e.g. preserved biological, geological, and paleontological items); and (2) living collections (e.g. zoos, aquaria, botanical gardens, research colonies, and cells within tissue and germplasm repositories). Both types of collections have enabled countless scientific discoveries in biodiversity, taxonomy, genomics, biogeography, evolutionary biology, plant systematics, and medical improvements to human health (NASEM 2020). Moreover, these collections are the pillars of academic studies in botany, mammalogy, herpetology, ichthyology, ornithology, mycology, and microbiology, to name a few.

Specimens, and their associated metadata, have for centuries supported hypothesis-driven research across the natural and life sciences; these preserved, fossilised, and living specimens are a vast repository of data about Earth's biodiversity and how it has changed over time and space (Wildt 2000; Meineke *et al.* 2018). Recent estimates predict that natural history collections in the United States may hold between 800 million and 1 billion specimens in long-term repositories (Kemp 2015). If that is scaled across the planet, including living collections, the amount of accessible, stored, information on biodiversity in repositories will far exceed 1 billion specimens. For example, O'Connell *et al.* (2014) and Hedrick *et al.* (2020) extrapolated the limited global digitised content available and predicted that there may be roughly 2.5–3 billion specimens housed in natural history collections worldwide. While substantial, the number of specimens within biological collections represents only a fraction of the compendium of species on Earth. The most comprehensive catalogue of all known species on the planet, the Catalogue of Life (CoL; https://www.catalogueoflife.org/data/download), approximates that only ~80% of the 2.3 million extant species have been documented by biologists, from an estimated 8.7 million forms of life.

Although many biological collections have existed for centuries, accessing information within them was challenging and required physical access to the institutions where specimens were held or from which they had to be shipped, thereby risking damage or loss (Owens and Johnson 2019). Historically, archival reports called 'voucher specimens' were primarily used by biologists to provide credibility to their studies and resulting publications, understand species' distributions and delimitations, document morphological distinctions, and assist with taxonomic identification (Reynolds *et al.* 1996; Rocha *et al.* 2014; NASEM 2020). These voucher specimens typically were preserved in either formalin or alcohol (e.g. amphibians, reptiles, and fish), or processed as skins (e.g. birds and mammals), with the following basic information: locality, date and time of collection, taxonomic identification, standard morphological measurements, sex, and other limited data (e.g. maps) (NASEM 2020). However, as new technologies have emerged, high-quality ancillary data (e.g. genome sequences) can be added to the voucher specimen, thereby building capacity for a diverse set of disciplines to be joined in studying interactions among organisms, species, and communities (Schindel and Cook 2018). With the advancement of the digital age there has been a push to digitise information into online searchable databases, allowing worldwide accessibility to specimens in repositories and relatable metadata. In fact, entirely new fields of research in biodiversity have developed through data mining such digital repositories, particularly in the fields of paleoecology, isotope ecology, and bioinformatics. The omics2 revolution resulted in the development of a suite of new techniques in molecular biology that has expanded the utilisation of data from biological collections and that has become central to the field of conservation genetics. As more investigators work on these repositories, information from a broad array of specimens (with diverse preparations or taxonomic representation) can be combined within the databases through social networks and the integration of data.

Germplasm repositories

Germplasm repositories often referred to in the broader sense as genetic resource banks (GRBs), are a specialised form of living collections, whereby captive-type and wild-type isolates maintained as living cells (sperm, eggs, embryos, diploid totipotent cells, or somatic cells) can contribute to perpetual replication of organisms. As the name implies, a GRB allows for the deposit and withdrawal of gametes or embryos from the cryobank for the genetic management, research, and sustainability of amphibian captive assurance colonies. Moreover, these living amphibian cells can be used to generate new animals for recovery programs, exchange genetics between *in situ* and *ex situ* populations, and facilitate reintroductions. In general, GRBs for wildlife conservation are maintained within zoological institutions as part of their living collections and there are dozens of these repositories globally. Probably the three most well-known wildlife GRBs, due to the scope of their holdings, are the San Diego Frozen Zoo* with over 10 000 samples representing nearly

1000 wildlife species (https://science.sandiegozoo.org/resources/frozen-zoo), the Center for Species Survival's Genome Resource Bank with 108 000 samples from 1500 species (https://nationalzoo.si.edu/center-for-species-survival/cryo-initiative), and the South African National Biodiversity Institute (SANBI) Wildlife Biobank with over 180 000 samples representing 880 different species (Labuschagne 2020; https://www.sanbi.org/biodiversity). The majority of samples in these banks are of tissue and blood, which are much more difficult to contribute to species' replication, yet are incredibly valuable for a wide range of biological studies detailed above. After nearly half a century of research into cryopreservation, the Smithsonian contains 2500 specimens of sperm from a small number of wildlife species ($n = 100$), highlighting the challenges associated with banking large numbers of species within GRBs. These three biobanks have accessions from a wide suite of charismatic vertebrates and would not be considered 'specialised' as they do not have a narrow focus on a particular taxon. One example of a specialised repository of germplasm is the Mississippi State University (MSU) National Amphibian Genome Bank (NAGB). The NAGB was established in 2009 and contains > 1500 samples of sperm (in 0.25 cc straws) from 13 different amphibian species, many of them threatened with extinction (Table 12.1).

The biobank was initially established as a unique partnership between academia (MSU), the USA Association of Zoos and Aquariums (AZA) Amphibian Taxonomic Group (ATAG), the USA Fish and Wildlife Service (USFWS), state wildlife agencies, Amphibian Ark, and the Institute for Museum and Library Services (IMLS). For more than a decade, IMLS has provided generous support to establish a multi-institutional, national leadership program for research on the care and sustainability of collections. The university has laboratory space, specialised equipment, and graduate students available to conduct research on collections, whereas AZA and USFWS have the living collections, curatorial expertise, field biologists, and access to wild-type specimens under federal protection. Hence, the NAGB team has taken advantage of the long-term, established, USA network of biologists with access to well-managed living collections of amphibians in order to collect germplasm representing North America's natural amphibian heritage. Although the NAGB was officially established in 2009, research on cryopreservation began more than 20 years ago at the Cincinnati Zoo and

Table 12.1:   Mississippi State University's National Amphibian Genome Bank (NAGB) inventory of frozen sperm and the production of offspring.

Data current as of 21 July 2021.

Species (common name)	IUCN Red Lst status	Total straws	Produced F1 offspring	Produced F2 offspring
Dusky gopher frog	CR Endangered	194	Yes	Yes
Puerto Rican crested toad	CR Endangered	345	Yes	Yes
Houston toad	Endangered	137	Yes	No
Black spotted newt	Endangered	24	No	No
Chiricahua leopard frog	Vulnerable	232	Yes	No
Southern Rocky Mountain boreal toad	Vulnerable	419	Yes	No
Kweichow newt	Vulnerable	14	No	No
Red-legged salamander	Vulnerable	3	No	No
Fowler's toad	Least Concern	118	Yes	No
Ornate chorus frog	Least Concern	1	No	No
Black-bellied salamander	Least Concern	4	No	No
Yonahlossee salamander	Least Concern	1	No	No
Tiger salamander	Least Concern	21	Yes	Yes
Totals	N = 8 threatened	N = 1530	N = 7	N = 3

Botanical Garden, Ohio, USA. In addition to developing the protocols for cryopreservation, a suite of assisted reproductive technologies (ARTs), such as hormonal therapy for the collection of gametes, ultrasonography, and *in vitro* fertilisation (IVF; also referred to as artificial fertilisation, AF), had to be developed for all 13 species mentioned in Table 12.1. Over the past two decades, nearly 100 peer-reviewed articles have been published by scientists from around the world developing these technologies for amphibian conservation; the previous chapters are excellent reviews showing the evolution of amphibian ART from its early stages. Globally, we are now at a pivotal point, where the research needs to move from predominantly basic science to applied conservation, beginning with saving as much of the amphibian tree of life as possible in GRBs.

The purpose of this chapter is to outline the fundamental ways amphibian GRBs can contribute to new biological discoveries and scientific innovations, conserve biodiversity, genetically manage small populations, and contribute to recovery of declining species. Not covered will be the application of biobanks for de-extinction and the authors refer the reader to other reviews on this subject (e.g. Richmond *et al.* 2016; Shapiro 2017). Throughout the manuscript examples are provided of lessons learned from the past decade of operating the NAGB that will hopefully assist similar initiatives in the future. The chapter also will touch on best practices for collecting, curating, and measuring the impact of these frozen collections on conservation of threatened species.

WHY AMPHIBIAN GRBs ARE NEEDED

Humans' activities, both directly and indirectly, have caused loss or alteration of habitats across much of the planet, leading experts to describe the current era as the 'Anthropocene' or sixth great mass extinction (Lewis and Maslin 2015; Waters *et al.* 2016; Spalding and Hull 2021). Paleontological records show we are experiencing a higher rate of extinction than any previous transition between geological eras (Waters *et al.* 2016), with a recent report to the United Nations showing that a quarter of all species on the planet face extinction within decades (Díaz *et al.* 2019). Unless immediate actions are taken, this catastrophic loss of biodiversity will be irreversible as each geological era took millions of years to recover the same number of species as previously were lost during similar cataclysmic events. Victims of this sixth mass extinction are amphibians, with nearly 200 having become extinct and several hundred more on the brink and in rapid decline (Stuart *et al.* 2004). Of the 8345 amphibian species currently known (AmphibiaWeb 2021), 43% have declining populations and are in immediate need of conservation (IUCN 2021).

During this time of rapid biological change and extinction, GRBs are a particularly valuable way to save a good portion of the amphibian biodiversity about to be lost. More importantly, these GRBs need to focus on preserving sperm, eggs, embryos, and totipotent or somatic cells that can be used to reconstitute/reanimate species while preserving an amount of their genetic heterozygosity sufficient to sustain healthy populations. Although few would argue that the highest priority should be to preserve biodiversity and their associated ecosystem services in nature, humans are failing dramatically in these efforts (Ceballos *et al.* 2020). Nearly 30 years ago, Soulé (1991) recommended a spatial hierarchy design for conserving biodiversity that ranged from whole ecosystems, communities, species, populations, and eventually down to *ex situ* germplasm repositories and seed banks. Very few resources have been placed into this final strategy or safety net for protecting biodiversity and we are now at a critical juncture where if we don't act soon to increase the capacity for GRBs in a range of countries around the world, much of our amphibian heritage will disappear forever. Thus, it is not far-fetched to suggest that cryobanks may be the only, and last, resort for cataloguing and saving threatened amphibian species for potential retrieval in the future.

The value of amphibian GRBs for conservation

The theoretical value of GRBs for preserving wildlife in germplasm repositories has been proposed for decades (Holt *et al.* 1996; Wildt *et al.* 1997; Kouba and Vance 2009; Kouba *et al.* 2013; Wisely *et al.* 2015; Howard *et al.* 2016) with one of its principal drivers being the prevention of genetic erosion over time. In this capacity, the goal of GRBs is simple and direct, effectively lengthening the genetic lifespan of valuable individuals who can continue being part of managed programs even after that individual's death (Holt and Pickard 1999). Although there are few examples of GRBs being actively employed for genetic management and for reintroductions of threatened species, there are two 'proofs of concept' to date for the black-footed ferret, *Mustella nigripes* (Howard *et al.* 2016), and the Puerto Rican crested toad, *Peltophryne lemur* (Burger *et al.* 2021). These two examples utilised sperm held in cryostorage to recover lost genetics from the population by using ARTs (e.g. IVF) to generate offspring and subsequently release them into the wild. Although the word 'repository' is regularly used throughout this article, a repository suggests that the germplasm deposited in it has

a static, dormant role that carries biological samples into some uncertain future. Rather, the NAGB is rapidly expanding the use of banked amphibian germplasm to backcross and breed animals, followed by releases into the wild. Hence, the NAGB is serving as more of an 'investment bank' where the gametes are actively reworked and revived to build new futures. Kouba and Vance (2009) originally proposed 13 benefits for establishing amphibian GRBS, which included addressing reproductive failure, minimising inbreeding, extending generation times, and preventing genetic drift, to name a few (Table 12.2). This updated table now includes a role for amphibian GRBs in de-extinction, along with captive breeding and reintroductions needed for rewilding natural landscapes; this transdisciplinary field also includes replacement of megafauna and other taxa, retrobreeding, abandonment of land, restoration of habitats, and spontaneous rewilding (Carver *et al.* 2021).

GRBs have had an enormous impact on biomedically important research models (e.g. transgenic rats, mice, pigs, zebrafish and non-human primates) by: (1) reducing the cost of resources and of housing for animals; (2) eliminating the need for direct shipment of live animals; (3) preventing catastrophic loss due to natural or man-made disasters; (4) removing the need for quarantine of live animals; (5) decreasing the workload of staff; (6) downgrading threats from genetic drift, instability, or contamination; and (7) extending generation times (Agca 2012; Martínez-Páramo *et al.* 2016). Recently, biomedically important models of aquatic amphibians, such as *Xenopus* spp. and *Axoltl*, have expanded research into cryopreservation and genome banking, as the expenses associated with managing all their transgenic lines has become cost-prohibitive (Buchholz *et al.* 2004; Grainger 2012; Pearl *et al.* 2017; Tilley *et al.* 2021). Theoretically, these same benefits should be realised within amphibian GRBs as they are put into practice. As an example, recent modelling of the economic cost benefits of biobanks for a USA amphibian, *Rana pretiosa* (Howell *et al.* 2021a), and two Australian amphibians, *Geocrinia vitellina* and *Geocrinia alba* (Howell *et al.* 2021b), showed a substantial reduction in expenses and inbreeding coefficients, compared to managing collections of live animals.

The original concept of a specialised amphibian GRB was first proposed in 1994 by the USSR group of Uteshev and Gakhova (1994) and again nearly a decade later (Uteshev and Gakhova 2005). Since then, their team has produced numerous research and review papers on the

Table 12.2: Justification for establishing a genome resource bank for amphibians.

Reason	Justification
Reproductive failure	Incompatible breeding in toads and frogs could be overcome through the use of *in vitro* fertilisation
Increased security	Provides some protection against outbreaks of chytridiomycosis causing local extirpations of amphibian populations
Unlimited space	Due to limited space for live animals, cryobanking offers a large amount of space to conserve diversity
Increased gene flow	Transportation of frozen gametes between zoos has advantages over moving live amphibians
Minimise introgression	Secures the integrity of a gene pool against the threat of hybridisation when introduced to new ponds
Extend generation times	The genetic lifespan of a toad or frog is extended, thereby reducing loss of alleles (genetic drift)
Maximise genetic diversity	Storage of unrepresented founder amphibians, under-represented descendants, and deceased animals
Minimise inbreeding	Restoring germplasm to unrelated or more distantly related amphibians from different breeding ponds
Manage effective population size (Ne)	Equalise family size by manipulating age-specific fertility rates and sex ratios
Minimise selection	Detailed analysis of pedigree combined with GRBs can reduce genetic drift and increase genetic diversity for small amphibian assurance colonies
Mutation	Extending the lengths of generations assists in decreasing the load of harmful mutations in small populations of amphibians
Preservation of cell lines	Transgenic models in *Xenopus* spp. and *Axoltl* for studies of human health
Rewilding	Utilising the genome bank to return wild-type alleles back into the wild as part of reintroduction and restoration programs
De-extinction	Possibility of restoring lost species through nuclear transfer or experiments in parthenogenesis (gynogenesis and androgenesis)
Medical benefits	Discovery of medicinal compounds important to human health and wellness
Studies of disease	Establishment of unique allelic diversity from amphibian species that are resistant to disease

topic of freezing amphibian sperm, which led to the development of sperm cryopreservation protocols for five species housed in their genome bank (Uteshev *et al.* 2019). Currently, there are only a few amphibian GRBs in existence with large numbers of curated frozen sperm for research or genetic management: the USSR amphibian genome bank at the Institute of Cell Biophysics, the Australian amphibian genome bank at the Taronga Zoo and University of Newcastle, and the USA National Amphibian Genome Bank (NAGB) at MSU. However, there are several other institutions beginning to develop cryopreservation protocols for new species and likely have a limited number of sperm samples banked down that ideally will be made publicly available in the future. While the majority of these banks have focused on anurans, the NAGB is actively researching and cryopreserving sperm from caudates as well; to our knowledge no one is working with caecilians. Although there are only a limited number of institutions with established genome banks, there are many more researchers building capacity for

their use by developing the amphibian ARTs that are needed to fully utilise and realise a GRB's potential. The reader is referred to other chapters on ART within the present book, such as hormonal therapy, collection of sperm and eggs, analysis of gametes, ultrasonography, *in vitro* fertilisation, cryopreservation, and nuclear transfer. In addition to the preceding chapters, there are also several excellent reviews that have been published over the past several decades that provide additional information on development of ART for amphibians, with particular emphasis on cryopreservation of gametes (see Kouba and Vance 2009; Kouba *et al.* 2013; Browne *et al.* 2011; Clulow and Clulow 2016; Zimkus *et al.* 2018; Uteshev *et al.* 2019; Della Togna *et al.* 2020; and Strand *et al.* 2020). Zimkus *et al.* (2018) highlighted several species that have either shown viable sperm following cryopreservation or, on a more limited scale, produced offspring. Unfortunately, fewer than half of the species reviewed by Zimkus *et al.* (2018) are actually available for genetic management in GRBs; the species were primarily part of research projects developing protocols for cryopreservation. Currently, amphibian GRBs focus primarily on creating sperm banks as the technology for the cryopreservation of oocytes has not been developed yet. However, there are promising protocols being developed for the cryopreservation of oocytes and embryos in commercial and model aquatic organisms that may have application in the future to amphibians (reviewed in Chapter 10).

Conservation milestones: the NAGB's germplasm inventory

Table 12.1 shows the current holdings within the NAGB and some of the outputs generated by its work to date. In pursuit of developing the NAGB, a substantial amount of knowledge was collected both on anuran and caudate reproductive physiology, thereby leading to several milestones in conservation. For example, in order to collect gametes for cryopreservation it was necessary to learn as much as possible about the best practices for hormonally inducing spermiation and ovulation in targeted species. These standardised hormone protocols were subsequently incorporated into species' studbooks beginning as early as 1999 with the Wyoming toad, *Anaxyrus baxteri*, and soon after for the Puerto Rican crested toad, *Peltophryne lemur*. These scientific discoveries made by the team at the Cincinnati Zoo, Memphis Zoo, and MSU on amphibian reproductive physiology and hormonal therapy have led to the collective reintroduction of over one million fertilised eggs, tadpoles, metamorphs, and adult anurans belonging to several different North American species, including *Lithobates sevosus*, *Anaxyrus houstonensis*, *Anaxyrus baxteri*, *Peltophryne lemur*, *Anaxyrus boreas boreas*, and *Lithobates chiricahuensis*. A few success stories from the NAGB team's work include: (1) freezing sperm from wild male anurans and using this germplasm to link *in situ* and *ex situ* populations through the production of mixed wild/captive offspring (Chapter 11); (2) the production of viable F1 generation offspring utilising frozen–thawed sperm from seven of the species shown in Table 12.1 and subsequent production of F2 generations in three of the species generated by these cryotechnologies and ART (Vance *et al.* 2018); (3) the production of an adult *Peltophryne lemur* toad utilising sperm from a deceased parent; and (4) the release of more than 5000 tadpoles from a male *Peltophryne lemur* produced from frozen–thawed sperm and ART. These successes in breeding and reintroduction, using frozen–thawed sperm in cryostorage, highlight just a few of the applications and value

beginning to be realised from actively incorporating this technology into AZA captive breeding programs.

DESIGNING AN AMPHIBIAN ACTION PLAN FOR CRYOBANKING

The idea of zoos and aquaria serving as Arks for preserving all threatened wildlife is not a realistic solution in the 21st century as these organisations simply do not have enough space or resources to address all species in need of *ex situ* conservation. For example, a survey of space in AZA institutions conducted in 2019 by its Amphibian Taxonomic Group identified a total of 104 bio-secure rooms across all their institutions, which collectively manage only five species-survival programs. However, if several of these same zoological institutions combined the management of live animals with genome banks, space and resources could be freed up to manage a larger suite of at-risk-species in need of *ex situ* management. The Captive Breeding Specialist Group (CBSG) of the IUCN has provided dozens of workshops on the value of incorporating GRBs into captive-breeding and recovery programs and have developed specific guidelines on how to draft an 'action plan' for each new species targeted for biobanking (CBSG 1992; Wildt *et al.* 1997; Wildt 2000). We recommend using these guidelines for drafting amphibian-specific action plans to help guide the selection of species. Drafting an amphibian action plan for targeted species is beneficial for justifying the banked germplasm based on the following parameters: (1) *in situ* and *ex situ* status; (2) knowledge of life history and natural reproduction; (3) numbers of animals available for use in the wild and in captivity; (4) accessibility of the amphibian species for collection; (5) type and amount of samples to collect for genetic management; and (6) various governance issues, described in greater detail below. Upon the collection of samples, I strongly recommend that amphibian GRBs be divided into three repository categories: an 'In-Perpetuity Bank', an 'Active Management Bank', and a 'Research Bank'. When inputting data from voucher specimens, these designations can be entered into the database and separated via different LN2 dewars. The NAGB typically places samples in the 'In-Perpetuity Bank' last, depending on whether enough sperm was collected, as the project is more actively engaged with functional management and research.

Once an amphibian has been chosen as a priority for gene banking and an initial action plan has been developed, choosing how many males to use or how much sperm to bank from each individual can be challenging. Early seminal work by Johnston and Lacy (1995) evaluated different scenarios for selecting males whose genes should be banked. They found that although maximal allelic retention was achieved by banking 'all males' in the population, nearly optimal retention of genetic diversity could be achieved by banking males with minimal values of 'mean kinship' (Johnston and Lacy 1995). From their study, they proposed that an ideal GRB would contain biomaterials from as few individuals as possible, capture the majority of all of the allelic diversity within a population, capture as much genetic diversity as possible, utilise this banked sperm for future genetic management, and bank germplasm from easily identified individuals to reduce duplication. Follow-up research by Harnal et al. (2002) utilised computer simulations and analysis of pedigrees to determine the efficacy of different strategies for banking sperm. They found that 'wild banks' of at least 10 individual captive males, along with regular utilisation of the banked material for genetic management, were the most efficient at maintaining genetic diversity over the lifespan of the program. Moreover, they noted that an interactive bank updated every few years with wild genes can prove very beneficial, depending on the species' mean kinship (Harnal et al. 2002). Using these recommendations, the NAGB works closely with studbook keepers for each species to minimise mean kinship for all males banked from the captive population. Furthermore, the NAGB partners with federal and state agencies with a plan to bank a minimum of 10 wild males from each fragmented population every four years. The advantages and benefits of linking these *ex situ* and *in situ* populations are described in Chapter 11.

PRIORITISATION OF SPECIES WITHIN THE NAGB

Realising that banking every globally threatened amphibian is not realistic given the geographical challenges and limitation of resources, we designed a program meant to attain modest and pragmatic goals. The initial focus of the USA NAGB was centred on North American species in partnership with AZA and the USFWS. To help direct the GRB efforts, the partnership focused on addressing biobanking priorities within the AZA Regional Collection Plan (RCP). The RCP divided amphibian species of concern into three categories, with priority-1 species having the greatest need for immediate biobanking (8/11 species; Table 12.3). Focusing on national species of concern has several advantages, with the most prominent being that federal and state financial resources typically are allocated for country-level conservation, rather than on out-of-country species. Given the large financial investment in operating a GRB, we strongly recommend that institutions establishing germplasm repositories concentrate their efforts on national at-risk species in order to secure resources for their long-term management. Although precedence is given to banking the RCP priority-1 species identified by the zoo and aquarium industry, as members of the NAGB visit zoological institutions to work with their collections we

Table 12.3: Association of Zoos and Aquariums priority-1 species from the Regional Collection Plan.

Common name	Scientific name	IUNC status[1]	A[2]	B[3]	C[4]	D[5]	E[6]	F[7]
Reticulated flatwoods salamander	*Ambystoma bishopi*	VU	✓		✓	✓		
Flatwoods salamander	*Ambystoma cingulatum*	VU	✓		✓	✓		
Wyoming toad	*Anaxyrus baxteri*	EW	✓		✓		✓	✓
Houston toad	*Anaxyrus houstonensis*	EN	✓	✓	✓		✓	✓
Panamanian golden frog	*Atelopus zeteki* and *A. varius*	CR	✓	✓	✓	✓	✓	✓
Ozark hellbender	*Cryptobranchus bishop*	NL	✓	✓	✓	✓	✓	
Cricket coqui	*Eleutherodactylus gryllus*	EN	✓		✓		✓	✓
Upland coqui	*Eleutherodactylus portoricensis*	EN	✓		✓		✓	✓
Richmond's coqui	*Eleutherodactylus richmondi*	CR	✓		✓	✓	✓	✓
Dusky gopher frog	*Lithobates sevosus*	CR	✓		✓		✓	✓
Puerto Rican crested toad	*Peltophryne lemur*	CR	✓		✓	✓	✓	✓

[1] VU: Vulnerable; EW: Extinct in the Wild; EN: Endangered; CR: Critically Endangered; NL: not listed; TH: Threatened. [2] Ark/rescue role (A). [3] *In situ* conservation role (B). [4] *In situ* research role (C). [5] *Ex situ* research role (D). [6] Conservation education role (E). [7] Biobanking role (F).

opportunistically bank down germplasm from other native and non-native threatened species. The NAGB's prioritisation schema and RCP list are a refinement of the Amphibian ARK prioritisation tool (Johnson *et al.* 2020), which also incorporates AZA-specific conservation actions and predominantly addresses species held in collections.

Chapter 2 of the present book provides an excellent summary of how to prioritise conservation actions, including *ex situ* breeding, and highlights the many applications of the Conservation Needs Assessment tool produced by the Amphibian Ark (https://conservation needs.org). Within the Conservation Needs Assessment tool, conservation actions are summarised for each species, with six species from the United States listed as high priority for biobanking: *Lithobates sevosus, Anaxyrus houstonensis, Anaxyrus baxteri, Ambystoma cingulatum, Ambystoma bishopi*, and *Ambystoma califoriniense*.

However, expanding this assessment tool for the United States with a focus on threatened IUCN categories of amphibians (Extinct in the Wild, Critically Endangered, Endangered, and Vulnerable), 81 species of anurans and caudates should be classified as priority-1 targets for biobanking germplasm. The number of priority-1 USA species needing immediate acquisition in GRBs represents ~25% (81/326 species) of all amphibians in the USA with a greater number of caudates needing a GRB safety net than do anurans (82% [66/81] versus 18% [15/81], respectively). As the NAGB completes acquisition of the smaller species list within the AZA-RCP priority group, it is likely to begin to include additional animals identified across the larger platform within the Conservation Needs Assessment tool.

Figure 12.1 shows a series of categories and situations in which collection of samples of amphibian germplasm should be prioritised to provide a basic overview/

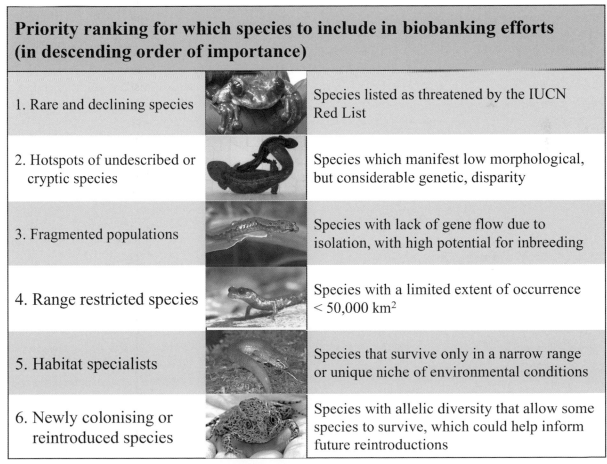

Priority ranking for which species to include in biobanking efforts (in descending order of importance)

1. Rare and declining species		Species listed as threatened by the IUCN Red List
2. Hotspots of undescribed or cryptic species		Species which manifest low morphological, but considerable genetic, disparity
3. Fragmented populations		Species with lack of gene flow due to isolation, with high potential for inbreeding
4. Range restricted species		Species with a limited extent of occurrence < 50,000 km²
5. Habitat specialists		Species that survive only in a narrow range or unique niche of environmental conditions
6. Newly colonising or reintroduced species		Species with allelic diversity that allow some species to survive, which could help inform future reintroductions

Figure 12.1: Priority ranking for which species to include in biobanking efforts (in descending order of importance). Photo-credits: Dustin Smith and Andy Kouba.

justification for each classification. These recommendations are a simplified roadmap and the sampling designs for each will depend ultimately on the goals related to specific research questions, the amphibians targeted, access to animals, permits, availability of collectors, and costs associated with acquisitions. The fundamentals for this prioritization are based on several published works for benchmarking genetic diversity within specific taxa (García and Robinson 2021) and have been modified by the present author for amphibians with an emphasis on: (1) rare and declining species; (2) undescribed or cryptic species; (3) fragmented populations; (4) range-restricted species; (5) habitat specialists; and (6) newly colonising or reintroduced amphibians (Figure 12.1). Each country or individual institution may have its own set of criteria for how to prioritise species within their GRB. For example, Mahony and Clulow (2011) recommended prioritising threatened species first and utilised a three-tiered system for determining the acquisition of Australian species into a GRB, including national or state-listed species of concern for specific ecoregions, IUCN conservation categories from the global amphibian assessment (https://www.globalamphibians.org), and the FrogsAustralia project (https://www.frogs.wwf.org.au), based on a broader set of biological variables.

BEST PRACTICES FOR GOVERNANCE, MANAGEMENT, SUSTAINABILITY, AND DATA-SHARING

When establishing an amphibian GRB there are several important considerations relative to 'best practices' that can impact the success and usefulness of the product. The lessons learned by the NAGB and its partners over the past decade regarding GRB governance have strengthened its value as a legitimate contributor towards the conservation of native amphibians among zoological, state, and federal agencies. An excellent resource recommended for review during the early stages of designing an amphibian GRB is the document by the International Society for Biological and Environmental Repositories (ISBER) on best practice for repositories. This ISBER document can be found online at https://www.isber.org/page/BPR and provides numerous resources on standardised methods for collection, handling, storage, retrieval, and distribution of biological specimens that can be modified for germplasm repositories; a thorough summary on new topics, updates, and expanded areas can be

found in Campbell *et al.* (2018). Next we discuss some of these governance, regulatory, data-sharing, and sustainability issues relative to managing an amphibian GRB.

Governance

The overarching purpose of having a governance structure in place is to ultimately provide good stewardship of the collection. High-quality governance entails addressing the practical, legal, and ethical issues surrounding the use of banked gametes for replicating threatened species, which may involve self-regulation, oversight by an advisory board, or state/federal legal parameters. Governance oversight of the NAGB currently is under the direction of MSU as the primary designee of the bank, with an advisory committee made up of AZA and USFWS partners. The group currently is writing the governance framework of the NAGB that will include a description of the governing body and advisory committee's composition and responsibilities. The procedures and policies that come out of this charter ultimately will guide processes related to the entire life cycle of the amphibian germplasm from collection, disposition, utilisation, and access by AZA institutions or USFWS. Once the governing body and responsibilities are established, and a business plan written, the NAGB document will be made publicly available for all stakeholders. Aspects of this charter will include: (1) maintenance, security, and integrity of the sperm samples; (2) paperwork signifying agreement about the deposition of material and access to it; (3) open-access website for sharing information with stakeholders; (4) a plan for prioritising specimens; (5) a business plan to include financial management; and (6) a plan for physical and electronic transfer of the GRB should constraints on resources, lapse in funding, retirements, withdrawal of donors, dissolution of the governing authority, or unanticipated disasters occur, such that the resource needs to be moved from MSU to another institution. For security purposes, the NAGB utilises both electronic and battery-powered low-temperature alarms for each tank; these are based on the volume of LN2 kept in the dewars. Moreover, there is a dedicated laboratory supervisor who monitors the biobank's security on a near-daily basis and who is responsible for management of its maintenance and inventory. Should any of the above scenarios described in number 6 occur, several of the AZA collaborators maintaining other frozen collections in zoos have agreed to take over responsibility for managing the NAGB. Ultimately, the NAGB repository should be split in two for security purposes, with a satellite storage facility at either a natural history museum or zoo (e.g. San Diego Zoo, Cincinnati Zoo, Omaha Henry Doorly Zoo, or Smithsonian National Zoo) that has the infrastructure and personnel in place to manage a dedicated collection of frozen germplasm.

Regulatory and legal matters and the issuing of permits

On a broader scale, international treaties such as the Cartagena Protocol on Biosafety or the Nagoya Protocol on Access and Benefit Sharing (https://www.cbd.int/abs) may impact the future availability of amphibian voucher specimens within some countries. For example, the Nagoya protocol mandates that the benefits from utilising genetic resources, including for non-commercial conservation research, be returned to the country or community of origin and calls upon each country to enact national legislation. However, many countries have yet to ratify or join these international treaties; thus, most regulatory activities related to the acquisition of germplasm falls to in-country federal or state agencies, especially when dealing with listed amphibians. Regulatory guidelines for issuing permits to collect specimens are different for each country, and sometimes within a country. Types of regulatory documents needed may include those for collecting in protected areas, collecting or working with federally listed or state-listed species, export and import permits for moving biomaterials between countries or across state/provincial boundaries, veterinary/disease certificates, or internal Institutional Animal Care and Use (IACUC) approvals for the ethical treatment of targeted amphibians. These documents constitute important relational database information and should be associated with the gametes when they enter the collection, or become subject to research, as the regulatory compliance issues will persist with long-term storage. Many of these permits take months to acquire, require annual or semi-annual reports on the use of specimens, and demand a significant amount of time and resources by the repository's personnel. In the USA, federal permits are required in order to work with federally listed species, whereas non-federally listed species may or may not require a state permit. Fortunately, as part of the partnership with AZA, the collections work by the NAGB has been added as an amendment to existing regulatory permits obtained by the collective zoological partners for the RCP priority-1 species.

For legal protection, we highly recommend having a 'Biomaterials Transfer Agreement (BTA)' in place with collaborators, state, or federal agencies in order to define the rights of the provider and the recipient with respect to the disposition of amphibian germplasm and any of its derivatives (such as offspring produced). Ownership and disposition of the germplasm and any resulting offspring (tadpoles, metamorphs, or adults) are usually the largest point of contention these BTAs attempt to mitigate. Typically, for incoming amphibian germplasm, the zoo or USFWS sending out the material will require the use of its own BTA form if they have one. In the USA, most zoos have moved towards a standardised BTA agreement that the NAGB completes and signs before working with their animals and before depositing voucher specimens. An incoming amphibian BTA protects a researcher's ability to conduct and publish research or generate intellectual property, and provides for the protection of confidential information. Utilising an incoming BTA also ensures that the terms of the agreement do not conflict with rights granted by other agreements associated with the research on amphibian germplasm. Biomaterials Transfer Agreements for outgoing gametes (e.g. NAGB sending frozen sperm to another AZA institution) typically prevent the provider of the material from losing control over it and its use in research. If no agreement exists, then the recipient of the germplasm has no legal restrictions on the use of the material, or on transferring it. As an example of a generic Material Transfer Agreement for genomic material, which could be modified for their own institution, the reader is recommended to visit the Global Genome Biodiversity Network (https://www.ggbn.org).

Curation and digitisation of data in amphibian genomic resource bank collections

Similar to any other biological collection, the value of a repository of germplasm is only as good as the information put into it during the accession phase. The input and accessibility of data in the amphibian GRB databases for scientific inquiry, genetic management, and recovery of species is the most critical component for use in establishing these resources for the conservation community. As advanced molecular-biology technologies (e.g. gene drives such as CRISPR-Cas9) and new software programs that utilise machine-learning augment the preserved biomaterials with a wealth of digitised metadata on genotypes and phenotypes, collections will become even more valuable for solving complex conservation problems resulting from amphibian declines. Webster (2017) best described this as the 'extended specimen' concept, whereby the species' aggregated data can be combined to form an information-rich network for exploring Earth's amphibian biota across taxonomic, temporal, and spatial scales (Thiers *et al.* 2019). Conceptually, this is more manageable now, as only a few amphibian GRBs are in existence globally and are still in their infancy. Establishing amphibian GRBs in the present allows for accessions of specimens to be 'born' digitally with GPS locations, images, digital spreadsheets, genomic sequences, and other relevant information, whereas older samples often need to be reloaded from field/laboratory notebooks; such tasks require substantial expenditure of time and resources.

There are several coordinated worldwide efforts to digitise biological collections that could serve as amphibian GRB species-specific data-aggregators, including: (1) the USA Integrated Digitized Biocollections (iDigBio; https://www.idigbio.org); (2) the European Distributed System of Scientific Collections (DiSSCo; https://www.dissco.eu); (3) the Global Biodiversity Information Facility (GBIF; https://www.gbif.org); and (4) the Atlas of Living Australia (ALA; https://www.ala.org.au). These aggregators provide the cyberinfrastructure that gathers and compiles data from individual collections; as new amphibian GRBs become established they should establish the framework to utilise one of these platforms to catalogue their acquisitions. Although the NAGB is not yet associated with any of these platforms, it is hoped that it will join iDigBio by 2022 as the digitised library is prepared for online access. While the two platforms mentioned above are incredibly valuable for linking a global research network, individual institutions often still need a specific software database, or local portal, that manages their collections internally. Several of these software programs exist around the world for purchase, or could be developed in-house using something like Microsoft Access. Currently, the NAGB utilises the program 'Freezerworks' (https://freezerworks.com), which has input fields specifically related to reproductive cells and tissues. This software works very well for organising inventories and managing semen straws; the downside is the very expensive upfront cost and the annual licence fees. Such large financial expenses can jeopardise long-term infrastructure of GRBs and is cost-prohibitive for smaller start-ups. If long-term cost is prohibitive for using programs like 'Freezerworks', then open-source databases

like 'Specify' (https://www.specifysoftware.org) are freely available for digitising collections. While many of the open-source, cloud-based databases are designed for natural history museums, and not germplasm repositories *per se*, they boast of highly configurable frameworks that collectors could modify to meet their needs. The migration to cloud-based IT infrastructures from traditional methods will only grow over time and offer many advantages, including enhanced security, dynamic allocation of services, prevention of disasters, flexible costs, scalability, quick deployment, and easily shared datasets in real time (Paul *et al.* 2017). A standard type of workflow for collecting and digitising data on amphibian germplasm can be seen in Figure 12.2.

Sustainability

The topic of sustainability is receiving its own section to highlight the importance of this subject for the planning of repositories, and will include additional information beyond the aforementioned good governance, regulatory issues, and curation of data that are drivers for a long-term successful collection. In addition to these three organizational standards, several other key points related to sustainability are worth mentioning, including: (1) financial management; (2) personnel; and (3) the infrastructure of the facility. First, the financial cost of collecting specimens (e.g. hormones, disposables, equipment, travel), processing them (cryopreservation, data input, and management), quality control (curation of data, LN2 expenses, security protocols), distribution (personnels' time and specialised shipping costs for frozen straws), and GRB staff (director(s), technicians, or graduate students) is considerable, so it is important to establish a financial plan early in the design phase. Even though operating a GRB is expensive the benefits of maintaining collections in stasis versus live collections are substantial. Howell *et al.* (2021a) recently modelled the cost and genetic benefits achieved by including biobanking in the captive breeding program of the Oregon spotted frog (*Rana pretiosa*) and showed an overall 26-fold reduction in costs compared to maintaining live animals over the duration of the program. Locating dedicated sources of funding, in addition to external granting agencies, to keep the GRB in perpetuity is one of the highest priorities for newly established programs.

Second, dedicated personnel are critical for maintenance of a GRB to ensure it stays within budget, has adequate funding, serves as liaison with partner

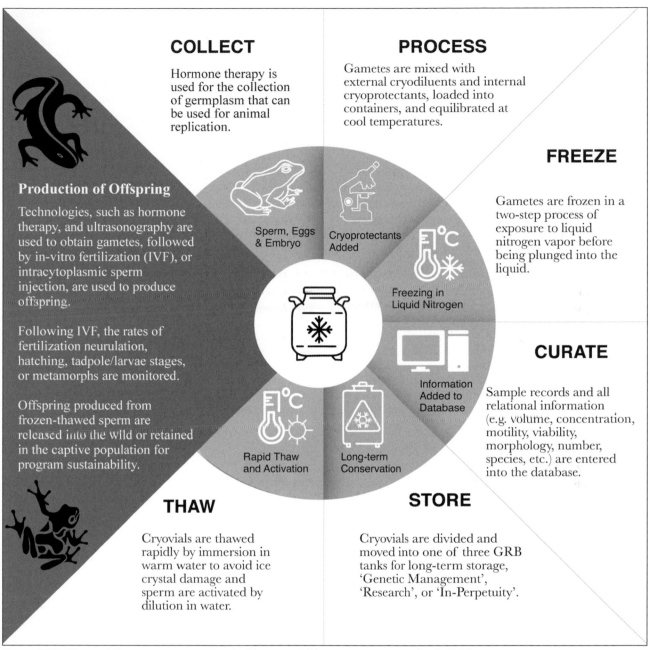

COLLECT

Hormone therapy is used for the collection of germplasm that can be used for animal replication.

PROCESS

Gametes are mixed with external cryodiluents and internal cryoprotectants, loaded into containers, and equilibrated at cool temperatures.

FREEZE

Gametes are frozen in a two-step process of exposure to liquid nitrogen vapor before being plunged into the liquid.

Production of Offspring

Technologies, such as hormone therapy, and ultrasonography are used to obtain gametes, followed by in-vitro fertilization (IVF), or intracytoplasmic sperm injection, are used to produce offspring.

Following IVF, the rates of fertilization neurulation, hatching, tadpole/larvae stages, or metamorphs are monitored.

Offspring produced from frozen-thawed sperm are released into the wild or retained in the captive population for program sustainability.

Sperm, Eggs & Embryo

Cryoprotectants Added

Freezing in Liquid Nitrogen

CURATE

Information Added to Database

Sample records and all relational information (e.g. volume, concentration, motility, viability, morphology, number, species, etc.) are entered into the database.

Rapid Thaw and Activation

Long-term Conservation

THAW

Cryovials are thawed rapidly by immersion in warm water to avoid ice crystal damage and sperm are activated by dilution in water.

STORE

Cryovials are divided and moved into one of three GRB tanks for long-term storage, 'Genetic Management', 'Research', or 'In-Perpetuity'.

Figure 12.2: The workflow processes associated with an amphibian genome resource bank (GRB).

institutions, meets all regulatory guidelines and reporting requirements, collects and cryopreserves gametes, conducts research on ART for the utilisation of specimens, and follows all standard operating procedures and best practices for the curation and the management of data. Finally, the infrastructure of the facility is critical for long-term sustainability and should take into consideration the availability of resources and the facility's environmental conditions. Good facilities for long-term storage of an amphibian GRB should have access to laboratories, research equipment, and personnel with expertise in reproductive physiology in order to collect, process, and utilise samples from the GRB. Environmental considerations within appropriate facilities would include: (1) heating, ventilation, and air conditioning; (2) lighting and electricity; (3) back-up power-generators; (4) restricted access and security; (5) fire-prevention systems; (6) protocols for dealing with emergencies; and (7) control of pests, disease, and contamination. University-based GRBs, like the NAGB, are uniquely staged to provide for long-term sustainability of biobanks as they meet many of the above criteria, whereas zoos, aquaria, and museums often are challenged by limited resources, do not receive taxpayer income and can be subjected to downsizing during economic hardships, as recently revealed by the COVID-19 pandemic. In contrast, public universities weather these types of hardships better and provide a more secure environment with greater access to research support, equipment, expertise, holding space, and security. Moreover, universities enjoy a high degree of public trust due to their institutionalised observance of research integrity, transparency, and open access to data. Thus, I strongly encourage more collaborations between academia, non-profit organisations (e.g. zoos and aquaria), and federal agencies that have yet to establish amphibian GRBs in-country to do so.

FUTURE DIRECTIONS, CHALLENGES AND HIGH PRIORITIES FOR AMPHIBIAN GRBS

Perhaps the most effective approach to establishing additional amphibian GRBs in the future is to take advantage of existing infrastructure, governance, expertise, and systems of management already in existence within museums, zoos, aquaria, and universities. These types of multi-institutional genome banking partnerships can be highly effective conservation groups, as observed through the establishment of the NAGB and a shared vision. That shared vision occurred because all partner organisations were working under the AZA umbrella of intensive management of species, within the amphibian taxon advisory group and species' survival plans. This intensive management ensures that the NAGB is an active, dynamic investment bank, rather than a static, dormant 'gene morgue' with no future or end goal. To finish, I recommend some future priorities, challenges, and opportunities for expanding the application of GRBs to the conservation of amphibians.

Priority 1: Establish more amphibian genome resource banks

Amphibian GRBs are all in their infancy and collectively hold frozen sperm for genetic management from no more than 20–25 species between them. On the conservative side, this represents ~0.06% of the amphibian species in global decline. More amphibian gene banks are in urgent need of being established if there is any hope of securing a future for amphibian biodiversity.

Priority 2: Develop a succession plan

The three largest germplasm GRBs currently in existence were all created as special projects by research scientists who are in their sunset years and close to retirement in the next decade or sooner. Creating a succession plan for the next generation to steward these incredibly valuable resources will be paramount to see that their value towards amphibian conservation is carried forward. A succession plan may also include where the collections could be moved, should the existing oversight institution no longer wish to manage the resource, given retirements and a change in mission.

Priority 3: Create more partnerships

The growth and sustainability of amphibian GRBs is absolutely dependent upon effective partnerships that respect and trust each other. Without buy-in from various stakeholders who will benefit from the GRB, they are short-lived and doomed to fail. Keeping these relationships going is time-consuming and requires significant investment on the host of the gene bank.

Priority 4: Establish linkages with other repositories

Every GRB should have a backup facility to curate duplicate samples as a hedge against catastrophic disasters. We recommend linking with other museums, zoos, or universities with existing repositories to create a second 'In Perpetuity Bank'. This way the host bank maintains control of the 'Genetic Management Bank' and the 'Research Bank'. The advantage to this strategy is that it addresses priorities 2 and 3 above.

Priority 5: Train the next generation

The accumulated knowledge of those working in cryobiology needs to be passed along to the next generation before institutional memory is lost. This type of 'research amnesia' can have serious consequences for the sustainability of GRBs. The future growth and application of amphibian GRBs will depend upon having a pipeline of enthusiastic new scientists ready to carry on their mission. The training of graduate students and post-doctoral fellows at universities can play an important role in establishing this pipeline.

Priority 6: Continue to develop the ART protocols that will be crucial for utilising a GRB

Although significant progress has been made in developing ART protocols for a limited number of species, very little is known about most amphibians' reproductive biology. Many of the protocols developed to date are species-specific or genera-specific with limited transferability to other taxonomic groups. Thus, research is needed to develop a wide ranging suite of biotechnologies that can be deployed when reviving germplasm from the repositories for the production of offspring. Genera-wide protocols will see the greatest future use for reintroducing or rewilding amphibians due to cost.

Challenges

If asked, most investigators would say that establishing and operating a successful GRB comes down to two things: money and resources. Those who have been successful at maintaining long-term programs have a responsibility to establish new programs and train the next generation on how to access private donors and external funding and to build partnerships that can provide resources (e.g. space and equipment). Establishing a three-tier collaborative system of public taxpayer-supported universities, zoos, and federal agencies is likely to provide the best chance for success. As detailed above, the number of amphibian species in need of biobanking is substantial, yet the species occupying space in existing global GRBs is less than 1% of the species in decline. Among the various reasons for these low numbers of species in GRBs is access to the animals. Collectively, zoos and aquaria maintain a relatively minor number of amphibians in their collection compared to the number of IUCN threatened species, with most of them in small numbers (e.g. two to four individuals) and primarily used for display. Moreover, for research projects like biobanking to be pushed through at many zoos, an internal champion is needed to coordinate agreements for the transfer of biological specimens, permits, IACUC approval, travel arrangements, access to animals, and keepers' time. With limited staff in many zoos, this drain on resources and time can seem substantial. For GRBs to be successful, they need to be raised to a higher priority by the World Association of Zoos and Aquariums (WAZA) in order to divert limited resources to saving as much biodiversity as possible by utilising a sustainable solution. Access to threatened wild amphibians can also be particularly challenging. In the USA, many of the threatened species targeted for a GRB are protected under the Endangered Species Act and activities outlined in their recovery plans rarely include biobanking as one of their rescue strategies. Because of legal ramifications if the federal agencies deviate from their specified recovery plan, getting buy-in from federal agencies is time-consuming and requires substantial regulatory paperwork. Similar to zoos, having a champion field biologist who is working with a species' recovery is one of the greatest assets when establishing a wild bank and is an essential part of a partnership team.

Opportunities

The single greatest opportunity for growth and development of existing and future amphibian GRBs centres on *education*, *extension*, and *outreach*. It is imperative that those currently in the field work together as a cohesive group to develop a step-by-step outreach plan to communicate the need and justification for GRBs with the public for mutual benefit. The goal of this plan should be to increase the knowledge and foster attitudes and behaviours of those in the public domain that garner support for GRBs; such an initiative could lead to greater resources and expansion of this tool for conservation. Similar to any education plan, the collective group needs to: (1) define the goals and objectives of the outreach; (2) identify the target audience to be reached; (3) design an attention-grabbing, specific message tied directly to what the loss of amphibians would mean to their personal health and wellness; (4) package and distribute the biobanking message through various social media, videos, and print platforms; and finally (5) evaluate the message's effectiveness over time and then adapt as necessary. The importance of this opportunity for outreach was discussed recently by the present author at a symposium called 'Reproductive Technologies and Biobanking for Amphibian Conservation' held at the 9th World Congress of Herpetology in New Zealand in 2020. If there was one overarching theme from this symposium it was that scientists working in the discipline of ART and GRB for amphibian conservation are serving as nature's security detail, first responders, and the last line of defence preventing imminent extinctions. As the final words in the last chapter, the message I would share with anyone not already in this field is: please be the next generation of biobankers, we need your passion!

ACKNOWLEDGEMENTS

First, I would like to thank all the graduate students, both past and present, that have collectively worked on building the resources and depositing the voucher specimens that support this work. Moreover, I would like to acknowledge the long-term partnership with Carrie Vance and Scott Willard who helped establish the NAGB and continue to support its growth. The collective work from the NAGB was supported by the Institute of Museum and Library Services (IMLS) National Leadership Grants (#LG-25-09-0064-09, #LG-11-0186-11 and #MG-30-17-0052-17) for sustaining collections of threatened amphibians; the Association of Zoos and Aquariums and Disney Conservation Grant Funds (#CGF-16-1396 and #CGF-19-1618); and the Morris Animal Foundation (#D14ZO-403). In addition, this manuscript is a contribution of the Mississippi Agricultural and Forestry Experiment Station USDA-ARS 58-6402-018 and the material is based upon work supported by the National Institute of Food and Agriculture, USA Department of Agriculture, Hatch project under accession number W3173. In addition, work was partially supported by the USDA-ARS Biophotonics Initiative grant #58-6402-3-018.

REFERENCES

Agca Y (2012) Genome resource banking of biomedically important laboratory animals. *Theriogenology* **78**, 1653–1665. doi:10.1016/j.theriogenology.2012.08.012

AmphibiaWeb (2021) <https://amphibiaweb.org>. University of California, Berkeley, CA, USA. Accessed 24 Jun 2021.

Browne RK, Li H, Robertson H, Uteshev VK, Shishova NV, McGinnity D, Nofs S, Figiel CR, Mansour N, Lloyd RE, Agnew D (2011) Reptile and amphibian conservation through gene banking and other reproduction technologies. *Russian Journal of Herpetology* **18**, 165–174.

Buchholz DR, Liezhen F, Yun-Bo S (2004) Cryopreservation of *Xenopus* transgenic lines. *Molecular Reproduction and Development: Incorporating Gamete Research* **67**, 65–69. doi:10.1002/mrd.20005

Burger I, Julien AR, Kouba AJ, Counsell KR, Barber D, Pacheco C, Kouba CK (2021) Linking in situ and ex situ populations of the endangered Puerto Rican crested toad (*Peltophryne lemur*) through genome banking. *Conservation Science and Practice* **3**(11), e525. doi:10.1111/csp2.525

Campbell LD, Astrin JJ, DeSouza Y, Giri J, Patel AA, Rawley-Payne M, Rush A, Sieffert N (2018) The 2018 revision of the ISBER best practices: summary of changes and the editorial team's development process. *Biopreservation and Biobanking* **16**, 3–6. doi:10.1089/bio.2018.0001

Carver S, Convery I, Hawkins S, Beyers R, Eagle A, Kun Z, Van Maanen E, Cao Y, Fsher M, Edwards SR, Nelson C, *et al.* (2021) Guiding principles for rewilding. *Conservation Biology* **35**, 1882–1893. doi:10.1111/cobi.13730

CBSG (Conservation Breeding Specialist Group) (1992) Guidelines for preparing a genome resource bank for wildlife conservation. *CBSG News* **3**, 29–31, <http://www.cbsg.org/sites/cbsg.org/files/Newsletters/CBSG_News.vol3.no3_Nov92.pdf>.

Ceballos G, Ehrlich PR, Raven PH (2020) Vertebrates on the brink as indicators of biological annihilation and the sixth mass extinction. *Proceedings of the National Academy of Sciences of the United States of America* **117**, 13596–13602. doi:10.1073/pnas.1922686117

Clulow J, Clulow S (2016) Cryopreservation and other assisted reproductive technologies for the conservation of threatened amphibians and reptiles: bringing the ARTs up to speed. *Reproduction, Fertility and Development* **28**, 1116–1132. doi:10.1071/RD15466

Della Togna G, Howell LG, Clulow J, Langhorne CJ, Marcec-Greaves R, Calatayud NE (2020) Evaluating amphibian biobanking and reproduction for captive breeding programs according to the Amphibian Conservation Action Plan objectives. *Theriogenology* **150**, 412–431. doi:10.1016/j.theriogenology.2020.02.024

Díaz SM, Settele J, Brondízio E, Ngo H, Guèze M, Agard J, Arneth A, Balvanera P, Brauman K, Butchart S, Chan K, *et al.* (2019) Summary for policymakers of the global assessment report on biodiversity and ecosystem services of the Intergovernmental Science-Policy Platform on Biodiversity and Ecosystem Services. IPBES Secretariat, Bonn, Germany. doi:10.5281/zenodo.3553579

García NC, Robinson WD (2021) Current and forthcoming approaches for benchmarking genetic and genomic diversity. *Frontiers in Ecology and Evolution* **9**, 622603. doi:10.3389/fevo.2021.622603

Grainger RM (2012) *Xenopus tropicalis* as a model organism for genetics and genomics: past, present and future. *Methods in Molecular Biology (Clifton, N.J.)* **917**, 3–15. doi:10.1007/978-1-61779-992-1_1

Harnal VK, Wildt DE, Bird DM, Monfort SL, Ballou JD (2002) Computer simulations to determine the efficacy of different genome resource banking strategies for maintaining genetic diversity. *Cryobiology* **44**, 122–131. doi:10.1016/S0011-2240(02)00013-5

Hedrick BP, Heberling JM, Meineke EK, Turner KG, Grassa CJ, Park DS, Kennedy J, Clarke JA, Cook JA, Blackburn DC, Edwards SV (2020) Digitization and the future of natural history collections. *Bioscience* **70**(3), 243–251. doi:10.1093/biosci/biz163

Holt WV, Pickard AR (1999) Role of reproductive technologies and genetic resource banks in animal conservation. *Reviews of Reproduction* **4**, 143–150. doi:10.1530/ror.0.0040143

Holt WV, Bennett PM, Volobouev V, Watwon PF (1996) Genetic resource banks in wildlife conservation. *Journal of Zoology* **238**, 531–544. doi:10.1111/j.1469-7998.1996.tb05411.x

Howard JG, Lynch C, Santymire RM, Marinari PE, Wildt DE (2016) Recovery of gene diversity using long-term cryopreserved spermatozoa and artificial insemination in the endangered black-footed ferret. *Animal Conservation* **19**, 102–111. doi:10.1111/acv.12229

Howell LG, Frankham R, Rodger JC, Witt RR, Clulow S, Upton RM, Clulow J (2021a) Integrating biobanking minimises inbreeding and produces significant cost benefits for a threatened frog captive breeding programme. *Conservation Letters* **14**, e12776. doi:10.1111/conl.12776

Howell LG, Mawson PR, Frankham R, Rodger JC, Upton RM, Witt RR, Calatayud NE, Clulow S, Clulow J (2021b) Integrating biobanking could produce significant cost benefits and minimise inbreeding for Australian amphibian captive breeding programs. *Reproduction, Fertility and Development* **33**, 573–587. doi:10.1071/RD21058.

IUCN (2021) *The IUCN Red List of Threatened Species. Version 2021–1*. <https://www.iucnredlist.org>.

Johnson KJ, Baker A, Buley K, Currillo L, Gibson R, Gillespie GR, Lacy RC, Zippel K (2020) A process for assessing and prioritizing species conservation needs: going beyond the Red List. *Oryx* **54**, 125–132. doi:10.1017/S0030605317001715

Johnston LA, Lacy RC (1995) Genome resource banking for species conservation: selection of sperm donors. *Cryobiology* **32**, 68–77. doi:10.1006/cryo.1995.1006

Kemp C (2015) Museums: the endangered dead. *Nature* **518**, 292–294. doi:10.1038/518292a

Kouba AJ, Lloyd RE, Houck ML, Silla AJ, Trudeau VL, Clulow J, Molinia F, Langhorne C, Vance CK, Arregui L (2013) Emerging trends for biobanking amphibian genetic resources: the hope, reality and challenges for the next decade. *Biological Conservation* **164**, 10–21. doi:10.1016/j.biocon.2013.03.010

Kouba AJ, Vance CK (2009) Applied reproductive technologies and genetic resource banking for amphibian conservation. *Reproduction, Fertility and Development* **21**, 719–737. doi:10.1071/RD09038

Labuschagne K (2020) Frozen in time: a biological back-up of species. *Quest* **16**, 16–17.

Lewis SL, Maslin MA (2015) Defining the Anthropocene. *Nature* **519**, 171–180. doi:10.1038/nature14258

Mahony MJ, Clulow J (2011) Appendix 2. Cryopreservation and reconstitution technologies: a proposal to establish a genome resource bank for threatened Australian amphibians. In *Guidelines for Minimising Disease Risks Associated with Captive Breeding, Raising and Restocking Programs for Australian Frogs*. (Eds K Murray, *et al.*) pp. 1–62. Australian Government, Department of Sustainability, Environment, Water, Population and Communities, Canberra.

Martínez-Páramo S, Horváth Á, Labbé C, Zhang T, Robles V, Herráez P, Suquet M, Adams S, Viveiros A, Tiersch TR, Cabrita E (2017) Cryobanking of aquatic species. *Aquaculture* **472**, 156–177. doi:10.1016/j.aquaculture.2016.05.042

Meineke EK, Davis CC, Davies TJ (2018) The unrealized potential of herbaria for global change biology. *Ecological Monographs* **88**, 505–525. doi:10.1002/ecm.1307

National Academies of Sciences, Engineering, and Medicine (NASEM) (2020) *Biological Collections: Ensuring Critical Research and Education for the 21st Century*. National Academies Press, Washington DC. doi:10.17226/25592

O'Connell AF, Jr, Gilbert AT, Hatfield JS (2004) Contribution of natural history collection data to biodiversity assessment in national parks. *Conservation Biology* **18**(5), 1254–1261.

Owens I, Johnson K (2019) One world collection: the state of the world's natural history collections. *Biodiversity Information Science and Standards* **3**, e38772. doi:10.3897/biss.3.38772

Paul S, Gade A, Mallipeddi S (2017) The state of cloud-based biospecimen and biobank data management tools. *Biopreservation and Biobanking* **15**, 169–172. doi:10.1089/bio.2017.0019

Pearl E, Morrow S, Noble A, Lerebours A, Horb M, Guille M (2017) An optimized method for cryogenic storage of *Xenopus* sperm to maximise the effectiveness of research using genetically altered frogs. *Theriogenology* **92**, 149–155. doi:10.1016/j.theriogenology.2017.01.007

Reynolds RP, Crombie RJ, McDiarmid RW, Yates TL (1996) Voucher specimens. In *Measuring and Monitoring Biological Diversity: Standard Methods for Mammals*. (Eds DE Wilson, FR Cole, JD Nichols, R Rudran and MS Foster) pp. 63–68. Smithsonian Books, Washington DC, USA.

Richmond DJ, Sinding SMH, Gilbert MTP (2016) The potential and pitfalls of de-extinction. *Zoologica Scripta* **45**, 22–36. doi:10.1111/zsc.12212

Rocha LA, Aleixo A, Allen G, Almeda F, Baldwin CC, Barclay MVL, *et al.* (2014) Specimen collection: an essential tool. *Science* **344**, 814–815. doi:10.1126/science.344.6186.814

Schindel DE, Cook JA (2018) The next generation of natural history collections. *PLoS Biology* **16**, e2006125. doi:10.1371/journal.pbio.2006125

Shapiro B (2017) Pathways to de-extinction: how close can we get to resurrection of an extinct species? *Functional Ecology* **31**, 996–1002. doi:10.1111/1365-2435.12705

Soulé ME (1991) Conservation: tactics for a constant crisis. *Science* **253**(5021), 744–750. doi:10.1126/science.253.5021.744

Spalding C, Hull PM (2021) Towards quantifying the mass extinction debt of the Anthropocene. *Proceedings of the Royal Society B* **288** (1949), 2020–2332.

Strand J, Thomsen H, Jensen JB, Marcussen C, Nicolajsen TB, Skriver MB, Søgaard IM, Ezaz T, Purup S, Callesen H, Pertoldi C (2020) Biobanking in amphibian and reptilian conservation and management: opportunities and challenges. *Conservation Genetics Resources* **12**, 709–725. doi:10.1007/s12686-020-01142-y

Stuart SN, Chanson JS, Cox NA, Young BE, Rodrigues AS, Fischman D, Waller RW (2004) Status and trends of amphibian declines and extinctions worldwide. *Science* **306**(5702), 1783–1786. doi:10.1126/science.1103538

Thiers B, Monfils AK, Zaspel J, Ellwood E, Bentley A, Levan K, Bates J, Jennings D, Contreras D, Lagomarsino L (2019) Extending US biodiversity collections to promote research and education. *Biodiversity Collections Network* 1–28, <https://bcon.aibs.org/2019/04/04/bcon-report-extending-u-s-biodiversity-collections-to-promote-research-and-education>.

Tilley L, Papadopoulos SC, Pende M, Fei JF, Murawala P (2021) The use of transgenics in the laboratory axolotl. *Developmental Dynamics* **2021**, 1–15. doi:10.1002/dvdy.357

Uteshev VK, Gakhova EN (1994) The prospects of amphibian gene cryobank creation. *Biophysics of the Living Cell* **6**, 28–34.

Uteshev V, Gakhova EN (2005) Gene cryobanks for conservation of endangered amphibian species. In Ananjeva N, Tsinenko O (Eds.), *Herpetologica Petropolitana. Proceedings of the 12th Ordinary General Meeting of the Societas Europaea Herpetologica, Russian Journal of Herpetology* **12**(Suppl.), 233–234.

Uteshev VK, Gakhova EN, Kramarova LI, Shishova NV, Kaurova SA, Browne RK (2019) Cryobanking of amphibian genetic resources in Russia: past and future. *Russian Journal of Herpetology* **26**, 319–324. doi:10.30906/1026-2296-2019-26-6-319-324

Vance CK, Julien A, Counsell K, Marcec R, Agcanas L, Tucker A, Kouba A (2018) Amphibian art over the generations: frozen sperm offspring produce viable F2 generation. *Cryobiology* **85**, 178. doi:10.1016/j.cryobiol.2018.10.220

Waters CN, Zalasiewicz J, Summerhayes C, Barnosky AD, Poirier C, Gałuszka A (2016) The Anthropocene is functionally and stratigraphically distinct from the Holocene. *Science* **351**, aad2622. doi:10.1126/science.aad2622

Webster MS (2017) The extended specimen: emerging frontiers. In *Collections-based Ornithological Research* (Ed. MS Webster). CRC Press, Boca Raton FL, USA. doi.org/doi:10.1201/9781315120454

Wildt DE (2000) Genome resource banking for wildlife research, management, and conservation. *ILAR Journal* **41**(4), 228–234. doi:10.1093/ilar.41.4.228

Wildt DE, Rall WF, Critser JK, Monfort SL, Seal US (1997) Genome resource banks. *Bioscience* **47**, 689–698. doi:10.2307/1313209

Wisely SM, Ryder OA, Santymire RM, Engelhardt JF, Novak BJ (2015) A road map for 21st century genetic restoration: gene pool enrichment of the black-footed ferret. *The Journal of Heredity* **106**, 581–592. doi:10.1093/jhered/esv041

Zimkus BM, Hassapakis CL, Houck ML (2018) Integrating current methods for the preservation of amphibian genetic resources and viable tissues to achieve best practices for species conservation. *Amphibian & Reptile Conservation* **12**, 1–27.

Appendix

Natalie Calatayud and Gina Della Togna

Researched and compiled by the co-chairs of the International Union for the Conservation of Nature, Species Survival Commission, Amphibian Specialist Group's Assisted Reproductive Technologies and Biobanking Working Group and published here with permission of the co-chairs, Natalie Calatayud and Gina Della Togna. This is the 2021 version. It should be emphasised that this is an evolving document and will be periodically updated as institutions change their priorities and resources. Please consult the co-chairs at the following email addresses: drncalatayud@gmail.com or gina.dellatogna@uip.pa with any queries about specific items. Another source of relevant information is the IUCN specialist group's website at: https://www.iucn-amphibians.org/working-groups/thematic/genome-resources/.

Principal institutions recently participating in the development and application of reproductive technologies and biobanking for amphibians.

Aalborg University, Department of Chemistry and Bioscience, Randers, Denmark [a]	Museum Victoria, Ian Potter Biobank, Melbourne, Australia [b]
Agricultural Research Service, U. S. Department of Agriculture, National Animal Germplasm Program, Fort Collins, USA [b]	Nashville Zoo, Nashville, USA [a,b]
	Nature's Safe, Shropshire, UK [b]
Alberta Museum, Alberta, Canada [b]	Omaha's Henry Doorly Zoo and Aquarium, Omaha, USA [a,b]
Central Zoo Authority, Bangalore, India [a,b]	Royal Zoological Society of Antwerp, Centre for Research and Conservation, Antwerp, Belgium [b]
Centro Nacional de Recursos Geneticos, Jalisco, Mexico [a,b]	Royal Zoological Society of Scotland, Edinburgh, Scotland [b]
Cincinnati Zoo, Center for Conservation and Research of Endangered Wildlife, Cincinnati, USA [a,b]	San Diego Zoo Wildlife Alliance, Beckman Center for Conservation research, San Diego, USA [a,b]
Copenhagen Zoo, Copenhagen, Denmark [b]	Smithsonian Institution, National Museum of Natural History, Washington, USA [b]
Detroit Zoological Society, National Amphibian Conservation Center, Detroit, USA [a]	Smithsonian's National Zoo and Conservation Biology Institute, Washington, USA [b]
Fort Worth Zoo, Texas, USA [a]	Smithsonian Tropical Research Institute, Panama Amphibian Rescue and Conservation Project, Panama [a,b]
Harvard University, Museum of Comparative Zoology, Cryogenic Collection, Cambridge, UK [b]	South African National Biodiversity Institute, Pretoria, South Africa [b]
Institute of Cell Biophysics, Russian Academy of Sciences, Moscow, Russia [a,b]	Taronga Conservation Society Australia, CryoDiversity Bank, Taronga Zoo, Mosman, Australia [a,b]
Liebniz Institute for Zoo and Wildlife Research, Department of Evolutionary Genetics Leibniz, Berlin, Germany [b]	University of Newcastle, Conservation Biology Research Group, School of Environmental and Life Sciences, Callaghan, Australia [a,b]
Louisiana State University, Aquatic Germplasm and Genetic Resources Center, Baton Rouge USA [a]	University of Ottawa, Department of Biology, Ottawa, Canada [a]
Memphis Zoological Society, Department of Research and Conservation, Memphis, USA [a,b]	University of Portsmouth, School of Biological Sciences and European Xenopus Resource Centre, Portsmouth, UK [a,b]
Mississippi State University, Department of Biochemistry, Molecular Biology, Entomology and Plant Pathology, Starkville, USA [a,b]	University of Wollongong, Evolution and Assisted Reproduction Laboratory, School of Earth, Atmospheric and Life Sciences, Wollongong, Australia [a,b]
Monash University, Australian Frozen Zoo, Melbourne, Australia [b]	

[a] Institutions developing and applying reproductive technologies (hormone therapies, artificial fertilisation and biobanking)
[b] Institutions housing genome resource banks that contain amphibian genetic material

Index

Note: Bold page numbers refer to Plates.